PROTEIN, LIPID AND MEMBRANE TRAFFIC: PATHWAYS AND TARGETING

NATO Science Series

A Series presenting the results of activities sponsored by the NATO Science Committee.
The Series is published by IOS Press and Kluwer Academic Publishers in conjunction with the NATO Scientific Affairs Division.

General Sub-Series

A Life Sciences	IOS Press
B Physics	Kluwer Academic Publishers
C Mathematical and Physical Sciences	Kluwer Academic Publishers
D Behavioural and Social Sciences	Kluwer Academic Publishers
E Applied Sciences	Kluwer Academic Publishers
F Computer and Systems Sciences	IOS Press

Partnership Sub-Series

1 Disarmament Technologies	Kluwer Academic Publishers
2 Environmental Security	Kluwer Academic Publishers
3 High Technology	Kluwer Academic Publishers
4 Science and Technology Policy	IOS Press
5 Computer Networking	IOS Press

The Partnership Sub-Series incorporates activities undertaken in collaboration with NATO's Partners the Euro-Atlantic Partnership Council - countries of the CIS and Central and Eastern Europe - in Prio Areas of concern to those countries.

NATO-PCO-DATA BASE

The NATO Science Series continues the series of books published formerly in the NATO ASI Series An electronic index to the NATO ASI Series provides full bibliographical references (with keywords and/or abstracts) to more than 50 000 contributions from international scientists published in all sections of the NATO ASI Series.
Access to the NATO-PCO-DATA BASE is possible in two ways:
- via online FILE 128 (NATO-PCO-DATA BASE) hosted by ESRIN, Via Galileo Galilei, I-00044 Frascati, Italy;
- via CD-ROM "NATO-PCO-DATA BASE" with user-friendly retrieval software in English, French an German (© WTV GmbH and DATAWARE Technologies Inc., 1989).

The CD-ROM of the NATO ASI Series can be ordered from PCO, Overijse, Belgium.

Series A: Life Sciences - Vol. 322 ISSN: 1387-6686

Protein, Lipid and Membrane Traffic: Pathways and Targeting

Edited by

Jos A.F. Op den Kamp

Centre for Biomembranes & Lipid Enzymology, Utrecht University, Utrecht, The Netherlands

IOS Press

Ohmsha

Amsterdam • Berlin • Oxford • Tokyo • Washington, DC

Published in cooperation with NATO Scientific Affairs Division

Proceedings of the NATO Advanced Study Institute on
Protein, Lipid and Membrane Traffic: Pathways and Targeting
Cargèse, France
7–19 June 1999

ISBN 1 58603 008 6 (IOS Press)
ISBN 4 274 90333 8 C3045 (Ohmsha)
Library of Congress Catalog Card Number: 99-68998

Publisher
IOS Press
Van Diemenstraat 94
1013 CN Amsterdam
Netherlands
fax: +31 20 620 3419
e-mail: order@iospress.nl

Distributor in the UK and Ireland
IOS Press/Lavis Marketing
73 Lime Walk
Headington
Oxford OX3 7AD
England
fax: +44 1865 75 0079

Distributor in the USA and Canada
IOS Press, Inc.
5795-G Burke Centre Parkway
Burke, VA 22015
USA
fax: +1 703 323 3668
e-mail: iosbooks@iospress.com

Distributor in Germany
IOS Press
Spandauer Strasse 2
D-10178 Berlin
Germany
fax: +49 30 242 3113

Distributor in Japan
Ohmsha, Ltd.
3-1 Kanda Nishiki-cho
Chiyoda-ku, Tokyo 101
Japan
fax: +81 3 3233 2426

PRINTED IN THE NETHERLANDS

Preface

The NATO Advanced Study Institute/FEBS Advanced Course on "Protein, Lipid and Membrane Traffic: Pathways and Targeting" which was held in June 1999 in Cargèse, France has concentrated on the following specific topics:

- the dynamic character of lipids and proteins in biological membranes,
- the existence of specific domains in membranes including their visualisation,
- the molecular mechanisms of intracellular transport of membrane constituents and the involvement of lipid-protein interactions in these processes,
- protein assembly, structure and folding and transport through membranes and
- the intracellular sorting and targeting of individual membrane components as well as different organelles.

The intention of the course was to discuss these topics starting from several disciplines, including biochemistry, cell biology, genetics, and biophysics.

The proceedings presented here give a good overview of these topics. It starts with a detailed review of the biosynthesis of phosphoplipids, enravelled by advanced genetic approaches and is followed by reviews of the specific intracellular routing of the various phospholipids, the arrangements in micro-domains in the membrane and the visualisation of these structures by Atomic Force Microscopy. The role of lipids in protein sorting, routing, insertion and translocation through membranes is presented in several contributions. Up to date information on membrane protein structure and assembly can be found in two articles followed by a series of contributions dealing with the molecular mechanisms of protein insertion and translocation. Both the prokaryotic systems, dealing with insertion and translocation of proteins in relation to export, as well as eukaryotic mechanisms of intracellular translocation are discussed in relation to organelle biogenesis and in particular the mitochondrion caught attention. Finally, intracellular protein and organelle traffic is summarised in detail. This complex phenomenon involves a large number of cell compartments and different tracks along which compounds can traffic into or out of the cell interior.

It was obvious from the various contributions to this ASI that the application of a scala of modern biochemical, cell biological and genetic techniques has led to surprisingly detailed information about the principles and mechanisms involved in the biological processes discussed here.

Jos A.F. Op den Kamp

Contents

Preface v

Phospholipid Synthesis and Turnover in Yeast: Regulation and Coordination with other Cellular Pathways, *S.A. Henry, J. Patton-Vogt, S.R. Dowd, A. Sreenivas, J.A. Graves, V. Bruno, J. Yang, M. Ruiz-Noriega, H.J. Chang, C.M. Almaguer and M. Villa-Garcia* 1

Phosphatidylserine Decarboxylation and CDP-Ethanolamine Pathway Contribute to the Supply of Phosphatidylethanolamine to Mitochondria of the Yeast, *M. Bürgermeister, R. Birner, C. Hrastnik and G. Daum* 19

Structure and Function of Sphingolipid and Cholesterol-Rich Membrane Domains and of GPI-Anchored Proteins, *D. Brown* 27

Near Field Microscopy Approach to the Lateral Heterogeneity of Artificial and Biological Membranes, *V. Vié, N. Van Mau, M.-C. Giocondi, E. Lesniewska, J.-P. Goudonnet, F. Heitz and C. Le Grimellec* 39

Lipid Heterogeneity in Mammalian Cells: Domains and Translocators, *G. van Meer, H. Sprong and R. Raggers* 55

Transmembrane Domain Length and Protein Sorting in the Secretory Pathway: An Examination of the Exit of Axl2p from the Yeast ER, *T.P. Levine and S. Munro* 63

Lipid Microdomains within the Apical Plasma Membrane of Polarized Epithelial Cells. A Potential Role of the Lipid Bilayer in the Retention of Prominin in Microvilli, *K. Röper, D. Corbeil and W.B. Huttner* 73

Spectrofluorometric Characterization of Lipid-Peptide Interactions, *S. Gangl, E. Staudegger, K. Lohner, S. Blondelle, G. Grabner and G. Köhler* 85

Membrane Protein Assembly and Structure, *G. von Heijne* 95

Helix Interactions in Membrane Proteins, *D.M. Engelman, L.F. Fisher, K.R. MacKenzie, J.-L. Popot, W.P. Russ and G.K. Sonoda* 105

Chaperones and Osmolytes in Protein Folding, *S.L. Lindquist* 121

Crystallization of Membrane Proteins in Lipidic Cubic Phases, *E.M. Landau* 137

Mitochondrial Targeting Signals, *C.M. Lee, W. Neupert and R.A. Stuart* 151

Protein Translocation into the Mitochondrial Matrix and Inner Membrane, *A. Geissler, N. Pfanner and K. Truscott* 161

Membrane Traffic Extravaganza, *T. Kirchhausen* 175

Overexpression of rabGDI Induces Vesiculation of the Golgi Apparatus, *G. Bode and T. Mayer* 183

The SRP and SecA/SecB Targeting Pathways in *E. coli* Exhibit Distinct Specificities in Substrate Recognition, *K. Beck, H.-G. Koch, U. Schäfer and M. Müller* 195

Sphingosine Base Requirement for Endocytosis in Yeast, *S. Friant, B. Zanolari, A. Munn and H. Riezman* 203

Endocytosis Meets Exocytosis, *V. Laird* 211

Budding of Retrograde Transport Vesicles from the cis-Golgi in Yeast *Saccharomyces Cerevisiae*, *A. Spang and R. Schekman* 221

Investigating the Roles of SecB Chaperone on HasA Secretion via the ABC (ATP Binding Cassette) Dependent Pathway, *G. Sapriel and P. Delepelaire* 227

Glucose Transport in the Extremely Thermoacidophilic *Sulfolobus solfataricus*, *S.-V. Albers, M.G.L. Elferink, A.J.M. Driessen and W.N. Konings* 235
Towards the Localization and Characterization of the Ligand Binding Sites of the Lactose Transport Protein from *Streptococcus thermophilus*, *L.M. Veenhoff, P.J.R. Spooner, A. Watts and B. Poolman* 245
Homotypic Vacuole Fusion: A New Window on Interorganelle Traffick, *W. Wickner* 255
Membrane Targeting in Pea Root Nodule Cells, *E. Butelli and N.J. Brewin* 261

Subject Index 269
Author Index 277

Protein, Lipid and Membrane Traffic: Pathways and Targeting
J.A.F. Op den Kamp (Ed.)
IOS Press, 2000

Phospholipid synthesis and turnover in yeast: Regulation and coordination with other cellular pathways

Susan A. HENRY, Jana PATTON-VOGT, Susan R. DOWD, Avula SREENIVAS, J. Anthony GRAVES, Vincent BRUNO, Jun YANG, Monica RUIZ-NORIEGA, Hak J. CHANG, *Claudia M. ALMAGUER, *Manuel VILLA-GARCIA
Carnegie Mellon University,
Department of Biological Sciences,
4400 Fifth Avenue,
Pittsburgh, PA 15213, USA
**Centro de Biotecnologia, ITESM Monterrey, Mexico 64849*

1. Introduction

In the yeast *Saccharomyces cerevisiae*, the processes of phospholipid biosynthesis and turnover are highly regulated and tightly coordinated with each other and with other aspects of cellular metabolism. In this paper, we will review recent progress in elucidating the mechanisms of regulation of phospholipid biosynthesis and turnover. We will also discuss the role of major signal transduction pathways in regulating phospholipid metabolism in yeast.

Regulation of phospholipid metabolism in yeast has been intensively studied over the last decade [1-7]. Progress in this field has profited enormously by the completion of the *Saccharomyces cerevisiae* genome in 1996 [8], the first eukaryotic genome to be completed. The public availability of the yeast genome database has spurred functional genomics, leading to the rapid assignment of gene-enzyme relationships and to the discovery of unanticipated gene functions. It has also led to rapid progress in describing interactions among diverse regulatory networks. This progress and the powerful molecular genetics of *S. cerevisiae* present extraordinary opportunities, including the possibility of producing a complete functional description of metabolic control in a eukaryotic cell.

2. Regulation of phospholipid biosynthesis in yeast

Phospholipid biosynthesis in *S. cerevisiae* is regulated on multiple levels [1, 4, 7]. Here we will deal primarily with transcriptional control of a large number of structural genes encoding enzymes of phospholipid biosynthesis. These genes contain the repeated element, UAS_{INO} (consensus 5' CATGTGAAAT 3'), in their promoters. Maximal expression of UAS_{INO}–containing genes occurs during logarithmic growth in media lacking the phospholipid precursors, inositol and choline. When inositol is present, the UAS_{INO}–containing genes are repressed during logarithmic growth. The addition of choline alone has little effect, but when it is added to medium already containing inositol, it causes further repression of UAS_{INO}–containing genes [4, 9]. The structural gene for inositol-1-phosphate synthase, *INO1*, is the most highly regulated of the UAS_{INO}–containing genes [1, 4]. The transcription of the UAS_{INO}–containing genes is influenced by the positive regulatory factors, Ino2p and Ino4p, and by the negative regulatory protein, Opi1p. The *OPI1* gene encodes a protein that contains a leucine zipper motif and polyglutamine stretches [10]. Deletion of the *OPI1* gene leads to constitutive expression of *INO1* [10] and *opi1Δ* mutants over-produce inositol and excrete it into the growth medium, a phenotype known as Opi^- which can be detected in a bioassay (Figure 1).

The regulatory genes, *INO2* and *INO4*, encode basic helix loop helix (bHLH) proteins, which interact to form a heterodimer that binds to UAS_{INO} and activates transcription [11-14]. The *ino2Δ* and *ino4Δ* mutants are inositol auxotrophs (Ino^-) due to failure to derepress *INO1* [9].

Changes in the ongoing status of phospholipid biosynthesis and turnover have immediate effects on expression of UAS_{INO}–containing genes. For example, increased turnover of phosphatidylcholine (PC), via a phospholipase D (PLD)-mediated route, leads to immediate derepression of UAS_{INO}–containing genes, even in stationary phase [15, 16]. A model for this regulation, including the hypothesis that production of excess phosphatidic acid (PA) produces a signal for derepression of UAS_{INO}–containing genes, is presented in Henry and Patton-Vogt [4]. This model is consistent with the repression of UAS_{INO}–containing genes observed in response to inositol or in stationary phase. PA levels drop when inositol is added to the growth medium because PA serves as a precursor to phosphatidylinositol (PI). The levels drop during stationary phase as energy, phosphate, and carbon sources are consumed.

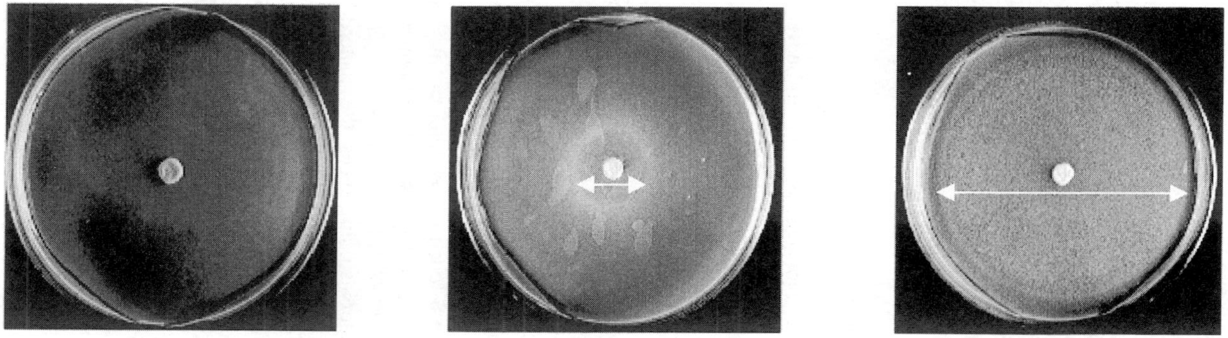

WT W303 1 x *INO1* **_Opi 1Δ_ 1 x _INO1_** **_Opi 1Δ_ 10 x _INO1_**

Fig.1. Opi^- test of yeast strains with increasing copies of the *INO1* gene. To test for the Opi^- phenotype, the strains were patched onto plates with media lacking inositol and incubated at 30°C for two days. These plates were sprayed with a tester strain (*ade1 ino1*) auxotrophic for inositol and grown for another two days. The excretion of inositol was detected by the growth of the tester strain as observed by a halo surrounding the patch of the tested strain. The arrows indicate the diameter of the halo.

Recently, our laboratory has investigated the relative roles of the Ino2p, Ino4p, and Opi1p regulatory proteins in the expression and regulation of *INO1* and other UAS_{INO}–containing genes. We have determined that, in certain genetic backgrounds, the cell can respond to inositol and repress *INO1* when the *OPI1*, *INO2,* and *INO4* genes are deleted [17] (T. Graves and J. Yang, unpublished results). Whereas, Ino2p and Ino4p have been shown to bind to each other and to UAS_{INO} thereby activating transcription [13, 14], we find no evidence that Opi1p binds to Ino2p, Ino4p, or UAS_{INO} (T. Graves and J. Yang, unpublished results). We propose that Opi1p and the Ino2p/Ino4p complex have antagonistic, but additive, effects on expression of UAS_{INO}–containing genes, but that none of the three regulatory proteins is involved in transmitting the signal for repression in response to inositol.

3. Regulation of phospholipid turnover

Recently, we have shown that biosynthesis and turnover of phospholipids are tightly coordinated processes. For example, we have demonstrated that increased turnover of PC via a PLD-mediated route led to derepression of UAS_{INO}–containing genes, such as *INO1* [15, 16]. Furthermore, the route of turnover of phosphatidylinositol (PI) is regulated in response to inositol, but deletion of the *OPI1* gene eliminates this regulation, just as it does the repression of *INO1* and other UAS_{INO}–containing genes [18]. Patton et al. [18] demonstrated that glycerolphosphoinositol (GroPIns), a product of PLB-mediated turnover, is produced and exported from yeast cells during logarithmic growth in the presence of inositol. Both production and reutilization of GroPIns are regulated by inositol availability [18, 19]. When inositol is present in the growth medium, GroPIns is exported at a much higher rate, and it is not transported back into the cell until inositol levels drop. A transporter for GroPIns (*GIT1*) has been identified and cloned [19].

We have discovered that choline also influences the pattern of phospholipid turnover. We have used the techniques of radiolabeling and ^{31}P-NMR to monitor the production and flow of PC metabolites upon temperature elevation and choline supplementation. Under standard growth conditions (2% glucose, 30°C), the products of PLD-mediated PC turnover are rapidly recycled into PC via the CDP-choline pathway [19]. The choline produced is at least partially released into the medium, but then is immediately retaken up by the cell. The relatively small amount of free choline found in wild type cells suggests that choline is quickly phosphorylated by choline kinase (Cki1p). In order to follow *in vivo* PC turnover without introducing a mutation in the *CKI1* gene to block reutilization via the CDP-choline pathway, we have developed a protocol in which ^{14}C-choline labeled cells are chased with excess non-radioactive choline. In these experiments, the environmental stress of elevating the temperature from 30° to 37° was used to induce changes in PC catabolism.

Table 1. Distribution of ^{14}C-choline-containing counts in the medium (Extra) and intracellular water-soluble fraction (Intra) of wild type and *cki1Δ* strains. Strains were grown to uniform labeling in ^{14}C-choline, harvested, washed free of label, and inoculated in medium containing 10mM non-radioactive choline at either 30° or 37°C. Data is presented as a percentage of the total counts recovered following 6 hours of chase. Counts do not add up to 100%, because membrane-associated counts are not included.

	30°C		37°C	
	Extra	Intra	Extra	Intra
		wild type		
Choline	17	<1	13	2
Choline-P	<1	3	<1	2
GroPC	3	18	4	44
		cki1Δ		
Choline	23	<1	20	<1
Choline-P	<1	<1	<1	<1
GroPC	2	7	5	7

Table 1 shows the water-soluble choline metabolites produced when a wild type strain, uniformly labeled with ^{14}C-choline at 30°C, is washed free of label, and re-Inoculated in fresh non-radioactive medium containing 10 mM choline at both 30°C and 37°C. At various times, aliquots of the culture were removed and separated into three fractions: the medium, the membrane, and the water-soluble intracellular fraction as previously described [15]. The choline-containing compounds (choline, choline-P, GroPC, and CDP-choline) in the medium and in the intracellular water-soluble fractions were separated by column chromatography [15] or by TLC [20] and quantitated by liquid scintillation counting.

In the presence of unlabeled choline at 30°C, label is slowly chased out of the PC pool (data not shown) and slowly chased into the growth medium, primarily as free choline (Table I). This result is consistent with previous findings [15], and suggests that PC turnover, most likely via a PLD-mediated route, results in the production of extracellular choline. When the chase is performed at 37°C in the presence of choline, PC turnover is accelerated and the label lost from PC appears in the intracellular fraction as GroPC (Table I). Thus, the shift from 30° to 37°C seems to induce a deacylating phospholipase activity, possibly a phospholipase of type A or B that results in the production of intracellular GroPC. A more accurate description of this turnover awaits the identification of the phospholipase gene products responsible.

Surprisingly, a *cki1Δ* deletion mutant does not show this changed pattern of PC turnover at 37°C. In fact, the profile of PC turnover and the accumulation of choline-containing metabolites in the *cki1Δ* strain are virtually identical at 30° and 37°C in the presence of choline (Table 1). This suggests that either the accumulation of choline-P, another intermediate in the CDP-choline pathway, the metabolic flux through the pathway or the presence of Cki1p, itself, is necessary to induce the PLB-mediated turnover upon shifting to 37°C.

Using ^{31}P-NMR of cell extracts, we have found that choline supplementation, like temperature elevation, activates PLB-mediated PC turnover. The use of ^{31}P-NMR to compare the levels of the various water-soluble intracellular intermediates in different strains adds another dimension to these studies. While the radiolabeling experiments must be performed in the presence of choline, it is unnecessary with ^{31}P-NMR. Obviously, this feature is essential to determining the effect of choline supplementation on PLB-mediated PC turnover. In addition, while the pulse-chase experiments described above monitor PC turnover and resynthesis via the CDP-choline pathway, ^{31}P-NMR allows for a steady state picture of total phosphate-containing metabolites. The combination of these two methods provides a powerful means of analyzing phospholipid turnover and metabolite production. We have analyzed the water-soluble metabolites produced in wild type and *cki1Δ* mutants in the presence and absence of exogenous choline. Consistent with the temperature elevation studies, the *cki1Δ* strain was unable to induce PLB-mediated turnover in response to choline supplementation.

Thus, these experiments represent two different techniques (pulse-chase radiolabeling and mass measurement by NMR) and two different conditions (elevated temperature and choline availability) by which it has been demonstrated that deletion of a gene *(CKI1)* encoding an enzyme of the Kennedy pathway for PC biosynthesis results in an inability to induce PLB-mediated PC turnover. This finding is reminiscent of the observation that ongoing PC biosynthesis and turnover is necessary in order to maintain proper regulation of *INO1* transcription in response to inositol and growth phase [21]. Recent results suggest, as detailed in Patton-Vogt et al. [15] and Sreenivas et al. [16], that PLD-generated PA is the key metabolite in maintaining this regulation

4. Metabolic engineering of phospholipid metabolism in yeast

The availability of cloned structural and regulatory genes, coupled with extensive knowledge of regulatory mechanisms controlling phospholipid metabolism in yeast, makes it possible to use this organism to produce lipid metabolites for industrial applications. We have been exploring the use of yeast to produce inositol and inositol-containing metabolites by fermentation. Inositol is a vital component of the diet of humans and animals of economic importance [22-24]. It is also increasingly being tested in medicinal applications [25-27] and as an additive to infant formula [28].

Existing industrial methods for producing inositol are based on the chemical conversion of phytic acid (inositol hexaphosphoric acid ester) to *myo*-inositol and phosphate using corn mill factory steep water [29]. The first practical industrial method was used by the A. E. Staley Co., based on the 1938 patent of E. Bartow and W. W. Walker [29]. This process, which is still used in Asia today, involves the breakdown of phytic acid with high heat, very acidic conditions and high pressures. A reliable method for producing inositol by fermentation would be both economically and environmentally desirable. To address these requirements, we have produced genetically modified yeast [30, 31] that excretes inositol into the fermentation broth. This strain was constructed by deleting both copies of the *OPI1* gene in a diploid strain, and introducing multiple integrated stable copies of the *INO1* structural gene. This strain excretes increased levels of inositol compared to standard *opi1Δ* strains and is currently undergoing industrial trials. Additional products can be produced using yeast strains engineered based on knowledge of phospholipid regulatory circuits, including the role of inositol- and choline-containing phospholipids as well as water-soluble metabolites produced during phospholipid biosynthesis and turnover.

5. Role of phospholipids in membrane trafficking in yeast

Since phospholipids are major structural components of cellular membranes, regulation of phospholipid metabolism must be tightly integrated with major pathways of membrane trafficking. A first indication of a connection between the secretory pathway and phospholipid metabolism in yeast was provided by the discovery that the *SEC14* gene encodes a phospholipid transfer protein capable of binding PI or PC [32, 33]. Temperature-sensitive *sec14^{ts}* mutants arrest at the late Golgi stage of the secretory pathway when shifted to their restrictive temperature [34]. Mutants in the CDP-choline pathway for PC biosynthesis suppress the *sec14* secretory and growth phenotypes as do a number of other second-site suppressors, including *sac1* mutants [35]. Kearns et al. [36] have proposed that DAG is essential to the secretory process, and that excess consumption of DAG in *sec14^{ts}* mutants shifted to the restrictive temperature is responsible for the observed growth and secretory phenotypes. According to this hypothesis, the mechanism of suppression of *sec14* by mutants in the CDP-choline pathway and *sac1* mutants would involve either blocking excess DAG consumption or generating additional DAG [36]. However, recently we have shown that *sec14^{ts} cki1* strains, shifted to the restrictive temperature, exhibit accelerated PLD-mediated PC turnover [15]. Furthermore, Pld1p is required for *sec14* bypass suppression by CDP-choline mutants and other suppressors [16, 37]. In mammalian cells, it has been proposed that PLD-mediated turnover and its product, PA, are involved in vesicle budding [38-43].

In all eukaryotic cells, vesicles mediate transport between the membrane compartments of the secretory pathway. Vesicle budding involves the capture of specific cargo molecules into coated vesicle intermediates [44, 45]. To date, three classes of coated vesicles have been characterized: COPII-coated vesicles mediate anterograde transport from the ER to the Golgi apparatus [45, 46]; COPI-coated vesicles mediate intra-Golgi and endoplasmic Reticulum (ER)-Golgi bidirectional transport [44, 47]; clathrin-coated vesicles that mediate transfer among the plasma membrane, endocytic, and *trans*-Golgi compartments [48]. The essential membrane components that are required for the recruitment of coat subunits to membrane have been analyzed *in vitro* [49]. The stepwise formation of COPII coat vesicles *in vitro* from liposomal templates requires only COPII-coat protein, GTP and acidic phospholipids, such as PI(4,5)P_2, PI4P, and PA [49].

Because lipid metabolism can be manipulated in yeast using genetic approaches, it should be possible to use this organism to study the roles of specific phospholipids in membrane trafficking events *in vivo*. As yet, the involvement of specific lipids in most steps in the secretory pathway in yeast has not been determined *in vivo*. However, there is increasing evidence of participation of PI3P in vacuolar trafficking. Vesicular transport to the vacuole is a major membrane trafficking pathway in yeast. The vacuole makes up as much as 25% of the yeast cellular volume [50, 51]. Recent studies revealing the involvement of phosphoinositides and their kinases in the vacuolar protein sorting process [52-54] raise the possibility that membrane trafficking and phospholipid metabolism are integrated and coordinated processes. For example, PIP3 kinase, encoded by the *VPS34* gene, is required for sorting of proteins from the Golgi to the vacuole [52]. Furthermore, Vps27p and Pep7p/Vac1p, which are required for events at the endosome, contain RING FYVE domains that specifically bind to PIP3 [53]. Vps27p functions in vacuolar and endocytic traffic through the endosome [55]. Pep7p functions in the vesicle consumption step in vesicle-mediated transport to the endosome [56]. The *FAB1* gene also contains the RING FYVE domain. *fab1Δ* mutants fail to produce PI(3,5)P but contain PI3P and exhibit altered vacuolar size and acidification but normal Golgi to vacuole trafficking [54].

6. Role of the Unfolded Protein Response Signal Transduction Pathway in *INO1* Expression

In higher eukaryotes, phospholipids and their turnover products have been shown to play important cellular roles. In particular, the phosphorylated derivatives of PI interact with proteins to affect their activity or location, and they can be hydrolyzed by phospholipase C to generate inositol phosphates that act as second messengers in mammalian cells [57]. In yeast, even though PI3P, PI4P, PI(3,5)P and PI(4,5)P have all been shown to be present and to play important roles in membrane trafficking [49, 58] clear roles for phospholipids or their turnover products in specific signal transduction pathways have yet to be established [7].

We have shown that PLD-mediated turnover of PC, and production of PA produces a signal that leads to derepression of *INO1* and other UAS_{INO}–containing genes [15, 16]. How this signal is transmitted to the transcription apparatus controlling expression of the UAS_{INO}–containing genes is unknown at present. As described above, we have recently shown that the signal, produced by the presence of inositol leading to repression of UAS_{INO} containing genes, involves a lipid intermediate but is not transmitted via the known regulatory factors, Ino2p, Ino4p, or Opi1p. We have begun an extensive survey of known signal transduction pathways to determine which, if any, of these pathways are involved. This analysis, if successful, could simultaneously provide insight into the roles of phospholipids in signal transduction in yeast and the mechanisms of coordination of phospholipid synthesis with other cellular processes. For example, the recent discovery of involvement of the unfolded protein response (UPR) signal transduction pathway in *INO1* expression [59] provides evidence for coordination of synthesis of the protein and lipid components of the ER in yeast. The UPR is activated when cells are under certain conditions of stress involving changes in secretory output, generally known as "ER stress" [60, 61]. The UPR allows eukaryotic cells to respond to changing conditions in the ER by regulating the synthesis of ER-resident proteins such as Kar2p (BiP) and Pdi1p [60]. Two major components of the UPR pathway are Ire1p and Hac1p. Ire1p is an ER transmembrane kinase that initiates the UPR signal transduction from the ER [62, 63]. Under ER stress, such as the accumulation of unfolded proteins in the ER, Ire1p senses the event in the ER lumen, most likely with its N-terminal domain [Cox, 1993 #50; Mori, 1993 #51; [59, 61]. The UPR signal is transmitted across the ER membrane to the C-terminus of Ire1p that lies in the nucleus where oligomerization and autophosphorylation of Ire1p occur in response to the signal [64, 65]. The C-terminus of Ire1p contains two critical domains: the serine/threonine kinase domain and the site-specific endoribonuclease domain [66]. Upon sensing the UPR signal, the kinase domain autophosphorylates and activates the endoribonuclease domain of Ire1p to catalyses the splicing of *HAC1* mRNA. Since only the spliced form of *HAC1* transcript is translated at normal rate, this process is an effective form of translational control [67]. Hac1p is consequently detectable only in UPR-activated cells. Hac1p, a basic leucine zipper (bZIP) transcription factor, then binds to the upstream activator sequence of the target genes at the unfolded protein response element (UPRE) activating the unfolded protein response [68].

In addition to being defective in activating the UPR pathway, *hac1* and *ire1* mutants are inositol auxotrophs [59]. Furthermore, inositol depletion induces the UPR while *INO1* expression is elevated in UPR-induced cells [59], indicating a complex interplay between UPR and regulation of phospholipid metabolism. *hac1* and *ire1* null mutants express *INO1* at wild-type levels up to 3 hours after induction of the UPR, but expression declines drastically thereafter. This suggests that Hac1p and Ire1p are responsible for maintaining a high level of *INO1* expression, but not for controlling the initial expression of *INO1* under inositol-depleting conditions [59]. A further indication of the integration of the UPR pathway and *INO1* expression is the observation that *opi1Δ* suppresses the Ino$^-$ phenotype of hac1 mutants [59]. *opi1 hac1* cells are able to express the *INO1* gene, but transcription of UPRE-containing genes is not activated in this strain.

We are investigating the mechanism by which the UPR signal transduction pathway regulates *INO1* using the *sec14^{ts}cki1Δ genetic* background. In this strain, *INO1* expression can be induced by a shift to the *sec14^{ts}* restrictive temperature [15, 16]. By constructing triple mutants (*sec14^{ts} cki1 ire1* and *sec14^{ts}cki1 hac1*), we are determining whether Ire1p and Hac1p are responsible for mediating the signal that results in elevated *INO1* expression in the *sec14^{ts}cki1* strain. Detailed kinetic analysis of *INO1* expression will be performed to determine if the signal for *INO1* induction, produced when excess PA is generative by PLD-mediated turnover of PC in *sec14^{ts}cki1* strains, is transmitted via the UPR pathway.

7. Role of the Glucose Response Pathway in Expression of *INO1*

Two recent studies have revealed that the glucose response pathway also plays a role in the expression of *INO1* and other UAS_{INO}–containing genes [69, 70]. A screen for suppressors of the inositol auxotrophy of the *ino4-8* missense allele of *INO4*, led to the identification of *reg1* mutants. Mutations at the *REG1* locus result in constitutive expression of the *INO1* gene, suggesting that Reg1p is required for repression of *INO1* [69]. Shirra and Arndt [70] independently identified a recessive *reg1* mutation and a dominant mutation of *SNF4* as suppressors of the Ino$^-$phenotype of the *spt15-328* mutation in the TATA binding protein.

Reg1p has also been shown to be required for repression of glucose-repressed genes [71, 72] and functions as the regulatory subunit of the Glc7 protein phosphatase [73, 74]. The Glc7p/Reg1p complex regulates the activity of the Snf1p/Snf4p kinase, which is required to activate the expression of glucose-repressed genes (For reviews see: [75, 76]. We have shown that a *glc7* mutant that fails to interact with Reg1p is able to suppress the inositol auxotrophy of *ino4-8*, as is a dominant allele of *SNF4*. The Ino$^-$ phenotype of the *ino4-8* allele is also suppressed by growth on ethanol. We have found that when cells are grown on carbon sources other than glucose, particularly ethanol, the overall levels of *INO1* expression are significantly higher than when they are grown on glucose. However, independent of the carbon source used, the *INO1* gene is repressed in response to inositol. Even though the maximal levels of expression vary by carbon source, *INO1* expression is reduced in the presence of inositol and is maximally expressed in its absence, no matter what carbon source is employed. We do not believe that the glucose response pathway is responsible for transmitting the signal for repression in response to inositol. Examination of the kinetics of *INO1* derepression in the presence and absence of inositol show that *INO1* is repressed in the presence of inositol in media containing either 0.05% or 2% glucose. These results and other evidence collected in our laboratory suggest that even though regulatory proteins involved in glucose repression have an effect on *INO1* expression, the *INO1* gene is not subject to classical glucose repression. Nevertheless, the involvement of the glucose response pathway in expression of *INO1* and other UAS_{INO}–containing genes suggests that lipid and carbohydrate metabolism are linked in yeast as they are in mammals via the insulin response pathway [77]. It is interesting, in this regard, that there is evidence that acetyl-CoA carboxylase is regulated by Snf1p in yeast as it is in mammals [78, 79], and that the structural gene encoding this enzyme in yeast, *ACC1*, contains UAS_{INO} in its promoter and is regulated by inositol [78, 80].

8. Concluding Remarks

Recently, our laboratory has made progress in elucidating the mechanism controlling regulation of phospholipid biosynthesis. A great deal has also recently been learned through studies in a number of laboratories about the coordination of phospholipid metabolism with membrane trafficking and the major signal transduction pathways in yeast. Progress in these areas has been very rapid in the last several years, due in part to the availability of the sequence of the complete genome of *S. cerevisiae* coupled with powerful molecular genetic approaches available for the analysis of gene function in yeast. Because of the similarities of phospholipid membrane trafficking and signal transduction pathways in all eukaryotes, this progress in yeast is expected to provide major new insights into these processes in higher organisms.

References

[1] G. M. Carman, and S. A. Henry, Phospholipid biosynthesis in yeast, *Ann. Rev. Biochem.* **58** (1989) 635-669.

[2] F. Paltauf, and G. Daum, Phospholipid transfer proteins from yeast. In: (Dennis, E. A., and Vance, D. E., Eds.), Methods in Enzymology. 1992, pp. 514-522.

[3] M. L. Greenberg, and J. M. Lopes, Genetic regulation of phospholipid biosynthesis in yeast, *Microbiological Reviews* **60** (1996) 1-20.

[4] S. A. Henry, and J. L. Patton-Vogt, Genetic regulation of phospholipid metabolism: Yeast as a model eukaryote. In: (Cohn, W. E., and Moldave, K., Eds.), Progress in Nucleic Acid Research and Molecular Biology. Academic Press, Inc., San Diego, CA. 1998, pp. 133-179.

[5] M. Schweizer, Lipids and membranes. In: (Dickinson, J. R., and Schweizer, M., Eds.), The metabolism and molecular physiology of *Saccharomyces cerevisiae*. Taylor and Francis. Inc., Philadelphia, PA. 1999, pp. 79-155.

[6] G. Daum, *et al.*, Biochemistry, cell biology and molecular biology of lipids of *Saccharomyces cerevisiae*, *Yeast* **14** (1998) 1471-1510.

[7] G. A. Carman, and S.A. Henry, Molecular characterization and regulation of phospholipid metabolism in yeast. In: Progress in Lipid Research. Elsevier Science, Amsterdam, The Netherlands. 1999, pp. In press.
[8] A. Goffeau, *et al.*, Life with 6000 genes, *Science* **274** (1996) 546-567.
[9] J. P. Hirsch, and S. A. Henry, Expression of the *Saccharomyces cerevisiae* inositol-1-phosphate synthase (*INO1*) gene is regulated by factors that affect phospholipid synthesis, *Mol. Cell. Biol.* **6** (1986) 3320-3328.
[10] M. J. White, *et al.*, The *OPI1* gene of *Saccharomyces cerevisae*, a negative regulator of phospholipid biosynthesis, encodes a protein containing polyglutamine tracts and a leucine zipper, *J. Biol. Chem.* **266** (1991) 863-872.
[11] J. M. Lopes, and S. A. Henry, Interaction of *trans* and *cis* regulatory elements in the *INO1* promoter of *Saccharomyces cerevisiae*, *Nucl. Acids Res.* **19** (1991) 3987-3994.
[12] D. M. Nikoloff, and S. A. Henry, Functional characterization of the *INO2* gene of *Saccharomyces cerevisiae*, *J. Biol. Chem.* **269** (1994) 7402-7411.
[13] J. Ambroziak, and S. A. Henry, *INO2* and *INO4* gene products, positive regulators of phospholipid biosynthesis in *Saccharomyces cerevisiae*, form a complex that binds to the *INO1* promoter, *J. Biol. Chem.* **269** (1994) 15344-15349.
[14] S. Schwank, *et al.*, Yeast transcriptional activator *INO2* interacts as an Ino2p/Ino4p basic helix-loop-helix heteromeric complex with the inositol/choline-responsive element necessary for expression of phospholipid biosynthetic genes in *Saccharomyces cerevisiae*, *Nucl. Acids Res.* **23** (1995) 230-237.
[15] J. L. Patton-Vogt, *et al.*, Role of the yeast phosphatidylinositol/phosphatidylcholine transfer protein (Sec14p) in phosphatidylcholine turnover and *INO1* regulation, *J. Biol. Chem.* **272** (1997) 20873-20883.
[16] A. Sreenivas, *et al.*, A role for phospholipase D (Pld1p) in growth, secretion, and regulation of membrane lipid synthesis in yeast, *J. Biol. Chem.* **273** (1998) 16635-16638.
[17] J. A. Graves, Analysis of the role of the *OPI1* gene product in the negative regulation of the phospholipid biosynthetic pathway of *Saccharomyces cerevisiae*, Ph.D. Thesis, (1996) Biological Sciences, Carnegie Mellon University, Pittsburgh, PA.
[18] J. L. Patton, *et al.*, Production and reutilization of an extracellular phosphatidylinositol catabolite, glycerophosphoinositol, by *Saccharomyces cerevisiae*, *J. Bacteriol.* **177** (1995) 3379-3385.

[19] J. L. Patton-Vogt, and S. A. Henry, *GIT1*, a gene encoding a novel transporter for glycerophosphoinositol in *Saccharomyces cerevisiae*, *Genetics* **149** (1998) 1707-1715.
[20] T. W. Martin, Formation of diacylglycerol by a phospholipase D-phosphatidate phosphatase pathway specific for phosphatidylcholine in endothelial cells, *Biochimica et Biophysica Acta* **962** (1988) 282-296.
[21] P. Griac, *et al.*, The role of phosphatidylcholine biosynthesis in the regulation of the *INO1* gene of yeast, *J. Biol. Chem.* **271** (1996) 25692-25698.
[22] B. J. Holub, The nutritional importance of inositol and the phosphoinositides [editorial comment], *N Engl J Med* **19** (1992) 1285-1287.
[23] C. D. Berdanier, Is inositol an essential nutrient?, *Nutrition Today* **27(2)** (1992) 22 (5).
[24] Report: Nutrient requirements of fish, (1993) National Research Council Committee on Animal Nutrition, National Academy Press, Washington, D.C., pp. x, 114.
[25] J. Levin, Controlled trials of inositol in psychiatry, *Eur Neurophychopharmacol* **7** (1997) 147-155.
[26] L. Colodny and R. L. Hoffman, Inositol - clinical applications for exogenous use, *Altern. Med. Rev.* **3** (1998) 432-447.
[27] J. Corcoran, What are the molecular mechanisms of neural tube defects?, *Bioessays* **20** (1998) 6-8.
[28] K. E. Sharpless, L.M. Gill, et al., Preparation of standard reference material 2383 (Baby Food Composite) and use of an interlaboratory comparison exercise for value assignment of its nutrient concentrations, *J AOAC Int.* **82** (1999) 276-287.
[29] E. Bartow, and W. W. Walker, Manufacture of I-Inositol, (1938) US Patent Office 2,112,553, Iowa State University, USA
[30] S. A. Henry, and M. J. White, Inositol-excreting yeast, (1996) US Patent Office 5,529,912, Carnegie Mellon University, USA
[31] S. A. Henry, and M. J. White, Inositol-excreting yeast, (1997) US Patent Office 5,599,701, Carnegie Mellon University, USA
[32] J. F. Aitken, *et al.*, The gene encoding the phosphatidylinositol transfer protein is essential for cell growth, *J. Biol. Chem.* **265** (1990) 4711-4717.
[33] V. Bankaitis, *et al.*, An essential role for a phospholipid transfer protein in yeast for Golgi function, *Nature* **347** (1990) 561-562.
[34] P. Novick, *et al.*, Identification of 23 complementation groups required for post-translational events in the yeast secretory pathway, *Cell* **21** (1980) 205-215.

[35] A. E. Cleves, *et al.*, Mutations in the CDP-choline pathway for phospholipid biosynthesis bypass the requirement for an essential phospholipid transfer protein, *Cell* **64** (1991) 789-800.
[36] B. G. Kearns, *et al.*, Essential role for diacylglycerol in protein transport from the yeast Golgi complex, *Nature* **387** (1997) 101-105.
[37] Z. Xie, *et al.*, Phospholipase D activity is required for suppression of yeast phosphatidylinositol transfer protein defects, *Proc. Natl. Acad. Sci. USA* **95** (1998) 12346-12351.
[38] A. Siddhanta, and D. Shields, Secretory vesicle budding from the trans-Golgi network is mediated by phosphatidic acid levels, *J. Biol. Chem.* **273** (1998) 17995-17998.
[39] Y.-G. Chen, *et al.*, Phospholipase D stimulates release of nascent secretory vesicles from the *trans*-Golgi network, *J. Cell Biol.* **138** (1997) 495-504.
[40] K. Bi, *et al.*, Phosphatidic acid formation by phospholipase D is required for transport from the endoplasmic reticulum to the Golgi complex, *Curr. Biol.* **7** (1997) 301-307.
[41] J. H. Exton, New developments in phospholipase D, *J. Biol. Chem.* **272** (1997) 15579-15582.
[42] M. Liscovitch, Phospholipase D: Role in signal transduction and membrane traffic, *J. Lipid Mediators Cell Signalling* **14** (1996) 215-221.
[43] M. G. Roth, and P. C. Sternweis, The role of lipid signaling in constitutive membrane traffic, *Curr. Opin. Cell Biol.* **9** (1997) 519-526.
[44] J. E. Rothman, and F. T. Wieland, Protein sorting by transport vesicles, *Science* **272** (1996) 227-234.
[45] R. Schekman, and L. Orci, Coat proteins and vesicle budding, *Science* **271** (1996) 1526-1533.
[46] S. Y. Bednarek, *et al.*, Traffic COPs and the formation of vesicle coats, *Trends Cell Biol.* **6** (1996) 468-473.
[47] L. Orci, *et al.*, Bidirectional transport by distinct populations of COPI-coated vesicles, *Cell* **90** (1997) 335-349.
[48] J. S. Robinson, The role of clathrin, adaptors and dynamin in endocytosis, *Curr. Opin. Cell Biol.* **6** (1994) 538-544.
[49] K. Matsuoka, *et al.*, COPII-Coated vesicle formation reconstituted with purified coat proteins and chemically defined liposomes, *Cell* **93** (1998) 263-275.
[50] A. Wiemken, and M. Durr, Characterization of amino acid pools in the vacuolar compartment of *Saccharomyces cerevisiae*, *Arch. Microbiol.* **101** (1974) 45.
[51] E. J. Jones, *et al.*, Biogenesis and function of the yeast vacuole. In: (Broach, J. R., Pringle, J. R., and Jones, E. W., Eds.), The

Molecular and Cellular Biology of the Yeast *Saccharomyces cerevisiae*: Cell Cycle and Cell Biology. Cold Spring Harbor Laboratory Press, Plainview, NY. 1998.

[52] J. H. Stack, and S. D. Emr, Vps34p required for yeast vacuolar protein sorting is a multiple specificity kinase that exhibits both protein kinase and phosphatidylinositol-specific PI 3-kinase activities, *J. Biol. Chem.* **269** (1994) 31552-31562.

[53] C. G. Burd, and S. D. Emr, Phosphatidylinositol(3)-phosphate signaling mediated by specific binding to RING FYVE domain, *Molecular Cell* **2** (1998) 157-162.

[54] J. D. Gary, *et al.*, Fab1p is essential for PtdIns(3)p5-kinase activity and the maintenance of vacuolar size and membrane homeostasis, *J. Cell Biol.* **143** (1998) 65-79.

[55] R. C. Piper, *et al.*, *VPS27* controls vaculoar and endocytic traffic through a prevacuolar compartment in *Saccharomyces cerevisiae*, *J. Cell. Biol.* **131** (1995) 603-617.

[56] G. C. Webb, *et al.*, Pep7p provides a novel protein that functions in vesicle-mediated transport between the yeast Golgi and endosome, *Mol. Biol. Cell* **8** (1997) 871-895.

[57] D. A. Fruman, *et al.*, Phosphoinositide kinases, *Annu. Rev. Biochem.* **67** (1998) 481-507.

[58] J. C. Hay, *et al.*, ATP-dependent inositide phosphorylation required for Ca(2+)-activated secretion, *Nature* **374** (1995) 173-177.

[59] J. S. Cox, *et al.*, The unfolded protein response coordinates the production of endoplasmic reticulum protein and endoplasmic reticulum membrane, *Mol. Biol. Cell* **8** (1997) 1805-1814.

[60] Y. Kozutsumi, *et al.*, The presence of malfolded proteins in the endoplasmic reticulum signals the induction of glucose-regulated proteins, *Nature* **332** (1988) 462-464.

[61] C. E. Shamu, Splicing together the unfolded protein response, *Current Biology* **7** (1997) R67-R70.

[62] J. S. Cox, *et al.*, Transcriptional induction of genes encoding endoplasmic reticulum resident proteins requires a transmembrane protein kinase, *Cell* **73** (1993) 1197-1206.

[63] K. Mori, *et al.*, A transmembrane protein with a $cdc2^{+}$/CDC28-related kinase activity is required for signaling from the ER to the nucleus, *Cell* **74** (1993) 743-756.

[64] C. E. Shamu, and P. Walter, Oligomerization and phosphorylation of the Ire1p kinase during intracellular signaling from the endoplasmic reticulum to the nucleus, *EMBO J.* **15** (1996) 3028-3039.

[65] A. A. Welihinda, and R. J. Kaufman, The unfolded protein response pathway in *Saccharomyces cerevisiae*, *J. Biol. Chem.* **271** (1996) 18181-18187.
[66] C. Sidrauski, and P. Walter, The transmembrane kinase Ire1p is a site-specific endonuclease that initiates mRNA splicing in the unfolded protein response, *Cell* **90** (1997) 1031-1039.
[67] R. E. Chapman, and P. Walter, Translational attenuation mediated by a mRNA intron, *Curr. Biol.* **7** (1997) 850-859.
[68] J. S. Cox, and P. Walter, A novel mechanism for regulating activity of a transcription factor that controls the unfolded protein response, *Cell* **87** (1996) 391-404.
[69] Q. Ouyang, *et al.*, The *REG1* gene product is required for repression of *INO1* and other inositol-sensitive upstream activating sequence-containing genes of yeast, *Genetics* **152** (1999) 89-100.
[70] M. K. Shirra, and K. M. Arndt, Evidence for the involvement of the Glc7-Reg1 phosphatase and the Snf1-Snf4 kinase in the control of TBP function at the *INO1* promoter of *Saccharomyces cerevisiae*, *Genetics* **152** (1999) 73-87.
[71] K. Matsumoto, *et al.*, Recessive mutations conferring resistance to carbon catabolite repression of galactokinase synthesis in *Saccharomyces cerevisiae*, *J. Bacteriol.* **153** (1983) 1405-1414.
[72] D. Niederacher, and K. D. Entian, Characterization of Hex2 protein, a negative regulatory element necessary for glucose repression in yeast, *Eur J Biochem* **200** (1991) 311-9.
[73] J. Tu, and M. Carlson, The *GLC7* type 1 protein phosphatase is required for glucose repression in *Saccharomyces cerevisiae*, *Mol. Cell. Biol.* **14** (1994) 6789 6796.
[74] J. Tu, and M. Carlson, *REG1* binds to protein phosphatase type 1 and regulates glucose repression in *Saccharomyces cerevisiae*, *EMBO J.* **14** (1995) 5939-5946.
[75] J. M. Gancedo, Yeast carbon catabolite repression, *Microbiol Mol Biol Rev* **62** (1998) 334-61.
[76] M. Carlson, Regulation of glucose utilization in yeast, *Curr. Opinion in Genetics & Develop.* **8** (1998) 560-564.
[77] D. Wang, and H. S. Sul, Upstream stimulatory factor binding to the E-box at -65 is required for insulin regulation of the fatty acid synthase promoter, *J. Biol. Chem.* **272** (1997) 26367-26374.
[78] R. K. Beri, *et al.*, Molecular cloning, expression and chromosomal localisation of human AMP-activated protein kinase, *FEBS Lett.* **356** (1994) 117-121.
[79] A. Woods, *et al.*, Yeast *SNF1* is functionally related to mammalian AMP-activated protein kinase and regulates acetyl-CoA carboxylase *in vivo*, *J. Biol. Chem.* **269** (1994) 19509-19515.

[80] M. Hasslacher, *et al.*, Acetyl-CoA carboxylase from yeast is an essential enzyme and is regulated by factors that control phospholipid metabolism, *J. Biol. Chem.* **268** (1998) 10946-10952.

Protein, Lipid and Membrane Traffic: Pathways and Targeting
J.A.F. Op den Kamp (Ed.)
IOS Press, 2000

Phosphatidylserine decarboxylation and CDP-ethanolamine pathway contribute to the supply of phosphatidylethanolamine to mitochondria of the yeast

M. Bürgermeister, R. Birner, C. Hrastnik and G. Daum

Institut für Biochemie und Lebensmittelchemie
Technische Universität, Petersgasse 12/2
A-8010 Graz, Austria

Abstract

The contribution of the two phosphatidylserine decarboxylases of the yeast *Saccharomyces cerevisiae*, the mitochondrial Psd1p and Psd2p present in a Golgi/vacuolar compartment, to cellular phosphatidylethanolamine (PtdEtn) synthesis was compared to that of the CDP-ethanolamine pathway (Kennedy pathway). Psd1p is the major source of cellular PtdEtn, and deletion of *PSD1* resulted in a major decrease of mitochondrial PtdEtn as shown by quantitative analysis of lipids and labeling experiments. Lack of Psd2p affects synthesis of PtdEtn and its supply to mitochondria to a smaller extent. In a *Δpsd2* mutant the synthesis of phosphatidylcholine (PtdCho) by the methylation pathway is dramatically decreased indicating that PtdEtn formed by Psd2p is the preferred substrate for PtdEtn methyltransferases. Although mitochondria receive most of their PtdEtn from decarboxylation by Psd1p, import of PtdEtn produced by Psd2p or by the Kennedy pathway into mitochondria significantly contributes to the mitochondrial level of this phospholipid.

1. Introduction

Mitochondria are only partially autonomous regarding the supply of phospholipids to their membranes. Phosphatidylcholine (PtdCho) and phosphatidylethanolamine (PtdEtn) are the major phospholipid species of mitochondria. In the yeast, PtdEtn biosynthesis occurs by (i) decarboxylation of phosphatidylserine (PtdSer) through catalysis of phosphatidylserine decarboxylase 1 (Psd1p) in mitochondria, (ii) decarboxylation of PtdSer by phosphatidylserine decarboxylase 2 (Psd2p) in a Golgi/vacuolar compartment, and (iii) incorporation of ethanolamine (Etn) via CDP-ethanolamine pathway (Kennedy pathway), which is located in microsomes (Fig. 1).

The major pathway of PtdEtn formation in yeast appears to be the Psd1p route which involves formation of PtdSer in the endoplasmic reticulum, import of PtdSer into mitochondria and decarboxylation in the inner membrane of this compartment [for a review see ref. 1]. Although Psd1p catalyzes 90-95 % of PSD activity *in vitro*, a *Δpsd1* deletion strain grows like wild-type on standard media [2].

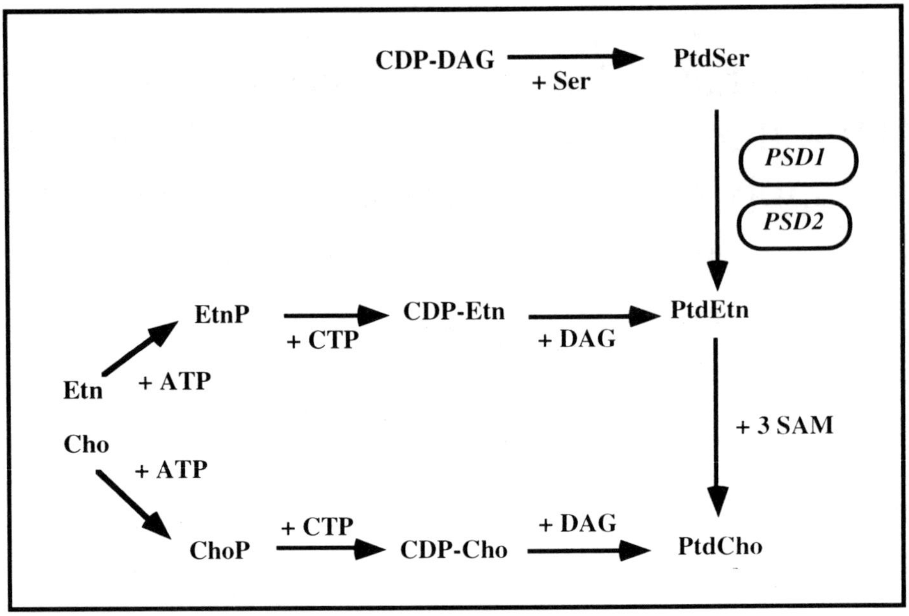

Fig. 1: Biosynthesis of aminoglycerophospholipids in the yeast.
CDP-DAG: CDP-diacylglycerol; Ser: serine; PtdSer: phosphatidylserine; PtdEtn: phosphatidylethanolamine; PtdCho: phosphatidylcholine; Etn: ethanolamine; EtnP: ethanolamine phosphate; CDP-Etn: CDP-ethanolamine; Cho: choline, ChoP: choline phosphate; CDP-Cho: CDP-choline; DAG: diacylglycerol; SAM: S-adenosyl methionine; *PSD1* and *PSD2*: phosphatidylserine decarboxylases 1 and 2.

Psd2p accounts for only 5-10% of the total cellular PSD activity *in vitro* [3] and loss of *PSD2* activity does not affect cell growth at all. Lack of both PtdSer decarboxylases Psd1p and Psd2p, however, turns the strain auxotrophic for ethanolamine (Etn) or choline (Cho).

PtdCho is the most abundant phospholipid of mitochondrial membranes, although mitochondria do not contain enzymes capable of phosphatidylcholine biosynthesis. Thus, all PtdCho must be imported into mitochondria. In yeast, PtdCho is mainly synthesized by a three step methylation of PtdEtn. The two methyltransferases catalyzing these reactions are localized to the endoplasmic reticulum. In addition, Cho can be incorporated into PtdCho in a reaction similar to the utilization of free Etn through the Kennedy pathway (see Fig. 1).

In this study the contribution of Psd1p, Psd2p and the Kennedy pathway to the biosynthesis of PtdEtn and to the supply of this phospholipid to mitochondria was investigated. Translocation of aminoglycerophospholipids between organelles as a prerequisite for balanced biosynthesis of PtdEtn and PtdCho is discussed.

2. Methods

The wild-type yeast strain *Saccharomyces cerevisiae* FY1679, *Δpsd1* and *Δpsd2* single mutants, and a *Δpsd1 Δpsd2* double mutant were used throughout this study. Strains deleted of *PSD1* and *PSD2* were constructed by standard procedures following the method described by Wach et al. [4]. All strains were cultivated in YPLac media (1% yeast extract,

2% peptone, 2% lactate, 0.1% glucose) under aerobic conditions at 30°C and harvested in the logarithmic phase.

Yeast mitochondria and microsomes were isolated by published procedures [5, 6]. Purity of isolated subcellular fractions was routinely tested by Western Blot analysis [7]. Lipids of the cell homogenate, mitochondria or microsomes were extracted by the procedure of Folch et al. [8]. Individual glycerophospholipids were separated by two-dimensional thin-layer chromatography (TLC) using chloroform/methanol/ 25% NH_3 (65:35:5; per vol.) as first, and chloroform/acetone/methanol/acetic acid/water (50:20:10:10:5; per vol.) as second developing solvent. Phospholipids were visualized on TLC plates by staining with iodine vapor, scraped off from the plate, and quantified by the method of Broekhuyse [9].

To distinguish between the contribution of Psd1p, Psd2p and the Kennedy pathway to the formation of mitochondrial PtdEtn and PtdCho a double labeling experiment was designed. Yeast spheroplasts were prepared by the method of Daum et al. [5], regenerated in minimal media buffered with 1 M sorbitol and 100 mM Na-citrate, pH 5.5, for one hour, followed by labeling with [^{3}H]serine and [^{14}C]ethanolamine or [^{14}C]choline, respectively, for 30 min. Subsequently, mitochondria and microsomal fractions were isolated. Lipids of these fractions and of the cell homogenate were extracted. Phospholipids were separated by TLC with chloroform/methanol/25% NH_3 (50:25:6; per vol.) as a developing solvent. Spots were visualized by iodine vapor, scraped off from the plate, and radioactivity in PtdEtn and PtdCho was measured by liquid scintillation counting.

3. Results and Discussion

3.1. Deletion of *PSD1* and *PSD2* affects cellular and mitochondrial phospholipid composition

To show the contribution of Psd1p, Psd2p and the Kennedy pathway to the maintenance of the level of glycerophospholipids we subjected mitochondria and cell homogenate of wild-type, *Δpsd1* and *Δpsd2* single mutants, and a *Δpsd1 Δpsd2* double mutant to glycerophospholipid analysis (Table 1).

Deletion of *PSD1* led to a significant depletion of PtdEtn to about one third of wild-type in mitochondria and whole cells. Thus, Psd1p is not only the PSD with highest activity *in vitro* but also the major source of total cellular and mitochondrial PtdEtn *in vivo*. PtdEtn formation via Psd2p or Kennedy pathway is less efficient. The decreased formation of PtdEtn in *Δpsd1* or *Δpsd1 Δpsd2* is compensated by an increased synthesis of PtdCho and PtdIns. In mitochondria of the *Δpsd1 Δpsd2* double mutant PtdEtn drops to a minimum level. Surprisingly, this decrease is not accompanied by a further accumulation of PtdSer as compared to the *Δpsd1* mutant. The surplus of CDP-DAG in the double mutant appears to be used rather for PtdIns production. Interestingly, the amount of cardiolipin (CL) decreased in mitochondria of *Δpsd1 Δpsd2*. This finding may indicate changes in the activity of mitochondrial enzymes involved in CL formation due to the altered lipid environment. Alternatively, a more general dysfunction of mitochondria may disturb formation of CL in the double mutant strain.

The phospholipid profile of the *Δpsd2* cell homogenate is similar to the wild-type. However, there are some intriguing differences in the mitochondrial phospholipid composition of this strain. In spite of an intact mitochondrial Psd1p, there is still an accumulation of PtdSer. The decrease of PtdCho in mitochondria of *Δpsd2* suggests that PtdEtn produced by Psd2p may preferentially be methylated to yield PtdCho. Thus, the role

Table 1: Phospholipid composition of homogenate and mitochondria of wild-type, *Δpsd1* and *Δpsd2* single mutants, and *Δpsd1 Δpsd2* double mutant grown on 2 % lactate media.

	Mol % glycerophospholipids in cell homogenate						
Strain	PA	PtdSer	PtdIns	PtdCho	PtdEtn	CL	Ly
Wild-type	2	6	17	53	19	3	0
Δpsd1	2	7	21	58	8	3	1
Δpsd2	1	7	15	55	17	3	2
Δpsd1 Δpsd2	2	7	27	60	1	2	1

	Mol % glycerophospholipids in mitochondria						
Strain	PA	PtdSer	PtdIns	PtdCho	PtdEtn	CL	Ly
Wild-type	1	1	13	51	25	7	2
Δpsd1	1	4	15	62	7	7	3
Δpsd2	1	4	12	45	27	9	1
Δpsd1 Δpsd2	0	3	25	66	1	4	1

PA: phosphatidic acid; PtdSer: phosphatidylserine; PtdIns: phosphatidylinositol; PtdCho: phosphatidylcholine; PtdEtn: phosphatidylethanolamine; CL: cardiolipin; Ly: lysophospholipids.

of Psd2p may not be primarily to supply the cell with large amounts of PtdEtn, but to form the substrate for PtdEtn methylation. Since Psd2p and enzymes of PtdEtn methylation are located to the same or at least related compartments, the site of synthesis pre-determines the pool of PtdEtn which is preferentially converted to PtdCho. Thus, Psd2p may also play an important role in the maintenance of the cellular phospholipid composition.

3.2. Contribution of Psd1p, Psd2p and the CDP-ethanolamine pathway to formation of mitochondrial PtdEtn *in vivo*

To further distinguish between the contribution of Psd1p, Psd2p and the Kennedy pathway to the formation of mitochondrial PtdEtn and PtdCho, double labeling experiments *in vivo* with subsequent cell fractionation were performed. Spheroplasts of *Δpsd1* and *Δpsd2* deletion mutants and wild-type were labeled with [^{3}H]serine and [^{14}C]ethanolamine or [^{14}C]choline, respectively, and the ratio of ^{3}H/^{14}C in PtdEtn and PtdCho was measured. ^{3}H in PtdEtn and PtdCho results from decarboxylation of PtdSer by Psd1p and Psd2p, whereas ^{14}C is incorporated into phospholipids through the Kennedy pathway. Thus, a high ^{3}H/^{14}C ratio is indicative of predominant formation of the lipids via decarboxylation.

The low ^{3}H/^{14}C ratio in all subcellular fractions of *Δpsd1* as compared to wild-type and *Δpsd2* confirmed that Psd1p-dependent decarboxylation is the major route of PtdEtn formation (Fig. 2). In wild-type, the higher ratio of ^{3}H/^{14}C in mitochondrial PtdEtn as compared to PtdEtn of the homogenate indicates that PtdEtn of mitochondria is mainly formed via decarboxylation by Psd1p. Surprisingly, deletion of *PSD2* also leads to a significant decrease of decarboxylation-derived PtdEtn in mitochondria versus the Kennedy pathway. Thus, supply of extramitochondrial PtdEtn, synthesized either by Psd2p or via Kennedy pathway, significantly contributes to the maintenance of the mitochondrial PtdEtn level, at least under conditions of Psd1p deficiency.

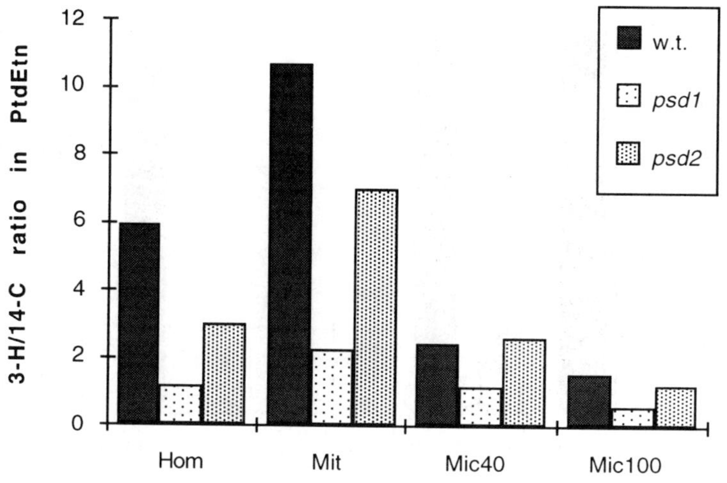

Fig. 2: $^{3}H/^{14}C$ ratio in PtdEtn isolated from cell homogenate (Hom), mitochondria (Mit), 40,000 x *g* microsomes (Mic40) and 100,000 x g microsomes (Mic100) of wild-type, *Δpsd1* and *Δpsd2*. Spheroplasts were labeled for 30 min with [^{3}H]serine and [^{14}C]ethanolamine, and lipids of isolated organelles were analyzed as described in the Methods section.

In microsomes, PtdEtn derived from the Kennedy pathway is more abundant than in mitochondria. This result is in line with the localization of Kennedy pathway enzymes in the microsomal fraction. When Psd1p is missing, the relative contribution of the Kennedy pathway to microsomal PtdEtn formation becomes even more pronounced. This result indicates that PtdEtn newly synthesized by Psd1p in mitochondria is efficiently exported to the endoplasmic reticulum. Only a minor effect on the formation of microsomal PtdEtn was observed when Psd2p was missing.

3.3. Contribution of Psd1p, Psd2p and the CDP-choline pathway to formation of mitochondrial PtdCho *in vivo*

The nearly constant $^{3}H/^{14}C$ ratio in PtdCho of cell homogenate, mitochondria and microsomes indicates that there is only one major pool of PtdCho in yeast cells from which this phospholipid is uniformly distributed among cellular membranes (Fig. 3). This observation may be result of the localization of both PtdCho biosynthesis pathways, methylation of PtdEtn and the Kennedy pathway, to related compartments. Certain steps of the latter pathway were detected in the Golgi [10] which is in line with regulation of PtdCho synthesis accomplished by Sec14p in this compartment [11]. Methylation of PtdEtn occurs preferentially in a microsomal fraction enriched for endoplasmic reticulum [6].

Surprisingly, the decrease of $^{3}H/^{14}C$ in PtdCho of total membranes and mitochondria is more pronounced with *Δpsd2* than with *Δpsd1*. We propose that PtdEtn formed by Psd2p is the preferred substrate for methylation to PtdCho. This finding is in agreement with the decrease of PtdCho in mitochondria isolated from *Δpsd2* (see Table 1).

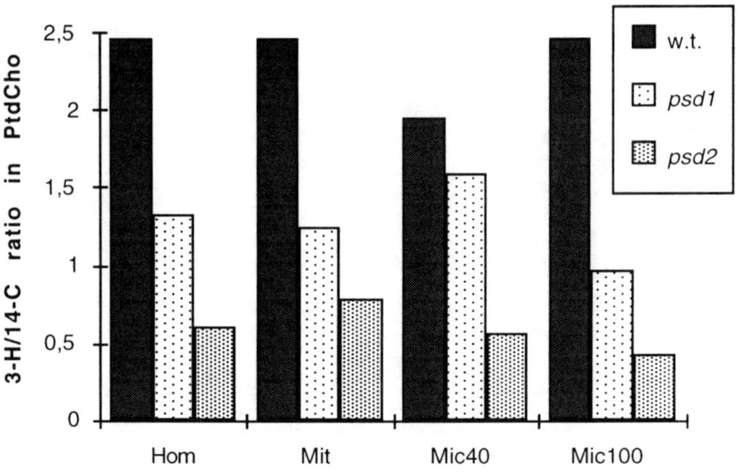

Fig. 3: $^3H/^{14}C$ ratio in PtdCho isolated from cell homogenate (Hom), mitochondria (Mito), 40,000 x *g* microsomes (Mic 40) and 100,000 x g microsomes (Mic 100) of wild-type, *Δpsd1* and *Δpsd2*. Spheroplasts were labeled for 30 min with [^{3}H]serine and [^{14}C]choline, and lipids of isolated organelles were analyzed as described in the Methods section.

3.4. Is the Kennedy pathway affected by deletion of *PSD1* or *PSD2 in vivo*?

Decrease of the $^3H/^{14}C$ ratio in PtdEtn or PtdCho as described above may be caused either by a decreased cellular PSD activity as assumed to occur in *Δpsd1* and *Δpsd2* deletion strains, or by an up-regulation of the activities of Kennedy pathway enzymes. To address this question absolute values of ^{3}H and ^{14}C in PtdEtn and PtdCho after labeling cells with [^{3}H]serine and [^{14}C]ethanolamine or [^{14}C]choline, respectively, were compared.

Table 2A: Incorporation of [^{3}H]serine and [^{14}C]ethanolamine into PtdEtn.

dpm in PtdEtn/10 mg protein of the cell homogenate		
Strain	^{3}H	^{14}C
Wild-type	1,192	207
Δpsd1	174	153
Δpsd2	531	176

Table 2B: Incorporation of [^{3}H]serine and [^{14}C]choline into PtdCho.

dpm in PtdCho/10 mg protein of the cell homogenate		
Strain	^{3}H	^{14}C
Wild-type	1,085	447
Δpsd1	684	493
Δpsd2	344	569

As can be seen from Table 2 the altered $^3H/^{14}C$ ratio in *Δpsd1* and *Δpsd2* deletion strains results mainly from the lowered incorporation of ^{3}H into phospholipids caused by a decreased decarboxylation rate. We conclude from this result that deletion of *PSD1* and *PSD2* does not affect the Kennedy pathway to a marked extent.

4. Conclusion

Three biosynthetic pathways of the yeast lead to the formation of the same product, PtdEtn, which is one of the major phospholipids of all yeast subcellular membranes. Different pools of PtdEtn, however, appear to fulfill different roles. PtdEtn formed by the mitochondrial Psd1p is retained in this compartment to a large amount. Psd2p derived PtdEtn formed in the microsomal fraction is a preferred substrate for methyltransferases yielding PtdCho. The Kennedy pathway appears to fill the gaps when decarboxylation of PtdSer is decreased, e.g., in mutants deficient in one or both of the PtdSer decarboxylases. It has to be noted, however, that the role of PtdEtn pools as described above is preferential but by far not strict. Thus, PtdEtn formed in mitochondria can be efficiently exported to the endoplasmic reticulum, and extramitochondrially synthesized PtdEtn can be imported into mitochondria when needed. The flexibility of these cellular systems which include communication between organelles and translocation of lipids between subcellular compartments guarantees efficient balance of lipid composition in organelle membranes. The spatial separation of redundant or equivalent pathways provides an additional regulatory tool to maintain cellular phospholipid levels.

Acknowledgment

This work was supported by the Fonds zur Förderung der wissenschaftlichen Forschung in Österreich (project 12076).

References

[1] Daum, G. and Vance, J. (1997) Prog. Lipid Res. 36, 103-130
[2] Trotter, P. J., Pedretti, J. and Voelker, D. R. (1993) J. Biol. Chem. 268, 21416-21421
[3] Trotter, P. J. and Voelker D. R. (1995) J. Biol. Chem. 270, 6062-6070
[4] Wach, A., Brachat, A., Pöhlmann, R. and Philippsen, P. (1994) Yeast 10, 1793-1808
[5] Daum, G., Boehni, P. C. and Schatz, G. (1982) J. Biol. Chem. 257, 13028-13033
[6] Zinser, E., Sperka-Gottlieb, C. D. M., Fasch, E.-V., Kohlwein, S. D., Paltauf, F. and Daum, G. (1991) J. Bacteriol. 173, 2026-2034
[7] Zinser, E. and Daum, G. (1995) Yeast 11, 493-536
[8] Folch, J., Lees, M. and Sloane-Stanley, G. H. (1957) J. Biol. Chem. 226, 497-509
[9] Broekhuyse, R. M. (1968) Biochim. Biophys. Acta 152, 307-315
[10] Leber, A., Hrastnik, C. and Daum, G. (1995) FEBS Lett. 337, 271-274
[11] Fang, M., Marcos, R. P. and Bankaitis, V. A. (1998) Biochim. Biophys. Acta 1404, 85-100

Structure and Function of Sphingolipid and Cholesterol-Rich Membrane Domains and of GPI-anchored Proteins

Deborah BROWN
Department of Biochemistry and Cell Biology, SUNY at Stony Brook
Stony Brook, NY 11794-5215, USA

Abstract. Lipids do not always mix uniformly in membranes, but can cluster to form microdomains. We will consider one type of microdomain that can form in cell membranes. These are enriched in cholesterol and sphingolipids, and are referred to as rafts. Rafts probably exist in membranes in the liquid-ordered phase or a phase with similar properties. We will briefly review membrane lipid phase behavior, and the differences between liquid-crystalline, liquid-ordered, and gel phase membrane bilayer domains. We will present evidence suggesting that phospholipid-rich liquid-crystalline phase domains and sphingolipid-rich liquid-ordered phase domains (rafts) can exist in equilibrium in biological membranes, especially the plasma membrane. Preferential partitioning of membrane proteins into rafts can affect function. Among the proteins that are targeted to rafts are those anchored in the outer leaflet of the membrane through covalent attachment to a special glycolipid, glycosyl phosphatidylinositol (GPI). GPI-anchored proteins are ubiquitous among eukaryotes, although they are expressed most abundantly by certain unicellular organisms. The GPI anchor core is synthesized as a unit, and then added *en bloc* to the protein on the inner surface of the endoplasmic reticulum. The protein is then transported via the secretory pathway to the cell surface. Although GPI-anchored proteins have a wide variety of functions, it has been difficult to define a general function for GPI anchorage, or to determine why certain proteins are GPI-anchored. We will focus on one property of GPI-anchors; the ability to target proteins to rafts. Targeting of GPI-anchored proteins and other proteins to rafts plays a role in signal transduction in hematopoietic cells, and possibly also in sorting in intracellular membranes and regulation of cell-surface proteolysis in other mammalian cells.

1. Introduction

The fact that membrane lipids have such heterogeneous structures suggests that they may mix non-randomly in the bilayer to form lipid microdomains. However, despite much interest in the subject, good evidence that lipids cluster in cell membranes has been slow to emerge. We will discuss one type of microdomain that has recently been described in cell membranes. These are rich in cholesterol and sphingolipids, and probably exist in membranes in the liquid-ordered (l_o) phase or a phase with similar properties. These domains have also been called rafts. The fact that rafts are in the l_o phase gives them a surprising but technically useful property; they are resistant to solubilization by non-ionic detergents such as Triton X-100 in the cold. Thus, after extraction of cells in detergent, insoluble rafts can be separated from fully-solubilized non-raft proteins and lipids and isolated for further analysis. We will use the term raft to refer to the domain in the intact membrane, and the term DRM to refer to the structure isolated by detergent insolubility.

Certain proteins, including those anchored in membrane by glycosyl phosphatidylinositol (GPI) and those linked to dual saturated acyl chains such as palmitate and myristate, are targeted to rafts. We will discuss GPI-anchorage, an unusual form of membrane anchorage, in detail, covering synthesis and the types of proteins that are GPI-anchored. We will also discuss the functions of GPI-anchorage, focusing on the ability of this modification to target proteins to lipid rafts.

2. Phase Behavior of Phospholipids, Sphingolipids and Cholesterol in Bilayers

Because rafts are probably in a separate phase from the rest of the bilayer, we will briefly review membrane lipid phase behavior. Artificial bilayers consisting of one pure phospholipid species exist in a "frozen", ordered, gel phase at low temperatures. Acyl chains of gel-phase lipids are extended and packed tightly together, and lipids have low rates of rotational mobility and of diffusion in the bilayer. These frozen acyl chains melt at a melting temperature (T_m) that is characteristic of each phospholipid species. Above this temperature, the bilayer is present in a phase, termed liquid-crystalline (l_c) or liquid-disordered (l_d), in which the lipid acyl chains are fluid and disordered. In the l_c phase, the acyl chains are disordered and do not pack tightly together. They also have high rates of rotational and lateral mobility in the bilayer. Binary mixtures of phospholipids with different T_m can be examined at temperatures between the T_m of the two lipids. When one component is present at low levels, the mixture is uniform, and generally remains in the phase favored by the major component. Above a threshold concentration of one component, cooperative phase separation occurs, and gel and l_c phase domains can coexist. In these two-phase systems, the gel phase is enriched in the high-T_m lipid, and the l_c phase is enriched in the low-T_m lipid. Phase separation in these mixtures is affected by temperature; high temperatures favor the l_c phase, while low temperatures favor the gel phase. Whenever two phases are present, the lipid composition of each phase remains constant. Thus, as the lipid composition and the temperature are changed, only the relative amount of the two different phases in the membrane changes.

Eukaryotic cell membranes contain mixtures of glycerolipids (in mammalian cells, all phospholipids except sphingomyelin), sphingolipids, and sterols. Biological glycerolipids generally have very low T_m, while sphingolipids (especially glycosphingolipids) have much higher T_m. This disparity has suggested that phase separation between glycerolipid- and sphingolipid-rich domains (possibly in the l_c and gel phases respectively) might occur in membranes in cells. Because lipids in the gel phase are insoluble in detergent (1), it might be imagined that DRMs are gel-phase domains. However, although phase separation between l_c and gel phases has been well characterized in model membranes, the gel phase does not appear to exist in biological membranes except in unusual cases (2). In addition, DRMs are rich in cholesterol (3), making it unlikely that they are present in the gel phase. However, another type of phase separation has also been described in model membranes (2). This is phase separation between two fluid phases in binary mixtures of a saturated-chain phospholipid (which has a high T_m) and cholesterol (4). In this case, above the T_m of the phospholipid, a liquid-ordered (l_o) phase can separate from the l_d phase as the amount of cholesterol is increased. Acyl chains of lipids in the l_o phase have properties that are intermediate between those of the gel and l_d phases. They are extended, ordered, and tightly packed together, as in the gel phase, but have high lateral mobility in the bilayer, as in the l_d phase. Phase separation between the l_d and l_o phases also occurs in model membranes containing ternary mixtures of high- and low-T_m lipids and cholesterol, although the process is more complex (5, 6). In particular, l_c/l_o phase separation occurs at 37°C in mixtures of phospholipids, sphingolipids, and cholesterol in proportions to those found in the plasma membrane (6). This result is important because it shows that formation of l_o phase domains in the plasma membrane is plausible from a biophysical standpoint.

3. Evidence that DRMs are in the l_o Phase

Several studies suggest that DRMs isolated from cells exist in intact membranes as domains in the l_o phase. First, DRMs are in the l_o phase when they are isolated (7). Second, control experiments have addressed the possibility that detergents might introduce artifacts. One such artifact might be the following. Lipids in membranes might be in a uniform phase, with a concentration of sphingolipids (high-T_m lipids) too low to induce phase separation. However, detergent might preferentially extract low-T_m phospholipids first. As phospholipids were gradually removed, it might be imagined that the concentration of sphingolipids in the membrane might gradually increase. This process might eventually lead to phase separation when a threshold concentration of sphingolipids was reached. We showed that detergent does not act in this manner in model membranes by making a series of liposomes whose phase behavior was determined by methods independent of detergent extraction (6, 8). Some of these liposomes contained sphingolipids and cholesterol, but at levels too low for phase separation. When these liposomes were extracted with detergent, no DRMs were produced. This showed that detergent does not rearrange membrane lipids to induce DRM formation in the absence of pre-existing phase separation.

Two different models of raft formation have been proposed. Our data suggest that rafts form by phase separation. Thus, structural features of lipids that affect T_m and promote phase separation will promote raft formation. One key difference between phospholipids and sphingolipids is the length and saturation of their acyl chains. Lipid head-groups can also contribute to T_m by altering the way lipids pack together. According to our model, acyl chain packing plays a key role in raft formation. Head group interactions might also be important, although these interactions may affect raft formation indirectly by affecting acyl chain packing. Cholesterol is recruited to rafts through its ability to pack tightly with lipids of high T_m when the l_o phase is formed. Another group has proposed a different model for raft structure (9, 10). According to this model, rafts are primarily clusters of glycosphingolipids held together through hydrogen-bonding or other interactions between glycosphingolipid head groups. Because cholesterol has a small head-group, it is recruited to rafts because it fits well in the gaps between bulky-headed glycosphingolipids.

We used several approaches to distinguish between these models. The first used a model membrane system that we have developed, called SCRL (sphingolipid and cholesterol-rich liposomes). SCRL contain biological phospholipids, sphingolipids (including both glycosphingolipids and sphingomyelin, which has the same phosphocholine headgroup as PC), and cholesterol. After extraction with Triton X-100, SCRL phospholipids are mostly solubilized, while substantial amounts of sphingolipids and cholesterol can be isolated in the form of DRMs. This suggests that SCRL contain both l_o and l_c phase domains. (Analysis of similar liposomes using a detergent-free method for detecting phase separation also supports the idea that phase separation occurs in SCRL (6).) To test our model of raft formation, we incorporated small amounts of two different [^{3}H]phosphatidylcholine (PC) species into SCRL. These were either dipalmitoyl PC (DPPC, with saturated acyl chains and a high T_m similar to that of sphingomyelin) or dioleoyl PC (DOPC, with unsaturated chains and a very low T_m). We then extracted the SCRL with Triton X-100 and asked whether the [^{3}H]PC species associated with DRMs. We found that [^{3}H]DOPC was mostly solubilized, while [^{3}H]DPPC was mostly associated with DRMs. Because both lipids have the same head group, the difference in their insolubility shows that acyl chain interactions can determine whether lipids are targeted to DRMs.

Although this experiment shows that lipids can associate with DRMs regardless of head group, it might be imagined that glycosphingolipid-specific head group interactions are required for the initial formation of rafts. Other lipids might then be able to associate with these rafts based on acyl chain interactions. However, two other experiments show that glycosphingolipids are not required for raft formation. The first takes advantage of similarities between the gel and l_o phases. Like lipids in DRMs isolated from cells and l_o-phase liposomes, lipids in model membranes in the gel phase are insoluble in cold Triton X-100. This is likely to result from the tight packing that is characteristic of lipids in both gel and l_o-phases. Because pure DPPC in the gel phase is insoluble (1) and forms DRMs (8), while other PC species that are in the l_c phase are detergent-soluble, glycosphingolipids are not required for DRM formation.

The second experiment to show that glycosphingolipids are not required for raft formation was done in cultured cells. A mutant melanoma cell line that lacks glycosphingolipids has been isolated (11). We showed that DRMs can be isolated efficiently from these cells (12). The protein yield and profile of DRMs from the mutant and parental lines were similar. The lipid composition of DRMs form the mutant was very similar to that of the parental line, except that sphingomyelin was upregulated to make up for the missing glycosphingolipids. This result showed that glycosphingolipids are not required for DRM formation in cells.

4. Proteins in DRMs

The physiological significance of raft formation is likely to stem from the preferential partitioning of certain proteins into rafts. It has been known for several years that proteins anchored in membranes by glycosyl phosphatidylinositol (GPI) are detergent-insoluble and are present in DRMs. Acylated proteins, which are covalently linked to fatty acids, can also be targeted to DRMs. In both cases, this is likely to occur through partitioning of order-preferring lipid modifications to rafts. These findings suggest that as a general rule, proteins that have an affinity for an ordered lipid phase will partition into rafts. We will now turn to an introduction to the structure and function of GPI-anchored proteins.

5. Introduction to GPI-anchored Proteins

Most integral membrane proteins are anchored in the lipid bilayer by one or more hydrophobic peptide segments, usually in the form of an α-helix, which span the bilayer. A subset of integral membrane proteins, however, has a different form of anchorage, and is covalently linked to glycosyl phosphatidylinositol (GPI). In almost all cases, the PI is the only hydrophobic moiety in the protein, and there is no hydrophobic membrane-spanning peptide. All eukaryotes synthesize GPI-anchored proteins, although their prevalence varies greatly between species. For instance, in many unicellular protists, most plasma membrane proteins are probably GPI-anchored. By contrast, mammalian cells typically express approximately 5-10 different GPI-anchored proteins. Nevertheless, the core anchor structure is conserved between species. Although the fact that GPI-anchorage has been conserved among all eukaryotes suggests that it plays an important function, this function is not known. Because membrane-spanning peptides can serve as perfectly good membrane anchors, why do all eukaryotes bother maintaining a complex alternate mechanism for anchoring a subset of their membrane proteins? Several species-specific possibilities have been proposed, but to date there is no satisfactory unifying explanation of the function of GPI-anchorage.

6. Synthesis of GPI-anchored Proteins

GPI-anchored proteins are synthesized by co-translational translocation across the endoplasmic reticulum (ER) membrane, exactly the same as conventionally-anchored proteins. Proteins destined for GPI-anchorage have a hydrophobic moiety of about 8-20 amino acids at the extreme C-terminus. (If you find this motif in the sequence of an unknown protein, it is a good clue that the protein may be GPI-anchored, as it is unusual in other proteins.) After translation, a transamidase in the ER membrane (with its active site on the lumenal side of the membrane) binds the protein and cleaves it roughly 10 amino acids upstream of the C-terminal hydrophobic peptide. (The cleavage sequence is not precisely conserved, although there are some general rules. It is somewhat difficult to predict the exact site of cleavage). In a concerted reaction, the transmidase covalently links the new C-terminus of the protein to a preformed GPI anchor, also located on the lumenal side of the ER membrane. I will next describe how this GPI anchor is synthesized.

GlcNAc (donated by UDP-GlcNAc) is linked in an α(1-6) linkage to the inositol moiety of membrane PI. The GlcNAc-PI is then de-acetylated, to yield one of the few examples of free GlcN known. A mannose (Man) (donated by dolichol-P-Man) is then added to the GlcN-PI in an α(1-4) linkage. Another Man is then added (also from dolichol-P-Man) to the first Man, in an α(1-6) linkage. Next, a third Man (from the same donor) is added to the second Man in an α(1-2) linkage. Finally, ethanolamine phosphate (donated by membrane phosphatidylethanolamine) is linked through the phosphate group to the 6 position of the third Man. This represents the core GPI anchor structure (conserved among eukaryotes) that is joined to the C-terminus of the protein in an amide linkage. At least the first few steps of GPI anchor synthesis occur on the cytoplasmic side of the ER membrane. At some point during synthesis, the anchor is flipped across the membrane to be added to the protein in the ER lumen.

In most cases, after linkage of the anchor to the protein, the GPI-anchored protein is further modified. Additions of other groups as side chains to the glycan are common. Many of these probably occur after transport to the Golgi. By contrast to the conserved structure of the core, the modifications show a wide heterogeneity between species and cell types. The lipid moiety is also generally modified. As one modification, a palmitate chain that is linked to the inositol of the PI and serves as an additional anchor during synthesis is usually removed in the mature anchor. (Occasionally, this chain is retained in the mature species). In addition, the PI diacylglycerol-linked acyl chains undergo a process of remodeling, or replacement of the existing acyl chains by new ones. For instance, the trypanosome T. *brucei* substitutes the rare fatty acid myristate for both fatty acids in the anchor. In higher eukaryotes, mature GPI-anchored proteins generally contain two saturated acyl chains, in contrast to bulk membrane PI, which generally has one saturated and one unsaturated chain. In yeast, some GPI-anchored proteins are remodeled to contain ceramide as their hydrophobic moiety, while others that keep the PI anchor remodel their acyl chains to unusually long, saturated ones. Thus, the acyl chains of essentially all mature GPI anchored proteins that have been examined are more highly saturated than those in bulk membrane PI.

After GPI-anchorage, GPI-anchored proteins are processed in the same manner as conventionally-anchored proteins. They can be modified with N-linked oligosaccharides, and are transported rapidly along the secretory pathway. Virtually all GPI-anchored proteins are expressed on the cell surface. The only exceptions are a few proteins that are targeted to stored secretory granules near the cell surface. GPI-anchored proteins are generally targeted to the apical surface of polarized epithelial cells. It should be noted from the topology of synthesis that all GPI-anchored proteins are present on the outer, non-cytoplasmic, surface of the membrane.

7. Classes of GPI-anchored Proteins

A wide variety of proteins are GPI-anchored (13). In some cases, the specific functions of these proteins are known. However, it is difficult to discern a feature that all these proteins have in common, that might suggest a general function for GPI-anchorage. GPI-anchorage has been thoroughly studied in protozoa, including pathogens such as *Trypanosoma*, *Leishmania*, and *Plasmodium*. As GPI-anchored proteins are essential in these organisms, efforts are underway to develop therapeutic agents that specifically target the synthetic pathways in the pathogens. A substantial number of GPI-anchored proteins have been identified in these organisms, although the functions of many of these are not known. One of the best-studied GPI-anchored proteins is the variant surface glycoprotein (VSG) of *T. brucei,* a pathogen that lives in the bloodstream. Although the *T. brucei* genome has genes for hundreds of VSGs, only one is expressed at any time. This one is expressed at very high abundance and forms a thick coat around the organism, largely shielding other cell-surface proteins. Thus, the immune response mounted by the host is largely directed against VSG. Because mounting an immune response to a particular antigen is a slow process in mammals, the pathogen can successfully "outrun" the immune response by switching to synthesis of a new VSG and discontinuing synthesis of the old one at various intervals. Saccharomyces cerevisiae makes several GPI-anchored proteins. One of these is an agglutinin, involved in adhesion of cells during the mating response (14).

Mammalian cells express a variety of GPI-anchored proteins. Several are involved in adhesion or cell-cell interaction (13). These include LFA-3, heparan sulfate proteoglycan, Thy-1, and one isoform of the neural cell adhesion molecule (N-CAM). An intriguing feature of N-CAM is that organisms as diverse as humans, *Drosophila*, and the sea slug *Aplysia* all express three isoforms of the protein, one GPI-anchored and two with transmembrane domains. This suggests that an important specific function for the GPI-anchored form is waiting to be discovered. Another broad class of GPI-anchored proteins in mammalian cells is made up of cell-surface hydrolytic enzymes. These include alkaline phosphatase, 5'-nucleotidase, acetyl-cholinesterase (expressed on the surface of erythrocytes to break down any of the neurotransmitter acetylcholine inadvertently released into the bloodstream during neurotransmission), trehalase, aminopeptidase P, and lipoprotein lipase. uPAR (the urokinase-type plasminogen activator receptor) serves as a cell-surface docking site for the protease uPA. The uPA/uPAR complex fits in two classes; it has hydrolytic activity and is also an adhesion molecule, as uPAR binds vitronectin (15). Although not all cell-surface hydrolytic enzymes are GPI-anchored, the substantial number that are suggests a possible role for GPI-anchorage in the activity or regulation of these proteins. Other "miscellaneous" GPI-anchored proteins of known function on mammalian cells include the following (13); CD14, the monocyte receptor for bacterial lipopolysaccharide; FcγRIII, a receptor for the Fc portion of IgG on human neutrophils; the scrapie prion protein; the folate receptor; one subunit of the receptor for ciliary neurotrophic factor on neurons (16); and decay-accelerating factor (DAF), which breaks down complement inadvertently released into the bloodstream during the immune response.

The ability to synthesize GPI-anchored proteins appears to be essential for development in humans, as no patients completely lack GPI-anchored proteins. Synthesis of GPI-anchors is also essential for the yeast *S. cerevisiae*. (By contrast, several mammalian cell types that cannot synthesize GPI-anchors can be grown in tissue culture, showing that GPI-anchors are not essential for the growth of individual mammalian cells.) However, there is a human disease that results from partial loss of GPI-anchor synthesis. In patients with paroxysmal nocturnal hemoglobinuria (PNH), somatic mutations in stem cell progenitors of erythrocytes block synthesis of GPI-anchored proteins in these cells by blocking the first step in the synthesis of GPI anchors (17). Because these cells lack DAF, they are lysed easily by complement, and patients suffer from anemia.

8. Functions of GPI-anchorage

As we have seen, GPI-anchored proteins are expressed in a wide variety of cell types in a wide variety of organisms. The specific functions of some GPI-anchored proteins are known. However, no overall pattern emerges from these specific functions to suggest a general function for GPI-anchorage. GPI-anchored proteins can diffuse in the membrane more rapidly than transmembrane proteins, and it was originally thought that this was a function of the modification. However, further studies showed that GPI-anchored proteins do not always diffuse rapidly (18, 19), and few examples of functional significance of rapid diffusion of GPI-anchored proteins have emerged. In some cases, individual organisms have adapted GPI anchorage to what appear to be organism-specific purposes. For instance, the unusually short myristic acid chains on VSG in *T. brucei* may allow these proteins to be shed from the membrane into the host bloodstream at a high rate, to slow the immune response. In addition, *S. cerevisiae* cleaves the GPI-anchor after transport of some GPI-anchored proteins to the cell surface, and use the products in cell wall biosynthesis (14). In mammalian cells, it has been suggested that GPI-anchored proteins may serve as precursors for downstream mediators of the insulin response (20, 21). Thus, GPI-anchored proteins can be cleaved and released from muscle and fat cells in response to insulin, and inositol phosphoglycans that may be formed from cleavage of the anchor can mediate signaling. In addition to these organism-specific examples, it is tempting to speculate that a more "universal" function of GPI-anchorage may also exist.

9. GPI-anchored Proteins in Membrane Rafts

GPI-anchorage can target proteins to rafts. It has become clear in the last few years that targeting to rafts is an essentially universal property of GPI-anchored proteins in mammalian cells. Proposed function(s) of this targeting, which are still being elucidated, include sorting in polarized epithelial cells, signal transduction, and regulation of cell-surface hydrolytic activity.

GPI-anchored proteins have long been known to be insoluble in non-ionic detergents such as Triton X-100 in the cold. The basis of this property was not understood until recently. Integral membrane proteins are generally detergent-soluble unless they are linked to cytoskeleton. However, GPI-anchored proteins do not have access to cytoskeleton, as they are located in the outer leaflet of the plasma membrane. Detergent-insoluble GPI-anchored proteins are enriched in DRMs (3). This finding led us to postulate that partitioning of the saturated acyl chains on GPI-anchored proteins into the l_o phase targets the proteins to rafts. Alternatively, it might be imagined that protein-protein contacts might be required for detergent-insolubility of GPI-anchored proteins. To test these possibilities, we purified a GPI-anchored protein, placental alkaline phosphatase (PLAP), and incorporated it into SCRL liposomes that (as discussed above) probably contain both detergent-soluble l_c and detergent-resistant l_o phase domains. When we incorporated PLAP into these liposomes, subjected them to detergent extraction, and separated DRMs and the Triton-soluble fractions, we found that PLAP associated with the DRMs. This result showed that the protein interacted directly with lipids in the l_o-phase (22). Other GPI-anchored proteins behaved the same way, while a control transmembrane protein was fully solubilized (8).

As discussed above, although our data suggested that acyl chain packing was the most important driving force in raft formation, another group had suggested that head-group interactions (possibly hydrogen bonds) between glycosphingolipid head groups were most important (10). According to this model, the glycan in the GPI anchor of PLAP might interact with glycosphingolipid head groups in order to target the protein to rafts. To distinguish between these possibilities, we incorporated PLAP into dipalmitoylphosphatidylcholine (DPPC) liposomes in the gel-phase, subjected them to detergent extraction, and isolated DRMs. (As a control, PLAP was incorporated into l_c phase dioleoyl phosphatidylcholine (DOPC) liposomes). We found that PLAP associated with gel-phase DPPC DRMs, but was solubilized from DOPC liposomes (8). To show that PLAP preferred the gel phase to the l_c phase if given a choice, the protein was next incorporated into two-phase gel/liquid-crystalline liposomes containing both DPPC and DOPC. As predicted, PLAP also associated with DPPC-rich DRMs isolated from these liposomes (8). These results showed that interactions between GPI-anchored proteins and glycosphingolipid head-groups are not required for partitioning of GPI-anchored proteins into ordered-phase lipid rafts. Experiments using the mutant glycosphingolipid-negative melanoma mutant cell line also supported this conclusion. PLAP stably expressed in these cells associated with DRMs as efficiently as it did in the parental line (12).

10. Functions of Rafts and Functions of Raft-Targeting of GPI-anchored Proteins

Rafts have been proposed to form in the *trans*-Golgi network, and to function during sorting of apical and basolateral proteins into separate transport vesicles in polarized epithelial cells (9, 10). In what appears to be an analogous process, rafts have also been implicated in axonal versus dendritic sorting of proteins in neurons (23). The finding that GPI-anchored proteins are generally targeted to the apical membrane in epithelial cells (24) suggested that the GPI anchor might act as an apical targeting signal. Finding that PLAP was targeted to rafts during transport through the secretory pathway was consistent with this idea (3). Studies with hybrid proteins also suggested that the GPI anchor was an apical targeting signal (25, 26). For instance, linking a GPI-anchored protein to the transmembrane and cytoplasmic domains of a basolaterally-targeted protein caused the hybrid protein to be targeted to the basolateral surface (26). By contrast, linking the ectodomain of a normally basolateral protein to a GPI anchor resulted in apical targeting (26). However, another possible explanation of these results was that the cytoplasmic domain of basolateral proteins contains "dominant" sorting information that is required for basolateral targeting. (It is now generally recognized that basolateral proteins contain cytoplasmic sorting signals (27).) In this case, proteins (such as GPI-anchored proteins) that lack these signals might be targeted apically by default, or through recognition of other apical targeting signals, such as glycosylation (28). Hybrid proteins in which PLAP was linked to transmembrane and cytoplasmic domains from apical proteins were consistent with this idea; they were still targeted apically, although they did not associate with rafts (29, 30). Because the nature of apical targeting signals is obscure, and the default pathway, or fate of proteins without targeting signals, is still not known, there is no clear evidence that GPI-anchors act as apical targeting signals (31). While the potential role of rafts in apical targeting is exciting, it remains somewhat controversial (32).

Association with rafts may regulate two cell-surface proteases. The first is the urokinase-type plasminogen activator (uPA). uPA acts as a cell-surface protease by binding to its GPI-anchored receptor, uPAR. Treatment of cells with agents that disrupt rafts inhibits cell-surface plasmin generation by uPA, suggesting that association of uPAR with rafts may regulate uPA enzymatic activity (33). A second example is the coagulation cascade on endothelial cells. This involves tissue factor (TF), a transmembraneous protease receptor. Briefly, a proteolytically active complex formed by TF and the serine protease factor VIIa can be inhibited by binding of other components (34). During this downregulation process, the complex was shown to gain an affinity for DRMs by associating with an unidentified GPI-anchored protein (34). The complex in DRMs was inhibited more than would be expected from binding of inhibitory proteins, suggesting that the membrane environment itself has an additional inhibitory effect. Thus, regulating the activity of cell-surface proteases such as uPA and the TF complex may be an important function of rafts.

Clustering or ligation of cell-surface GPI-anchored proteins can trigger transmembrane signal transduction (31). This is surprising, because GPI-anchored proteins cannot interact directly with proteins on the cytoplasmic side of the bilayer. In at least some cases, signaling occurs through activation of Src-family non-receptor tyrosine kinases (35, 36). As these proteins are restricted to the cytoplasmic face of the plasma membrane, communication between them and GPI-anchored proteins is a puzzle.

Signaling through GPI-anchored proteins process has been best studied in T cells (31). Several factors suggest a role for rafts in this process. First, antibody-mediated crosslinking of virtually any GPI-anchored protein on the surface of T cells activates the cells, in a process that requires the GPI anchor (37) and functional membrane rafts (38). In addition, GPI-anchored proteins and the Src-family non-receptor tyrosine kinases downstream of them in the signaling pathway are in the same rafts (35, 36). Thus, targeting of both GPI-anchored proteins and Src-family kinases to rafts appears to be important in coupling them during signal transduction. It is likely that other proteins participate in this coupling as well. For instance, GPI-anchored proteins may bind in *cis* (in the plane of the membrane) to one or more unidentified transmembrane proteins. These proteins could transmit signals across the membrane, for instance by binding directly or indirectly to Src-family kinases. An alternative, highly speculative idea is that clustering of GPI-anchored proteins in the outer leaflet of the bilayer might organize or stabilize rafts in the inner leaflet, through interactions between lipids in opposite leaflets. If so, this might cause clustering and activation of Src-family kinases in the inner leaflet directly, without the need for a transmembrane "linker" protein (31).

Further work has shown that signaling through the conventional antigen receptors in both T cells and basophils (39) also occurs in rafts. This shows that the role of rafts in signal transduction is not limited to specific effects of GPI-anchored proteins. For instance, in T cells, the T cell receptor (40), associated kinases Lck and Fyn (41), the downstream adaptor protein LAT (42), and the costimulatory protein CD28 (43) are present in rafts during signaling. Mutations in Lck (44) and LAT (45) that prevent association of these proteins with rafts affect signaling through the TCR, as do agents that disrupt rafts (46).

These results show that rafts play an important role in signaling in T cells. This novel mechanism of regulating interactions between proteins by selective partitioning into membrane domains is unlikely to be restricted to one cell type. Further work is likely to demonstrate the importance of rafts in signaling in other cell types as well.

References

[1] A. A. Ribeiro, and E. A. Dennis, Effect of thermotropic phase transitions of dipalmitoylphosphatidylcholine on the formation of mixed micelles with Triton X-100, *Biochim. Biophys. Acta* **332** (1973) 26-35.

[2] D. A. Brown, and E. London, Structure and origin of ordered lipid domains in biological membranes, *J. Membr. Biol.* **164** (1998) 103-114.

[3] D. A. Brown, and J. K. Rose, Sorting of GPI-anchored proteins to glycolipid-enriched membrane subdomains during transport to the apical cell surface, *Cell* **68** (1992) 533-544.

[4] M. B. Sankaram, and T. E. Thompson, Cholesterol-induced fluid-phase immiscibility in membranes, *Proc. Natl. Acad. Sci. USA* **88** (1991) 8686-8690.

[5] J. R. Silvius, D. del Guidice, and M. Lafleur, Cholesterol at different bilayer concentrations can promote or antagonize lateral segregation of phospholipids of differing acyl chain length, *Biochemistry* **35** (1996) 15198-15208.

[6] S. N. Ahmed, D. A. Brown, and E. London, On the origin of sphingolipid-cholesterol rich detergent-insoluble domains in cell membranes: Physiological concentrations of cholesterol and sphingolipid induce formation of a detergent-insoluble, liquid-ordered lipid phase in model membranes., *Biochemistry* **36** (1997) 10944-10953.

[7] M. Ge, K. A. Field, R. Aneja, D. Holowka, B. Baird, and J. Freed, ESR characterization of liquid ordered phase of detergent resistant membranes from RBL-2H3 cells, *Biophys. J.* **In press.** (1999)

[8] R. J. Schroeder, S. N. Ahmed, Y. Zhu, E. London, and D. A. Brown, Cholesterol and sphingolipid enhance the Triton X-100-insolubility of GPI-anchored proteins by promoting the formation of detergent-insoluble ordered membrane domains, *J. Biol. Chem.* **273** (1998) 1150-1157.

[9] K. Simons, and G. van Meer, Lipid sorting in epithelial cells, *Biochemistry* **27** (1988) 6197-6202.

[10] K. Simons, and E. Ikonen, Functional rafts in cell membranes, *Nature* **387** (1997) 569-572.

[11] S. Ichikawa, N. Nakajo, H. Sakiyama, and Y. Hirabayashi, A mouse B16 melanoma mutant deficient in glycolipids, *Proc. Natl. Acad. Sci. USA* **91** (1994) 2703-2707.

[12] A. G. Ostermeyer, B. T. Beckrich, K. E. Grove, K. A. Ivarson, and D. A. Brown, Glycosphingolipids are not essential for formation of detergent-resistant membrane rafts in melanoma cells and do not determine their properties, *Submitted for publication*

[13] M. G. Low, and A. R. Saltiel, Structural and Functional Roles of Glycosyl-Phosphatidylinositol in Membranes, *Science* **239** (1988) 268-275.

[14] H. de Nobel, and P. N. Lipke, Is there a role for GPI's in yeast cell-wall assembly?, *Tr. Cell Biol.* **4** (1994) 42-45.

[15] Y. Wei, D. A. Waltz, N. Rao, R. J. Drummond, S. Rosenberg, and H. A. Chapman, Identification of the urokinase receptor as an adhesion receptor for vitronectin, *J. Biol. Chem.* **269** (1994) 32380-32388.

[16] N. Stahl, and G. D. Yancopoulos, The tripartite CNTF receptor complex: activation and signaling involves components shared with other cytokines, *J. Neurobiol.* **25** (1994) 1454-1466.

[17] T. Miyata, J. Takeda, Y. Iida, N. Yamada, N. Inoue, M. Takahashi, et al., The cloning of PIG-A, a component in the early step of GPI-anchor biosynthesis, *Science* **259** (1993) 1318-1320.

[18] F. Zhang, B. Crise, B. Su, Y. Hou, J. K. Rose, A. Bothwell, et al., The lateral mobility of some membrane proteins is determined by their ectodomains, *Biophys. J.* **62** (1992) 92-94.

[19] F. Zhang, B. Crise, B. Su, Y. Hou, J. K. Rose, A. Bothwell, et al., Lateral diffusion of membrane-spanning and glycosylphosphatidylinositol-linked proteins: Toward establishing rules governing the lateral mobility of membrane proteins, *J. Cell Biol.* **115** (1991) 75-84.

[20] D. E. Misek, and A. R. Saltiel, An inositol phosphate glycan derived from a Trypanosoma brucei glycosyl-phosphatidylinositol mimics some of the metabolic actions of insulin, *J.Biol.Chem.* **267** (1992) 16266-16273.

[21] D. F. Lazar, J. J. Knez, M. E. Medof, P. Cuatrecasas, and A. R. Saltiel, Stimulation of glycogen synthesis by insulin in human erythroleukemia cells requires the synthesis of glycosyl-phosphatidylinositol, *Proc. Natl. Acad. Sci. USA* **91** (1994) 9665-9669.

[22] R. Schroeder, E. London, and D. Brown, Interactions between saturated acyl chains confer detergent resistance on lipids and GPI-anchored proteins: GPI-anchored proteins in liposomes and cells show similar behavior, *Proc. Natl. Acad. Sci. USA* **91** (1994) 12130-12134.

[23] M. D. Ledesma, K. Simons, and C. G. Dotti, Neuronal polarity: Essential role of protein-lipid complexes in axonal sorting, *Proc. Natl. Acad. Sci. USA* **95** (1998) 3966-3971.

[24] M. P. Lisanti, M. Sargiacomo, L. Graeve, A. R. Saltiel, and E. Rodriguez-Boulan, Polarized apical distribution of glycosyl-phosphatidylinositol anchored proteins in a renal epithelial cell line, *Proc. Natl. Acad. Sci. USA* **85** (1988) 9557-9561.

[25] M. P. Lisanti, I. W. Caras, M. A. Davitz, and E. Rodriguez-Boulan, A glycosphingolipid membrane anchor acts as an apical targeting signal in polarized epithelial cells, *J. Cell Biol.* **109** (1989) 2145-2156.

[26] D. A. Brown, B. Crise, and J. K. Rose, Mechanism of membrane anchoring affects polarized expression of two proteins in MDCK cells, *Science* **245** (1989) 1499-1501.

[27] M. Yokode, R. K. Pathak, R. E. Hammer, M. S. Brown, J. L. Goldstein, and R. G. W. Anderson, Cytoplasmic sequence required for basolateral targeting of LDL receptor in livers of transgenic mice, *J. Cell Biol.* **117** (1992) 39-46.

[28] K. Fiedler, and K. Simons, The role of N-glycans in the secretory pathway, *Cell* **81** (1995) 309-312.

[29] J. E. Casanova, G. Apodaca, and K. E. Mostov, An autonomous signal for basolateral sorting in the cytoplasmic domain of the polymeric immunoglobulin receptor, *Cell* **66** (1991) 65-75.

[30] G. Arreaza, and D. A. Brown, Sorting and intracellular trafficking of a glycosylphosphatidylinositol-anchored protein and two hybrid proteins with the same ectodomain in MDCK kidney epithelial cells., *J. Biol. Chem.* **270** (1995) 23641-23647.

[31] D. A. Brown, and E. London, Functions of lipid rafts in biological membranes, *Annu. Rev. Cell Devel. Biol.* **14** (1998) 111-136.

[32] T. Weimbs, S. H. Low, S. J. Chapin, and K. E. Mostov, Apical targeting in polarized epithelial cells: there's more afloat than rafts, *Tr. Cell Biol.* **7** (1997) 393-399.

[33] A. Stahl, and B. M. Mueller, The urokinase-type plasminogen activator receptor, a GPI-linked protein, is localized in caveolae, *J. Cell Biol.* **129** (1995) 335-344.

[34] J. R. Sevinsky, L. V. M. Rao, and W. Ruf, Ligand-induced protease receptor translocation into caveolae: A mechanism for regulating cell surface proteloysis of the tissue factor-dependent coagulation pathway, *J. Cell. Biol.* **133** (1996) 293-304.

[35] D. Brown, The tyrosine kinase connection: how GPI-anchored proteins activate T cells, *Curr. Opin. Immunol.* **5** (1993) 349-354.

[36] A. H. Zisch, L. D'Alessandri, K. Amrein, B. Ranscht, K. H. Winterhalter, and L. Vaughan, The glypiated neuronal cell adhesion molecule contactin/F11 complexes with src-family protein tyrosine kinase Fyn, *Molec. Cell. Neurosci.* **6** (1995) 263-279.

[37] P. J. Robinson, M. Millrain, J. Antoniou, E. Simpson, and A. L. Mellor, A glycosylphospholipid anchor is required for Qa-2-mediated T cell activation, *Nature* **342** (1989) 85-87.

[38] T. M. Stulnig, M. Berger, T. Sigmund, H. Stockinger, V. Horejsi, and W. Waldhausl, Signal transduction via glycosyl phosphatidylinositol-anchored proteins in T cells is inhibited by lowering cellular cholesterol, *J. Biol. Chem.* **272** (1997) 19242-19247.

[39] E. D. Sheets, D. Holowka, and B. Baird, Critical role for cholesterol in Lyn-mediated tyrosine phosphorylation in FceRI and their association with detergent-resistant membranes, *J. Cell. Biol.* **145** (1999) 877-887.

[40] C. Montixi, C. Langlet, A.-M. Bernard, J. Thimonier, C. Dubois, M.-A. Wurbel, et al., Engagement of T cell receptor triggers its recruitment to low-density detergent-insoluble membrane domains, *EMBO J.* **17** (1998) 5334-5348.

[41] T. Cinek, and V. A. Horejsí, The nature of large noncovalent complexes containing glycosyl-phosphatidylinositol-anchored membrane glycoproteins and protein tyrosine kinases, *J. Immunol.* **149** (1992) 2262-2270.

[42] W. Zhang, J. Sloan-Lancaster, J. Kitchen, R. P. Trible, and L. E. Samelson, LAT: The ZAP-70 tyrosine kinase substrate that links T cell receptor to cellular activation, *Cell* **92** (1998) 83-92.

[43] A. Viola, S. Schroeder, Y. Sakakibara, and A. Lanzavecchia, T lymphocyte costimulation mediated by reorganization of membrane microdomains, *Science* **283** (1999) 680-682.

[44] P. Kabouridis, A. I. Magee, and S. C. Ley, S-acylation of LCK protein tyrosine kinase is essential for its signalling function in T lymphocytes, *EMBO J.* **16** (1997) 4983-4998.

[45] W. Zhang, R. P. Trible, and L. E. Samelson, LAT palmitoylation: its essential role in membrane microdomain targeting and tyrosine phosphorylation during T cell activation, *Immunity* **9** (1998) 239-246.

[46] R. Xavier, T. Brennan, Q. Li, C. McCormack, and B. Seed, Membrane compartmentation is required for efficient T cell activation, *Immunity* **8** (1998) 723-732.

Protein, Lipid and Membrane Traffic: Pathways and Targeting
J.A.F. Op den Kamp (Ed.)
IOS Press, 2000

Near Field Microscopy Approach to the Heterogeneity of Artificial and Biological Membranes

V. Vié*, N. Van Mau**, M-C. Giocondi*, E. Lesniewska***, J-P. Goudonnet***, F. Heitz**, C. Le Grimellec*.

**CBS /INSERM U414, Faculté de Pharmacie, 15 ave. Charles Flahault, 34060 Montpellier cedex, France, **CRBM, CNRS-UPR 1086, 1919 route de Mende, 34293 Montpellier cedex 5, France, ***Laboratoire de Physique, CNRS-URA 5027, UFR Sciences et Techniques, 9 rue Alain Savary, BP 400, 21011 Dijon cedex , France*

Abstract. The characterization of the two-dimensional organization of biological membranes remains a major issue for the understanding of the cellular functions regulation. An essential reason for this situation is the limited access of usual detection techniques to the mesoscopic scale, where the membrane lateral heterogenity phenomenoms are likely to take place. Recent data show that atomic force microscopy (AFM) seems to be well suited for filling this gap between molecular and light microscopy resolutions. AFM imaging of phase-separated lipid mixtures allows to determine the size, the shape of domains and their kinetics of growth. It also provides a direct access to the structural arrangement of an amphipathic peptide interacting with lipids, to the localization of a protein between lipids domains, and to the molecular structure of proteins with a resolution better than 1 nm in liquid medium. AFM images of the surface of intact or Triton X-100 extracted cells reinforce the view that this new technique might be of help in the understanding of structure-function relationships in biological membranes.

1. Introduction

Influence of lipids on the activity of membrane proteins has been extensively studied in the 70's and 80's. In numerous physiological and pathological situations, modifications in the activity of enzymes and transport systems, the accessibility of receptors and signals transmission are associated or correlated with changes in the membrane lipid physical state and composition [1-3]. The key role for the vectorial function of epithelial cells of the segregation of both proteins and lipids in apical and basolateral macrodomains well illustrates the importance of the lipidic matrix on the function of the cell membrane [4, 5].

The concept of lipid microdomains in biological membranes was proposed only three years after the fluid mosaic model of membrane structure [6, 7]. Data accumulated over the last past 25 years on a variety of cell types have strenghtened the view that specialized lipid domains can exist in membranes. This holds true for intestinal and renal epithelial cells, where determination of membrane lipid composition and physical state early strongly supported the existence of an heterogeneous lateral distribution of membrane constituents. Thus fluorescence polarization, fluorescence lifetime, ESR, DSC experiments on isolated renal and intestinal apical membranes [4,5], and later on intact cells [8], all showed the existence of phase separations close to, or including, the physiological temperature. Such biophysical evidences were however generally not found in non-epithelial cells where the existence of domains has been essentially inferred from FRAP, digital fluorescence microscopy, optical tweezers, single particle tracking as well as from biochemical, including detergent extraction, and morphological (EM) approaches [9-20]. Numerous questions about these microdomains remain open, even in model systems, like their size(s) and homogeneity, the respective role of lipids and proteins in their formation and growth, the relationships between microdomains found on the external leaflet and the presence of particular proteins on the internal leaflet. One reason for the difficulties encountered in the microdomains characterization is their probable size which might vary from ~50 nm to the micrometer scale. This range of sizes makes the observation of domains unpractical with usual techniques.

A new generation of microscopes, the near field (or scanning probe) microscopes (SXM) devoted to the study of surface, was born in 1982 with the invention of the scanning tunneling microscope (STM) whose resolution allows the imaging of single atoms [21]. In near field microscopes, a sharp tip is brought into close proximity to the sample, such as a specific interaction between the tip and the sample occurs. The tip raster scans the sample surface and a three-dimensional map that reflects the variation of the local interaction between the tip and the object is generated. So far, the biological applications of the STM have been limited because of the tunneling current which practically imposes to work on conducting samples. Among the members of the expanding family of SXM, the atomic force microscope (AFM, ref 22), also known as the scanning force microscope (SFM) can visualize non conducting samples, albeit at a lower resolution (0.1 nm and 0.01 nm for lateral and vertical resolution, respectively, on hard samples), in both air and aqueous buffer. These properties explain the interest of biologists for this particular SXM and excellent reviews are now available that illustrate the numerous biological applications of the AFM [23-32]. Following a brief presentation of the apparatus, we will focus our attention on recent data which suggest that this new technique can be helpfull in the understanding of membrane domains.

2. The AFM

2.1. Principle and operating modes

The principle of operation of the AFM is rather simple. Like other SXM, the AFM uses piezo ceramics to position the tip or the sample to an accuracy of fractions of atomic diameters. A detection system monitors the motion of the tip, mounted at the end of a soft cantilever spring. In « contact » mode AFM, the contrast mechanism results from the hard-core repulsion between tip and sample. However, additional contributions stem from van der Waals, electrostatic interactions and hydration forces. Upon tip-sample repulsion, the cantilever bends. The amplitude of the bending is a function of the cantilever spring constant. Vertical movement of less than 0.1 nm of the cantilever can be recorded using an optical lever composed of a laser diode beam focused on the back of the cantilever whose light is deflected on a split photodiode (Fig.1).

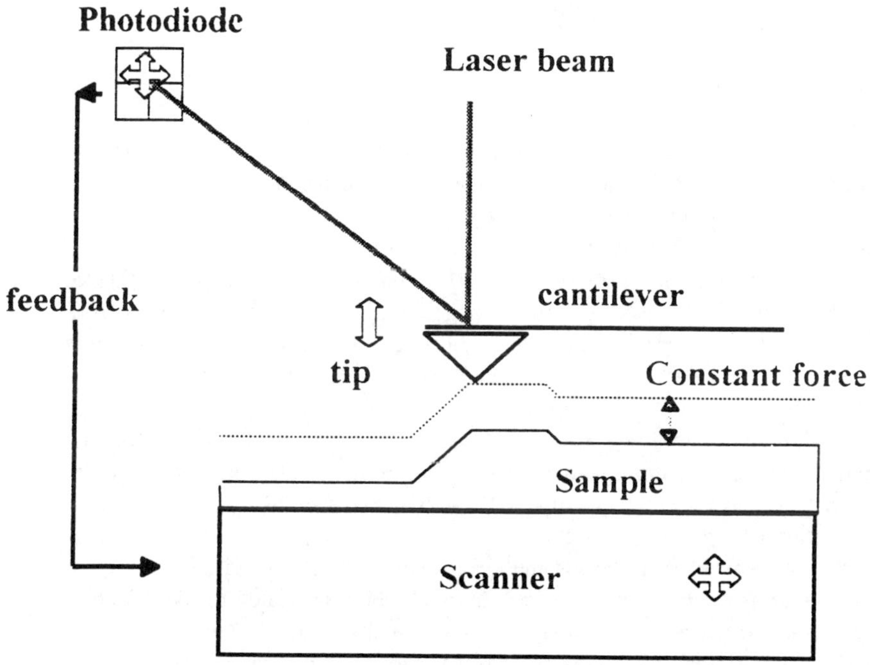

Figure 1. Schematics of an atomic force microscope.

Imaging of biological material requires the use of imaging forces as low as possible. For instance, for imaging 2D crystals of membrane proteins with minimal distortion, scanning forces below 200 pN have to be used. Even lower scanning forces, down to 20-50 pN, are recommended when imaging living cells [33]. This can be achieved with cantilevers whose spring constants are 0.1 N/m or below. Such cantilevers with integrated Si_3N_4 tips with spring constants as low as 0.01 N/m are microfabricated using semiconductor technology. The scanning force F can be estimated using F ≅ tip deflection x cantilever spring constant. Thus a 1 nm tip deflection using a 0.1 N/m cantilever will correspond to a 100 pN force. To maintain the imaging force constant, a feedback mechanism is used to adjust the sample height while scanning. The resulting images are isoforce reliefs of the sample. Lateral or frictional forces during scanning remain less well understood and can be deleterious when the sample does not firmly adhere to the support. It has to be reminded that AFM can only image objects that are deposited and adhere onto a flat support, like glass, mica, HOPG, etc. Unfortunately, many biological samples only weakly adsorb to the substrate surface and are easily whipped off during scanning. Tapping mode AFM (TMAFM, ref 34) has been developed to improve the imaging of such weakly adsorbed samples. In this mode, the cantilever is driven to oscillate at, or close to, its resonant frequency. As soon as the tip comes into contact with the surface, the amplitude of the oscillation is changed owing to the damping introduced by the surface and this amplitude change is used as the feedback signal. Thus, the oscillating tip touches the surface once each cycle, minimizing the friction forces present in contact mode [29].

2.2. AFM images

When scanning a sample, different signals related to different properties of the tip-sample interaction can be recorded. They provide the three dimensional images related to variations of the parameter chosen along the surface.

A) Height images. Basically, AFM provides an isoforce three dimensional map of surface. Considering a material with homogeneous surface properties, by keeping the force constant during scanning it is possible to obtain images with sample height subangstrom accuracy. For soft biological surfaces however, the local mechanical properties and electrostatic interactions between the AFM tip and the sample influence the measured height.

B) Deflection and amplitude images. Recording of the feedback signals, i.e deflection in contact mode and amplitude in TMAFM, provides sharper details of the surface features and gives images that resemble to those obtained with SEM, at the expense of the height information.

C) Lateral force (Friction) images. Using a four-elements photodiode, the torsional motion of the cantilever, in contact mode, can be recorded which is related to local friction properties of the surface. Interpretation of images on biological samples remains limited.

D) Phase. Upon tip-sample interaction in TMAFM, a change in the phase of oscillation occurs which reports on the energy dissipation due to the interaction and is related to local viscoelastic properties.

E) TMAFM deflection. This signal likely provides information on the local stiffness [35].

3. Phase separation and domains in lipid monolayers

Lipid-lipid immiscibility is one of the favored hypothesis to explain microdomains formation in biomembranes [18-20]. Because of the asymetric distribution of lipids between the outer and inner membrane leaflets and the common assumption that the two membrane leaflets are essentially uncoupled, lipid monolayers represent a good model when trying to define the parameters involved in microdomains properties. Fluorescence microscopy is an important tool for the study of the morphology and phase behavior of Langmuir monolayers, but its lateral resolution is limited and does not allow the detection of submicrometer size domains [36]. Much higher resolution images of monolayers are possible with the AFM which is now widely used by physicists and chemists involved in the processes of self-assembly in two-dimensions. Indeed, AFM observation of monolayers generally requires that the film is transferred to a solid substrate, by Langmuir-Blodgett (LB) or similar deposition techniques [37-39]. Whereas examination of POPC films (30 mN/m) reveals a homogeneous surface, images of brain SM transferred in identical conditions show the presence of branched domains 100-200 nm width, protruding by ~ 1 nm above the matrix (Fig. 2).

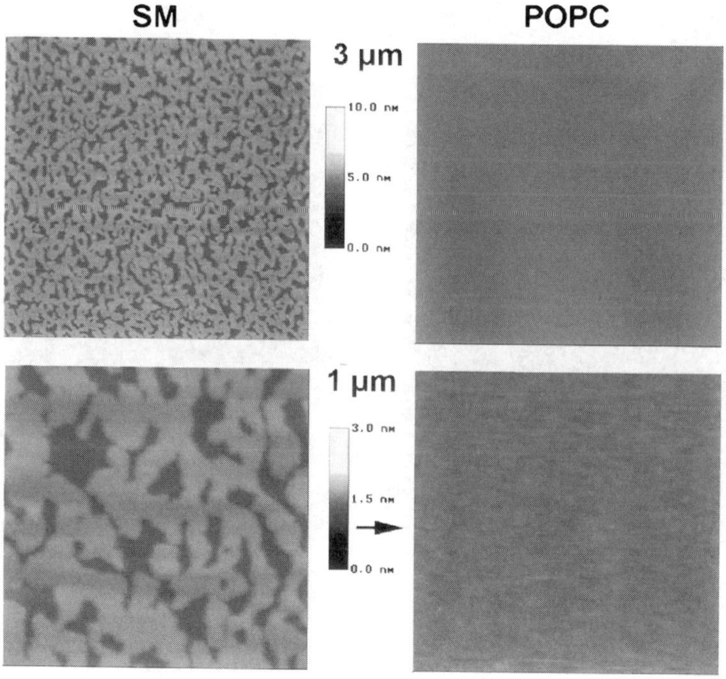

Figure 2. AFM topographical images of sphingomyelin (SM) and palmitoyl-oleoyl phosphatidylcholine (POPC). Note that for the POPC 1 µm scan the height scale has been reduced to 3.0 nm.

Studies on phase-separated thin films including mixed phosphatidyl ethanolamine and phosphatidylcholine [40,41], ceramide-cholesterol films also clearly show [42], besides the presence of large macrodomains, the existence of microdomains of mesoscopic size, undetectable by light microscopy. Both macro and microdomains can be detected on the basis of the height difference between aliphatic chains in the liquid condensed (Lc) and liquid expanded (Le, fluid) state: Lc domains emerge from the Le matrix by ~ 0.5 to 1.1 nm, according to the lipid mixture. It is worth noting that the presence of domains can be detected independently of the monolayer height, via the different frictional and energy dissipation properties of the two phases.

Sphingolipids and glycosphingolipids are likely to play an important role in microdomains formation. Their distribution between the gel and liquid-crystal phases of a two-components phospholipid mixture has been studied by indirect methods and has been a matter of debates. In a recent study, we have examined by AFM the distribution of the ganglioside G_{MI}, at low concentrations, in phase-separated 1:1 dioleoylphosphatidylcholine (DOPC)/dipalmitoyl phosphatidylcholine (DPPC) LB films transferred on mica [43]. For concentrations below 2 mole %, G_{MI} distributed exclusively in the liquid condensed DPPC phase where it formed nanometer scale patches of round and filamentous shapes both at the boundary and towards the middle of DPPC macrodomains (Fig. 3).

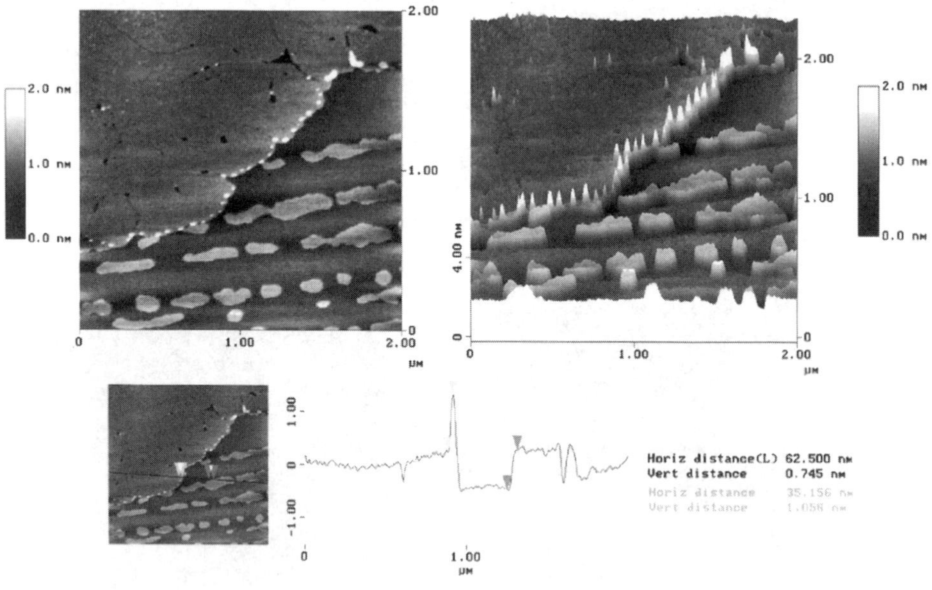

Figure 3. Localization of G_{MI} (1%) in phase separated DOPC/DPPC (1/1) mixture.

Raising the G_{MI} concentration up to 4% increased the number of patches and created fence-like structures at the DOPC/DPPC boundary whereas some G_{MI} clusters possibly emerged from the liquid expanded DOPC phase. Immiscibility of G_{MI} with DOPC was confirmed by examination of G_{M1}/DOPC LB films. These data were obtained by direct examination of the samples, without the help of G_{M1} headgroup markers, and support the view of the existence of G_{M1} enriched domains in biological membranes.

Such studies can be extended to mixed LB films made of lipids and other constituents. For instance, a phase separation has been observed, at a submicron scale, in dioleoyl phosphatidylcholine monolayers doped with increasing amounts of a primary amphipathic peptide containing a signal peptide associated with a nuclear localization sequence [44]. The AFM images further revealed that the peptide can be organized in superstructures, unpredictable from isotherms or spectroscopic analysis. Taken together, these experiments suggest that AFM constitutes an invaluable tool for the study of mesoscopic size lipid domains in LB films.

4. Examination of domains in lipid bilayers

The first images of a supported phospholipid bilayer, made of dimyristoylphosphatidyl ethanolamine by double transfer on mica were reported in 1991 [45]. Since that time few papers have been devoted to the study in aqueous medium of lipids bilayers under phase separation [46-48]. This likely results from the difficulties encountered in the obtention of stable samples, in particular when using a double tranfer technique [49]. Using this technique, Shao et al. [25] have described calcium-induced mesoscopic and micrometer scales domains in POPG/POPC (1/1) monolayer deposited on a diC15 PC first layer. Topography of bilayers obtained by the successive transfer of 2 monolayers made from phase separated liquid expanded-liquid condensed DPPC and DOPE/DPPE mixture were also recently reported [48].

An alternative to this procedure is the fusion of small unilamellar vesicles to a mica or a glass surface. Applied to a DPPC/POPC (1/1) mixture, the images again showed the presence of both mesoscopic and large domains [25]. Gel like domains appear thicker than the fluid domains by about 1nm, which suggests that the different mechanical properties of the two phases might contribute to the topographic images. On the other hand, isotropic gels domains of less than 10 nm were reported for mixtures of diarachinoyl PC (DAPC)/ DPPC (1/1) [47]. With the fusion method, however, the coverage of the substrate is often uncomplete. As far as the sample remains homogeneous, this does not hamper AFM analysis but allows to estimate the thickness of bilayers. For example, the thickness of a DPPC bilayer measured by AFM is close to 6 nm. In accordance with data obtained previously by other techniques, this indicates the presence of 1-2 nm of aqueous medium between the bilayer and the mica surface.

We have recently examined DOPC/DPPC (1/1) bilayers made by vesicle fusion. Following 1 hr incubation at 60°C , for the fusion step, the samples temperature was rapidly quenched to 25°C for AFM imaging. Both mesoscopic and microscopic size domains were observed. Gel phase DPPC enriched domains were still growing 1 hr after the temperature jump, emphasizing the pronounced metastability of this binary mixture. The data also strongly suggested that the two monolayers constituting the bilayer were coupled via the gel domains, i.e that gel domains at the surface of the accessible monolayer were coupled to gel domains of the monolayer facing the mica.

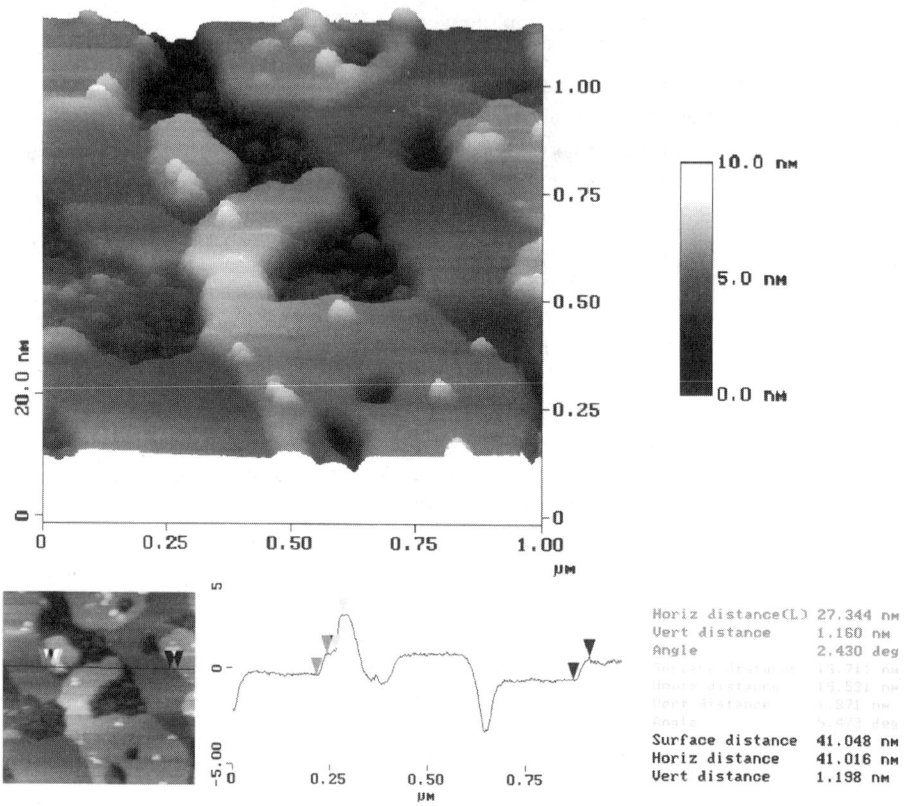

Figure 4. Imaging in liquid medium of G_{M1} (1%) clusters in DOPC/DPPC (1/1) bilayers made by vesicles fusion on mica.

Upon 1% GM1 addition to this mixture, clusters 20-50 nm diameter protruding by ~ 2 nm from the gel phase were visualized (Fig.4). This confirmed the propency of GM1, or at last of a part of it, to form microdomains that accumulate in the more ordered lipid phase. Additional evidences for this preferential localization of GM1 in gel phase domains were brought by adding , under the microscope, cholera toxin to the medium bathing the bilayer (GM1 is the toxin receptor in biological membranes). This resulted in a marked increase in the height difference between gel and liquid-crystal domains (Fig.5). 150 nm scans of the tall domains showed they were covered by small particles ~ 5nm diameter with a central pore (Fig.6). At higher magnification, the five subunits (12 kDa each) constituting the B oligomer were observed at a resolution down to ~ 1 nm [50], in accordance with the data of Mou et al. [51]. Interestingly, this resolution, in aqueous buffer, was obtained on a packed population of toxins and not on a 2-D crystal.

Numerous questions concerning the sizes of domains, their dependence upon the lipid composition, their kinetics of growth and melting, remain open. Localization of proteins in these domains and characterization of local lipid-protein interactions would also help in the understanding of their biological functions. The examples given above suggest that, in association with more classical techniques, the use of AFM can be fruitful in such studies.

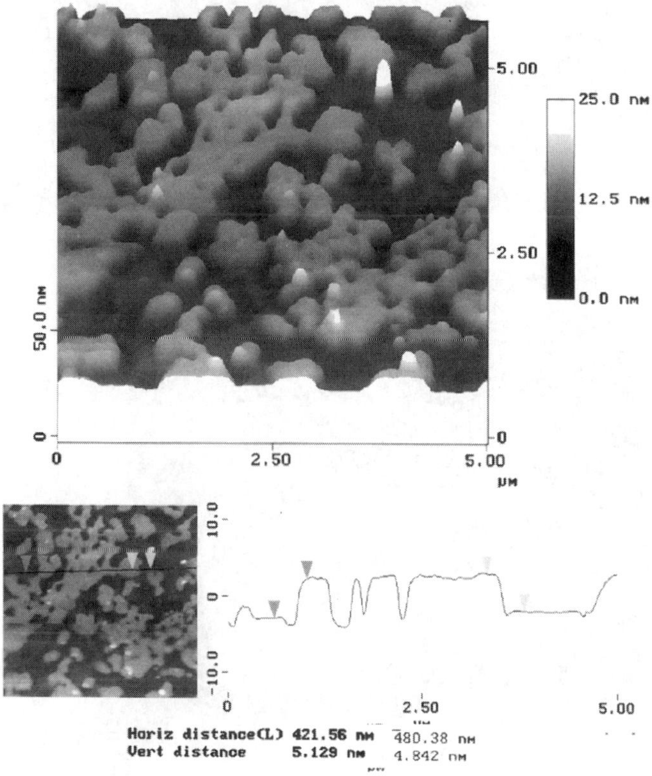

Figure 5. Low resolution image of DOPC/DPPC bilayer doped with 1% G_{M1} after cholera toxin addition. (Imaging in buffer)

5. Biological membranes in situ

Imaging of cells surfaces was early considered as an important biological application of the scanning probe microscopies [52,53], and more particularly of the AFM [32,54-57]. As for the other samples, good adherence to the support is a prerequisite for imaging of cells. Most of the experiments have been performed using the contact mode and

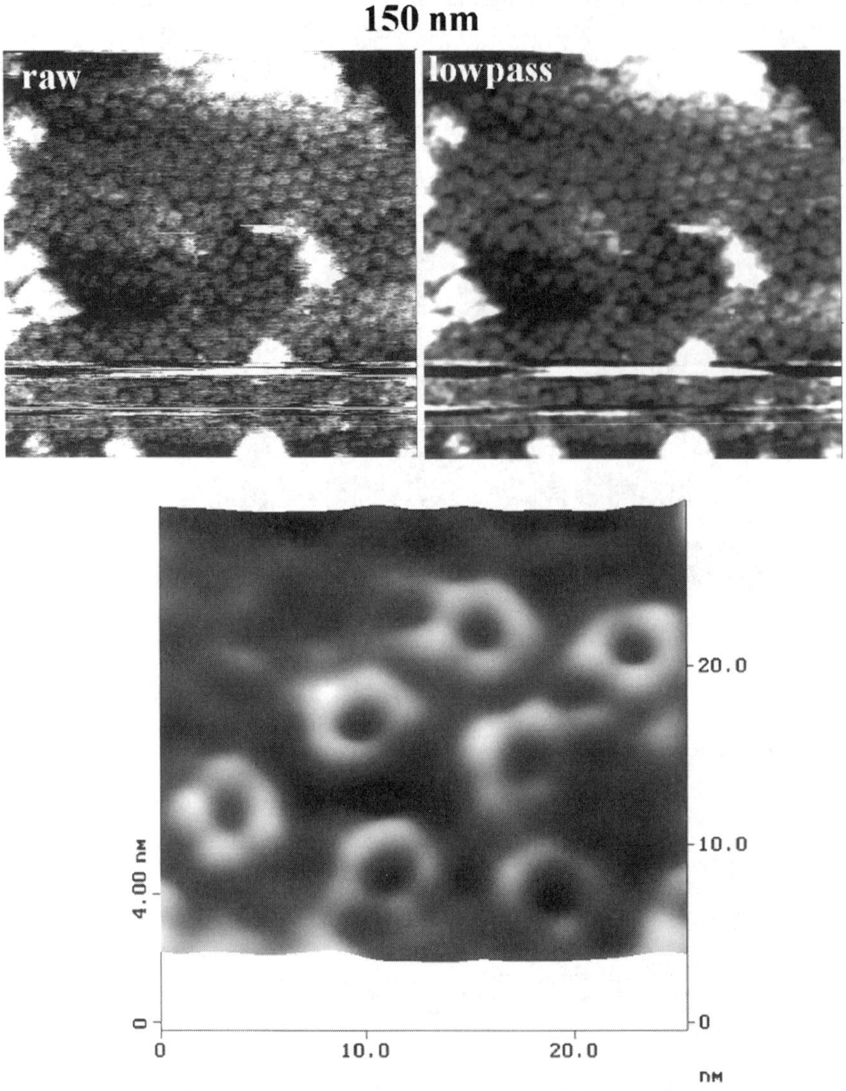

Figure 6. High resolution images of cholera toxin fixed on DOPC/DPPC bilayer doped with 1% G_{M1} . (Imaging in buffer).

the lateral resolution on cells, living or fixed, varies between ~ 10 and 500 nm, according mainly to the cell type and the force applied during scanning. Due to the softness of the material, the use of high scanning forces, i.e in the nN range, deforms the cell surface and limits the lateral resolution. Thus, loading forces as low as 20-50 pN is mandatory for high resolution imaging on intact cells. This allows to get three dimensional images of the cell surface with less than 10 nm local vertical deformation [33]. As shown by Fig. 7, at intermediate magnification, corresponding to 5-20 μm scan sizes, large and tall protrusions

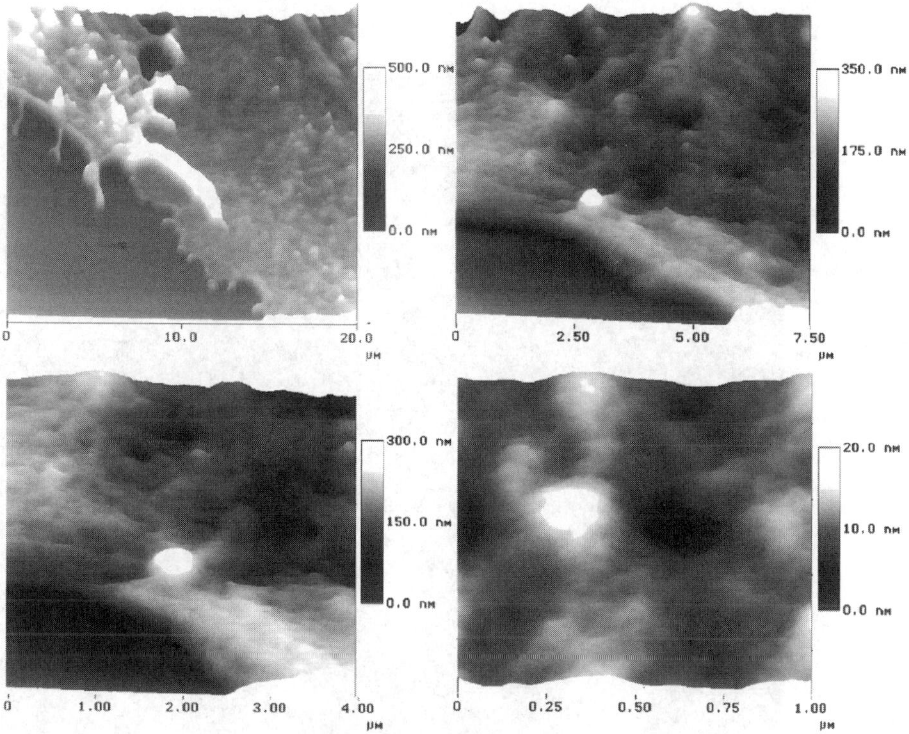

Figure 7. AFM contact mode imaging of the surface of living CV-1 cells.

are visualized that likely reflect the organization of submembrane and intracellular structures. Indeed it is impossible to conclude about the existence of plasma membrane domains only from such topographical images. Reducing the scan size to 1 µm or less, the surface appears granular with packed particles 5-30 nm xy size, but again, topography as viewed by AFM does not allow by itself to distinguish domains. In addition, small scans can only be successfully performed on particular regions of the cell surface which likely are poor in glycocalix [58]. The use of another SXM, the near field scanning optical microscope (NSOM, SNOM), in which the silicon nitride tip is replaced by an optical fiber thus giving access to fluorescence imaging at an enhanced resolution, might help in the identification of membrane domains and their constituents [48,59] .

Important informations on the constituents of membrane domains have been obtained through the use of detergent resistant membrane fractions (DRMs). We have recently examined in situ the surface of CV-1 cells, a fibroblast-like cell line, after Triton X-100 treatment. Because of the material fragility, imaging was performed using the tapping mode which also offers the advantage to detect the membrane skeleton organization. Fig. 8 gives an example of tapping mode images on untreated cells where both surface topography and membrane skeleton organization were simultaneously acquired [35].

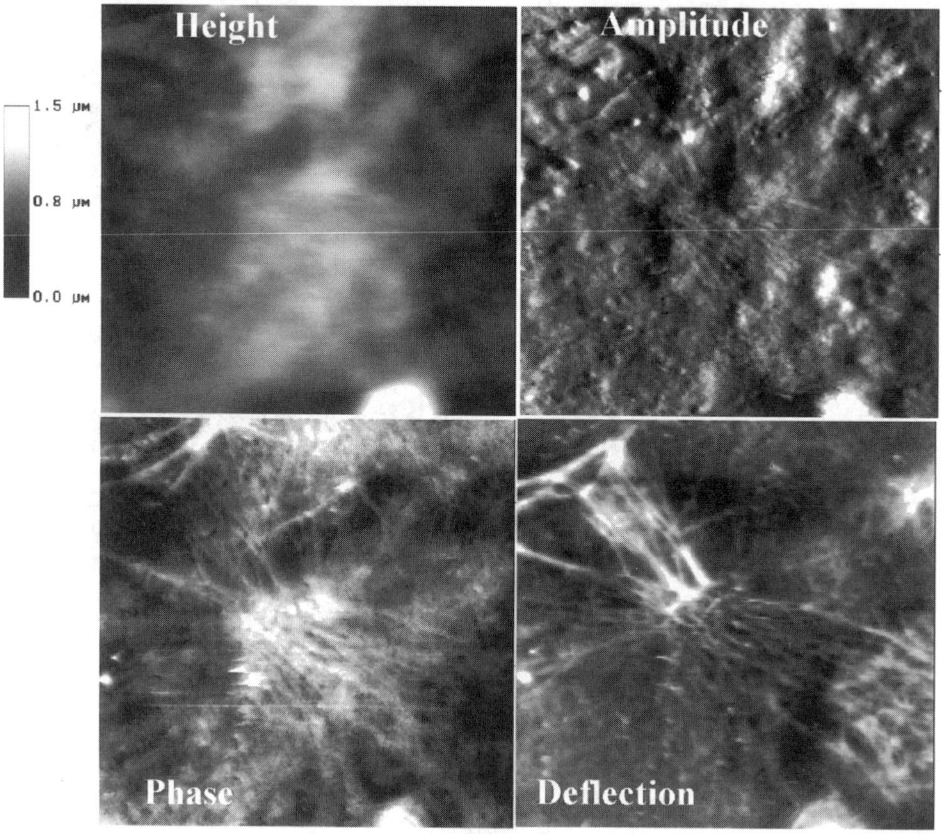

Figure 8. AFM tapping mode imaging of living CV-1cells. Scan size 25 µm.

In situ examination of Triton X-100 treated cells shows that DRMs can form domains whose minimal size often exceeds 15-20 μ^2 (Fig. 9). These results established the continuity of the membrane structures previously observed by optical microscopy [20,60]. They also indicate that, prior to their isolation for biochemical and biophysical analysis, DRMs are attached to the membrane skeleton. Current efforts to increase the lateral resolution on these samples would help in the characterization of their structural organization.

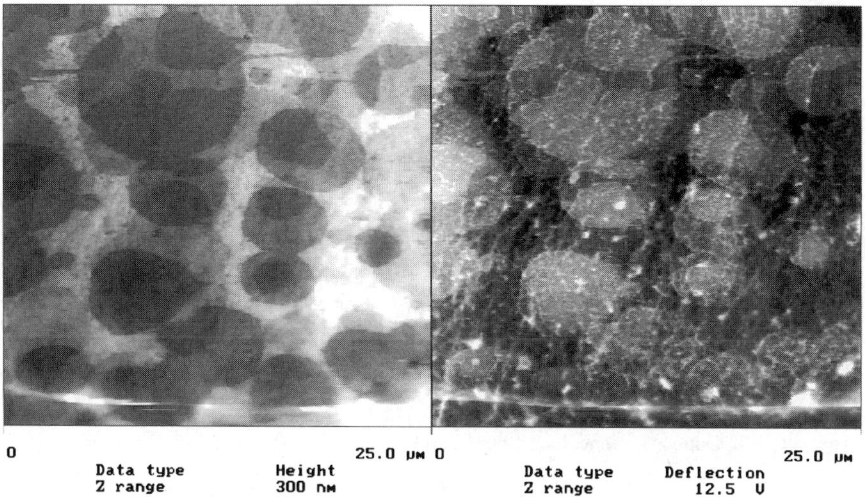

Figure 9. AFM tapping mode imaging of Triton X-100 treated cells (glutaraldehyde-fixed) in buffer. The left image is the height image. The right image is the corresponding deflection image.

6. Conclusion

Near field microscopies are new tools dedicated to the study of surface properties at high resolution. They constitute a rapidly growing family, with new applications emerging every year. Because of its possibility to work under physiological conditions, the AFM has been so far the privileged SXM among biologists. Its actual performances, from molecular to microscopic scales, fill the gap between light and electron microscopy. This direct access to the mesoscopic scale is likely to contribute to our understanding of the membrane microdomains properties.

References

1. Shinitzky, M. (1984) Membrane fluidity and cellular functions, in Physiology of Membrane fluidity, Shinitzky, M., ed, CRC Press, Boca Raton, pp 1-52.
2. Lipid Domains and the Relationship to membrane function (1988), Aloia, R.C., Curtain, C., Gordon, L.M., eds, Alan R. Liss, New York.
3. Vigh, L., Maresca, B. & Harwood, J.L. (1998) Does the membrane physical state control the expression of heat shock and other genes? *TIBS* **23**, 369-374.
4. Schachter, D. (1985) Lipid dynamics and lipid-protein interactions in intestinal plasma membranes. in Progress in Protein-lipid interactions, Watts, A., De Pont JJHHM, eds Elsevier, Amsterdam, pp 231-258.
5. Le Grimellec, C., Friedlander, G., El Yandouzi, E.H., Zlatkine, P. & Giocondi, M-C. (1992). Membrane fluidity and transport properties in epithelia. *Kidney Int.* **42**, 825-836.
6. Singer, S.J. & Nicolson, G.L. (1972) The fluid mosaic model of the structure of cell membranes . *Science* **175**, 1580-1582.
7. Morrisett, J.D., Pownall, H.J., Plumlee, R.T., Smith, L.C., Zehner, Z.E., Esfahani, M. & Wakil, S.J. (1975) Multiple thermotropic phase transitions in Escherichia coli membranes and membrane lipids. *J. Biol. Chem.* **250**, 6969-6976.
8. Giocondi, M-C & Le Grimellec, C. (1989) Temperature dependence of plasma membrane physical state in living Madin-Darby canine kidney cells. *Biochim. Biophys. Res. Comm.* **162**, 1004-1009.
9. Karnovsky, M.J., Kleinfeld, A.M., Hoover, R.L., & Klausner, R.D. (1982) The concept of lipid domains in membranes. *J. Cell Biol.* **94**, 1-6.
10. Yechiel, E., & Edidin, M. (1987) Micrometer-scale domains in fibroblast plasma membranes. *J. Cell Biol.* **105**, 755-760.
11. Glaser, M. (1993) Lipid domains in biological membranes. *Curr. Opin. Struct. Biol.* **3**, 475-481.
12. Tocanne, J-F., Dupou-Cézanne, L. & Lopez, A. (1994) lateral diffusion of lipids in model and natural membranes. *Prog. Lipid. Res.* **33**, 203-237.
13. Saxton, M.J. (1995) Single-particle tracking: effects of corrals. *Biophys. J.* **69**, 389-398.
14.Simson, R. Sheets, E.D., & Jacobson, K. (1995) Detection of temporay lateral confinement of membrane proteins using single-particle tracking analysis. *Biophys. J.* **69**, 989-993.
15. Kusumi, A. & Sako, Y. (1996) Cell surface organization by the membrane skeleton. *Curr. Opin. Cell Biol.* **8**, 566-574.
16. Anderson, R.G.W., Kamen, B.A., Rothberg, K.G. & Lacey, S. (1992) Potocytosis: sequestration and transport of small molecules by caveolae. *Science* **255**, 410-411.
17. Schnitzer, J.E., McIntosh, D.P., Dvorak, A.M., Liu, J. & Oh, P. (1995) Separation of caveolae from associated microdomains of GPI-anchored proteins. *Science* **269**, 1435-1439.
18. Simons, K. & Ikonen, E. (1998) Functional rafts in cell membranes. *Nature* **387**, 569-572.
19. Brown, D.A. & London, E. (1998) Structure and origin of ordered lipid domains in biological membranes. *J. Membrane Biol.* **164**, 103-114.
20. Brown, D.A. & London, E. (1998) Functions of lipid rafts in biological membranes. *Annu. Rev. Cell Dev. Biol.* **14**, 111-136.
21. Binnig, G. & Rohrer, H. (1982) Scanning tunneling microscope. *Helv. Phys. Acta* **55**, 726-735.
22. Binnig, G., Quate, C.F. & Gerber, C. (1986) Atomic force microscope. *Phys Rev Lett* **56**, 930-933.
23. Hoh, J.H. & Hansma, P.K. (1992) Atomic force microscopy for high-resolution imaging in cell biology. *Trends Cell Biol* **2**, 208-213.
24. Oberleithner, H., Brinckmann, E., Giebisch, G. & Geibel, J. (1995) Visualizing life on biomembranes by atomic force microscopy. *Kidney Int* **48,** 923-929.
25. Shao, Z., Mou, J., Czajkowsky, D.M., Yang, J. & Yuan, J-Y. (1996) Biological atomic force microscopy: what is acheived and what is needed. *Adv Physics* **45**,1-86.
26. Ikai, A. (1996) STM and AFM of bio/organic molecules and structures. *Surf Sci Rep* **26,** 261-332.
27. Engel, A., Schoenenberger, C-A. & Müller, D.J. (1997) High resolution imaging of native biological sample surfaces using scanning probe microscopy *Curr Opin Struct Biol* **7**, 279-284.
28. Czajkowsky, D.M. & Shao, Z. (1998) Submolecular resolution of single macromolecules with atomic force microscopy. *FEBS Lett.* **430,** 51-54.
29. Bustamante, C. & Keller, D. (1995) Scanning force microscopy in biology. *Physics Today* december, 32-38.
30. Bustamante, C., Rivetti, C. & Keller, D.J. (1997) Scanning force microscopy under aqueous solution. *Curr Opin Struct Biol* **7,** 709-716.
31. Müller, D.J., Fotiadis, D. & Engel, A. (1998) Mapping flexible protein domains at subnanometer resolution with the atomic force microscope. *FEBS Lett.* **430**, 105-111.
32. Radmacher, M., Tillmann, R.W., Fritz, M. & Gaub, H.E. (1992) *Science* **257**, 1900-1905.

33. Le Grimellec, C., Lesniewska, E., Giocondi, M-C, Finot, E., Vié, V., & Goudonnet, J-P (1998) Imaging of the surface of living cells by low-force contact-mode atomic force microscopy. *Biophys. J.* **75**, 695-703.
34. Zhong, Q., Inniss, D., Kjoller, K. & Elings, V.B. (1993) Fractured polymer/silica fiber surface studied by tapping mode atomic force microscopy. *Surface Sci. Lett.* **290**, 688-692.
35- Le Grimellec, C., Lesniewska, E., Giocondi, M.C., Finot, E., & Goudonnet, J.P. (1997) Simultaneous imaging of the surface and the submembraneous cytoskeleton in living cells by tapping mode atomic force microscopy. *C.R. Acad. Sci. Paris, Sciences de la vie / Life Sciences* **320**, 637-643.
36. Nag, K., Boland, C., Rich, N. & Keough, K.M.W. (1991) Epifluorescence microscopic observation of monolayers of dipalmitoylphophatidylcholine: dependence of domain size on compression rates. *Biochim. Biophys. Acta* **1068**, 157-160.
37. Zasadzinski, J.A., Viswanathan, R., Madsen, L.L. & Schwartz, D.K. (1994) Langmuir-Blodgett films. *Science* **263**, 1726-1733.
38. Schwartz, D.K. (1997) Langmuir-Blodgett film structure *Surface Science Reports*, **27**, 241-334.
39. Lee, K.Y.C., Lipp, M.M., Takamoto, D.Y., Ter-Ovaneysyan, E., Zasadzinski, J.A. & Waring, A.J. (1998) Apparatus for the continuous monitoring of surface morphology via fluorescence microscopy during monolayer transfer to substrates. *Langmuir* **14**, 2567-2572.
40. Soletti, J.M., Botreau, M., Sommer, F., Duc, T.M. & Celio, M.R. (1996) Characterization of mixed miscible and nonmiscible phospholipid langmuir-Blodgett films by atomic force microscopy. *J. Vac. Sci. Technol. B* **14**, 1492-1497.
41. Dufrêne, Y.F., Barger, W.R., Green, J-B. D. & Lee, G.U. (1997) Nanometer scale surface properties of mixed phospholipid monolayers and bilayers. *Langmuir* **13**, 4779-4784.
42. ten Grotenhuis, E., Demel, R.A., Ponec, M., Boer, D.R., van Miltenburg, J.C. & Bouwstra, J.A. (1996) Phase behavior of stratum corneum lipids in mixed Langmuir-Blodgett monolayers. *Biophys. J.* **71**, 1389-1399.
43. Vié, V., Van Mau, N., Lesniewska, E., Goudonnet, J.P., Heitz, F., & Le Grimellec, C. (1998) Distribution of ganglioside GM1 between two-component, two-phase phosphatidylcholine monolayers. *Langmuir* **14**, 4574-4583.
44. Van Mau, N., Vié, V., Chaloin, L., Lesniewsla, E., Heitz, F., & Le Grimellec, C. (1999) Lipid induced organization of a primary amphipathic peptide: a coupled AFM-Monolayer study. *J. membrane Biol.* **167**, 241-249.
45. Zasadzinski, J.A.N., Helm, C.A., Longo, M.L., Weisenhorn, A.L., Gould, S.A.C. & Hansma, P.K. (1991) Atomic force microscopy of hydrated phosphatidylethanolamine bilayers. *Biophys. J.* **59**, 755-760.
46.Czajkowsky, D.M., Allen, M.J., Elings, V. & Shao, Z. (1998) Direct visualization of surface charge in aqueous solution. *Ultramicroscopy* **74**, 1-5.
47. Gliss, C., Clausen-Schaumann, H., Günther, R., Odenbach, S., Randl, O. & Bayer, T.M. (1998) Direct detection of domains in phospholipid bilayers by grazing incidence diffraction of neutrons and atomic force microscopy. *Biophys. J.* **74**, 2443-2450.
48. Hollars, C.W. & Dunn, R.C. (1998) Submicron structure in L-∝-dipalmitoylphosphatidylcholine monolayers and bilayers probed with confocal, atomic force, and near-field microscopy. *Biophys. J.* **75**, 342-353.
49. Hui, S.W., Viswanathan, R., Zasadzinski, J.A. & Israelachvili, J.N. (1995) The structure and stability of phospholipid bilayers by atomic force microscopy . *Biophys. J.* **68**, 171-178.
50. Lesniewska, E., Giocondi, M-C., Vié, V., Finot, E., Goudonnet, J-P. & Le Grimellec, C. (1998) Atomic Force Microscopy of renal cells: limits and prospects. *Kidney. Int.* **53**, S42-S48.
51. Mou, J., Yang, J. & Shao, Z. (1995) Atomic force microscopy of cholera toxin B- oligomers bound to bilayers of biologically relevant lipids. *J. Mol. Biol.* **248**, 507-512.
52. Ruppersberg, J.P., J.K.H. Hörber, C. Gerber, & Binnig, G. (1989) Imaging of cell membraneous and cytoskeletal structures with a scanning tunneling microscope. *FEBS Lett* **257**,460-464.
53. Ito, E., T. Takahashi, K. Hama, T. Yoshioka, W. Mizutani, H. Shimizu, & Ono, M. (1991) An approach to imaging of living cell surface topography by scanning tunneling microscopy . *Biochem. Biophys. Res. Com.* **177**, 636-643 .
54. Butt, H.J., E.K. Wolff, S.A.C. Gould, B. Dixon Northern, C.M. Peterson, & Hansma, P.K. (1990) Imaging cells with the atomic force microscope. *J. Struct. Biol.* **105**, 54-61.
55. Häberle, W., J.K.H. Hörber, & Binnig, G. (1991) Force microscopy on living cells. *J. Vac. Sci Technol. B.* **9**, 1210-1213.
56. Henderson, E., P.G. Haydon, & Sakaguchi, D.S. (1992) Actin filaments dynamics in living glial cells imaged by atomic force microscopy. *Science* **257**, 1944-1946.
57. Hörber, J.K.H., W. Häberle, F. Ohnesorge, G. Binnig, H.G. Liebich, C.P. Czerny, H. Mahnel, & Mayr, A. (1992) Investigation of living cells in the nanometer regime with the scanning force microscope. *Scanning Microsc.* **6**, 919-930.
58. Le Grimellec, C., Lesniewska, E., Cachia, C., Schreiber, J.P., de Fornel, F. & Goudonnet, J.P. (1994) Imaging of the membrane surface of MDCK cells by Atomic Force Microscopy. *Biophys. J.* **67**, 36-41.

59. Hwang, J., Gheber, L.A., Margolis, L. & Edidin, M. (1998) Domains in cell plasma membranes investigated by near-field scanning optical microscopy. *Biophys. J.* **74**, 2184-2190.
60. Mayor, S. & Maxfield, F.R. (1995) Insolubility and redistribution of GPI-anchored proteins at the cell surface after detergent treatment. *Molec. Biol. Cell.* **6**, 929-944.

Lipid Heterogeneity in Mammalian Cells: Domains and Translocators

Gerrit VAN MEER, Hein SPRONG and René RAGGERS
Department of Cell Biology and Histology, Academic Medical Center, University of Amsterdam, P.O. Box 22700, 1100 DE Amsterdam, The Netherlands

Abstract. It has been realized for decades that cells contain multiple classes of lipids, that the various cellular membranes contain unique lipid mixtures, and that these mixtures are different for the two leaflets of most membranes. Still, only now we start to understand that membranes contain domains of lipids in a more ordered state. This domain organization is exploited by cells to bring proteins in contact or drive them apart in signal transduction events at the cell surface. However, it also constitutes an intricate part of the mechanism by which the unique protein and lipid compositions of membranes are established during vesicular traffic. The relation between the two bilayer leaflets is not understood at the molecular level, but an important role is expected for newly identified lipid translocators.

1. Unique Lipid Compositions

The various organelles in cells possess unique protein and lipid compositions. Differences in protein composition can often be directly coupled to defined functions. This has been more difficult in the case of membrane lipids, especially because, mostly, differences in lipid composition are not absolute. Most membranes contain more or less the same lipids but the ratio between the various lipids is different. An additional complication is the fact that the lipids are not evenly distributed over the two leaflets of the bilayer. To understand functions of lipids it is relevant to know the local density of a specific lipid on a specific membrane surface in time. This is determined by its sites of synthesis, modification and degradation, and by the mechanism of its transport through the cell. It is therefore relevant to identify the respective enzymes and molecular components of the transport machineries [1].

As one test system we, many years ago, adopted epithelial cells. The apical surface of epithelial cells along the gastro-intestinal and (distal) urinary tracts has been shown to contain high concentrations of glycosphingolipids. In intestinal cells, they constitute some 50% of the total polar lipids (see [2]). Because the (sparse) evidence indicates that glycolipids are confined to the outer leaflet of plasma membranes, the apical surface of the cells appears to be covered by glycolipids exclusively. Interestingly, the basolateral plasma membrane which is continuous with the apical membrane contains 2-3 times less glycolipids but is 2-4 times enriched in the glycero-phospholipid phosphatidylcholine (PC) and in the sphingo-phospholipid sphingomyelin (SM). Little difference was observed for the other phospholipids. In both plasma membrane domains cholesterol accounted for roughly 30 mol% of the total membrane lipids [3, 4]. As argued over 30 years ago by Forstner and colleagues [5], the latter parameter, percent of total lipids is probably more relevant for membrane properties than the cholesterol/phospholipid ratio. The high viscosity of the apical membrane [6] is therefore most likely due to the high concentration of glycolipids [7].

Madin Darby canine kidney (MDCK) cells in culture have maintained a polarized distribution of their membrane lipids, as was shown by antibodies [8] and cell fractionation [9], and by using viruses as probes for the plasma membrane lipids [10, 11].

The differences in lipid composition between the apical and the basolateral domains of the epithelial plasma membrane are maintained by a diffusion barrier situated at the tight junctions. The tight junctions connect neighboring epithelial cells and separate the apical from the basolateral domain by forming a continuous band around the apex of each cell. It has been found that lipids do not move between the exoplasmic (outer) leaflets of the apical and the basolateral plasma membrane [12-15], but are free to diffuse between the two domains on the cytosolic surface [15], suggesting the presence of the diffusion barrier in the exoplasmic but not the cytoplasmic leaflet of the plasma membrane lipid bilayer. Lipid models of the tight junction [16, 17] could not explain the observed lipid diffusion barrier between neighboring cells [14]. Models where the core of the tight junction consists of rows of proteins has gained tremendous support from the recent identification of two families of proteins that span the membrane four times [18].

The difference in lipid composition between the apical and basolateral surface implies that the membrane lipids do not exchange rapidly between the outer and the inner leaflet of the lipid bilayer, as in that case mixing between the two surface domains due to rapid diffusion via the continuous cytoplasmic leaflet of the plasma membrane would have been expected. This applies for the glycolipids in the outer leaflet of the apical plasma membrane domain, for the phospholipids SM and PC in the outer leaflet of the basolateral domain, and for the amino-phospholipids phosphatidylserine (PS) and phosphatidylethanolamine (PE) on the cytosolic surface.

2. Sites of Lipid Synthesis and Mechanisms of Transport

Glycolipids, SM and PC are not synthesized on the cell surface but in the endoplasmic reticulum (ER) and the Golgi complex. Thus, for the glycolipids there must be preferential transport from the site of synthesis to the apical surface and for the phospholipids PC and SM to the basolateral surface. When epithelial cells synthesized short-chain analogs of the glycolipid glucosylceramide (GlcCer) and SM, newly synthesized GlcCer was indeed delivered preferentially to the apical surface in MDCK cells and in human intestinal Caco-2 cells while SM was not [15, 19, 20]. The difference in polarity of delivery [19, 21] between the GlcCer and SM analogs in Caco-2 cells was sufficient to generate the apical enrichment of natural GlcCer over SM that has been observed in intestinal cells *in vivo*. Based on this and the fact that the complex glycolipids, like globosides which are enriched 5-fold in the apical membrane as % of total lipid [3], and SM are synthesized in the lumen of the Golgi apparatus, a model was formulated where glycosphingolipids aggregate into domains in the lumenal leaflet of the membrane of the trans Golgi network (TGN) analogous to the sorting of membrane proteins [2, 15, 22]. Through the years the model gained support from the findings that proteins linked to the lumenal aspect of membranes via a GPI-lipid-tail are sorted to the apical surface [23], that sphingolipids, an apical membrane protein and GPI-proteins share detergent-insolubility at low temperature [24, 25], and that one protein involved in apical sorting is caveolin, a protein in the cytosolic bilayer leaflet that underlies domains in the plasma membrane thought to be enriched in glycolipids and GPI-proteins (caveolae; review: [22, 26]). The physical basis of the preferential clustering of sphingolipids and cholesterol has been reviewed in great detail [27, 28], as has been the insight in functions of these domains at the cell surface [29].

3. Sphingolipid Translocators in Cellular Membranes

Roughly half of the glycolipids in the apical membrane of intestinal cells consists of the monohexosyl-ceramides glucosylceramide (GlcCer) and galactosylceramide (GalCer [3]). While the rat intestine contains exclusively GlcCer [30], four times more GalCer than GlcCer is present in the human intestine [31]. In contrast to the complex glycolipids and SM, these lipids are not synthesized in the lumen of the Golgi apparatus. GlcCer is synthesized on the cytosolic surface of the Golgi (see [32] and GalCer synthesis occurs in the lumen of the ER [33]. Although GlcCer can translocate across the Golgi membrane into the lumenal bilayer leaflet [34, 35] from where it would be transported to the apical surface via the TGN, its cytosolic synthesis opens the possibility that GlcCer reaches the cell surface not passing through the Golgi lumen. It would be transported to the cytoplasmic leaflet of the plasma membrane as a monomer [36], or on the cytoplasmic surface of transport vesicles followed by translocation across the plasma membrane. The enrichment on the apical surface could be caused by the selective apical localization of the translocator.

Evidence for the presence of an apical glycolipid translocator was obtained in a study of transport to the cell surface using sphingolipid analogs possessing a shortened acyl chain, which is essential for the transport assay. Short-chain GlcCer still reached the cell surface when vesicular lipid traffic to the plasma membrane was inhibited by brefeldin A [37]. This transport was inhibited by a series of known inhibitors of the multidrug resistance proteins, suggesting that such proteins might be responsible for the translocation across the plasma membrane. Direct proof that the multidrug transporter MDR1 P-glycoprotein can translocate short-chain lipids across the plasma membrane was obtained in transfected epithelial LLC PK1 cells [38]. MDR1 P-glycoprotein is expressed apically in MDCK and Caco-2 cells. MDR1 inhibitors had large effects on the apical transport of the short-chain GlcCer that had been used in our original transport studies [15, 19], demonstrating that 'sorting' in those studies was essentially due to the fact that the SM analog was transported by vesicles to both surfaces, while a large fraction of the GlcCer analog was transported selectively to the apical surface via monomeric exchange through the cytosol followed by translocation across the apical membrane. However, inhibitors had no effect on short-chain analogs possessing an α-hydroxy fatty acid (K. Burger et al., unpublished). The most simple interpretation of the sorting observed for those analogs [20] is still the formation of microdomains on the lumenal surface of the TGN. Such sorting by microdomains seems also the only possible mechanism for the apical enrichment of the native complex glycolipids. The assessment of the relative contribution of the TGN pathway and MDR1 P-glycoprotein to the apical enrichment of GlcCer awaits a transport assay for natural long-chain GlcCer.

To complicate the situation even further: also the multidrug resistance protein MRP1, an ABC-transporter that is only distantly related to MDR1 P-glycoprotein and is situated in the basolateral plasma membrane of epithelial cells, can translocate short-chain sphingolipids [39]. Triple knock-out mice for two types of MDR1 (Mdr1a and Mdr1b) and for MRP1 do not display a phenotype unless challenged with drugs (J. Wijnholds and P. Borst, Netherlands Cancer Institute, pers. commun.), which suggests either of three things: (a) Mdr1a, Mdr1b, and MRP1 are irrelevant for the polarized distribution of lipids in epithelia, (b) they are responsible for the polarized distribution of lipids but this surface polarity is not important for the well-being of the mouse, or (c) Mdr1a, Mdr1b, and MRP1 are responsible for the polarized distribution of lipids in epithelia, this surface polarity is important for the well-being of the mouse, but in the absence of these transporters other transporters (or mechanisms) take over. We are presently testing these possibilities.

4. Lipid Translocators and Domains

A significant fraction of the MDR1 P-glycoprotein has been assigned to detergent-insoluble domains on the cell surface [40]. Thus, this MDR1 would translocate GlcCer across the plasma membrane into a sphingolipid/cholesterol domain in the outer membrane leaflet. Thereby, MDR1 P-glycoprotein may actually regulate the formation and the localization of the domain. Of the three phospholipid translocators that have been identified, the MDR3 P-glycoprotein, a PC-specific ABC-transporter in specialized tissues (see [35]), the aminophospholipid translocator [41, 42] and the scramblase [43], the latter may translocate sphingolipids to the cytoplasmic leaflet. MDR1 P-glycoprotein and the scramblase may work in concert. Translocation of SM to the cytosolic surface may be relevant for signal transduction as it has been proposed [44, 45] that SM in the cytosolic leaflet of the plasma membrane is a substrate for the signaling neutral sphingomyelinase [46]. Interestingly, the presence of Glccer and possibly SM on the cytosolic surface of membranes opens the possibility that sphingolipid/cholesterol domains also exist on the cytosolic surface of cellular membranes.

Our recent findings suggest that part of the GlcCer is normally translocated across the Golgi membrane by an MDR1-independent unidentified translocator (R. Raggers, D. Sillence and G. van Meer, man. in prep.), that MDR1 P-glycoprotein acts as a back-up system and removes from the cytosolic surface of the plasma membrane a pool of GlcCer that may have served a function in protein sorting on the cytosolic surface of the Golgi (H. Sprong, P. van der Sluijs, and G. van Meer, man. in prep.). If this GlcCer is not properly removed by the MDR1 P-glycoprotein it is degraded (R. Raggers, D. Sillence and G. van Meer, man. in prep.).

In kidney cells, besides the aminophospholipid translocator, an inward translocating activity was observed for PC on both the apical and basolateral surface [47]. Whether the aminophospholipid translocator contributes to lipid domains in the cytosolic plasma membrane leaflet remains unclear. Remarkably, many proteins that are anchored to the cytosolic surface by lipid tails, notably saturated acyl chains [48], cluster in domains opposite the sphingolipid/cholesterol domains in the outer leaflet. Still, the existence of lipid domains in the cytosolic leaflet remains speculative, as does the nature of the coupling between domains in opposite leaflets.

Finally, lipids move across the ER membrane [49]. This is thought to be protein-mediated [50], and the system may have a preference for glycerophospholipids [51]. However, the ER is also the site where the glycolipid precursors for protein glycosylation must traverse the membrane into the lumen. These are dolichol-phosphoglucose and dolichol-phosphomannose, but, much more dramatically, dolichol-P-P-(N-acetylglucosamine)$_2$(mannose)$_5$. Also precursors for the glycosyl-phosphatidylinositol (GPI) membrane anchors must translocate across the ER membrane [52]. ER lipid translocators have so far not been identified. It is unclear whether glycolipids and SM that reach the ER [53] can use the machinery to flip to the cytosolic surface. In that case one might expect the need for an outward translocator that can correct this 'mislocalization'. Obviously our prime candidate would be MDR1 P-glycoprotein.

In summary, cells contain a number of lipid translocators of which only a few have been identified so far. Each of these may actually be a member of a family of transporters. The specificity of the transporters, their localization in the cell, their lateral organization in any particular membrane, and the regulation of their activity appear to be highly relevant for the lipid organization in cells. In conjunction with lipid domains, their activities are highly relevant for cellular physiology. In one particular case [54], the disruption of a translocator activity has been shown to result in disease. Undoubtedly, others will follow.

References:

[1] G. van Meer, Lipid traffic in animal cells. *Annu. Rev. Cell Biol.* **5** (1989) 247-275.

[2] K. Simons and G. van Meer, Lipid sorting in epithelial cells. *Biochemistry* **27** (1988) 6197-6202.

[3] K. Kawai, M. Fujita, and M. Nakao, Lipid components of two different regions of an intestinal epithelial cell membrane of mouse. *Biochim. Biophys. Acta* **369** (1974) 222-233.

[4] T.A. Brasitus and D. Schachter, Lipid dynamics and lipid-protein interactions in the rat enterocyte basolateral and microvillus membranes. *Biochemistry* **19** (1980) 2763-2769.

[5] G.G. Forstner, K. Tanaka, and K.J. Isselbacher, Lipid composition of the isolated rat intestinal microvillus membrane. *Biochem. J.* **109** (1968) 51-59.

[6] D. Schachter and M. Shinitzky, Fluorescence polarization studies of rat intestinal microvillus membranes. *J. Clin. Invest.* **59** (1977) 536-548.

[7] G. van Meer, How epithelial cells grease their microvilli. *Trends Biochem. Sci.* **13** (1988) 242-243.

[8] R.J. Turner, J. Thompson, S. Sariban-Sohraby, and J.S. Handler, Monoclonal antibodies as probes of epithelial membrane polarization. *J. Cell Biol.* **101** (1985) 2173-2180.

[9] G.E. Nichols, T. Shiraishi, and W.W. Young, Jr., Polarity of neutral glycolipids, gangliosides, and sulfated lipids in MDCK epithelial cells. *J. Lipid Res.* **29** (1988) 1205-1213.

[10] G. van Meer and K. Simons, Viruses budding from either the apical or the basolateral plasma membrane domain of MDCK cells have unique phospholipid compositions. *EMBO J.* **1** (1982) 847-852.

[11] G. van Meer and K. Simons, The function of tight junctions in maintaining differences in lipid composition between the apical and the basolateral cell surface domains of MDCK cells. *EMBO J.* **5** (1986) 1455-1464.

[12] P.R. Dragsten, R. Blumenthal, and J.S. Handler, Membrane asymmetry in epithelia: is the tight junction a barrier to diffusion in the plasma membrane? *Nature* **294** (1981) 718-722.

[13] S. Spiegel, R. Blumenthal, P.H. Fishman, and J.S. Handler, Gangliosides do not move from apical to basolateral plasma membrane in cultured epithelial cells. *Biochim. Biophys. Acta* **821** (1985) 310-318.

[14] G. van Meer, B. Gumbiner, and K. Simons, The tight junction does not allow lipid molecules to diffuse from one epithelial cell to the next. *Nature* **322** (1986) 639-641.

[15] G. van Meer, E.H.K. Stelzer, R.W. Wijnaendts-van-Resandt, and K. Simons, Sorting of sphingolipids in epithelial (Madin-Darby canine kidney) cells. *J. Cell Biol.* **105** (1987) 1623-1635.

[16] B. Kachar and T.S. Reese, Evidence for the lipidic nature of tight junction strands. *Nature* **296** (1982) 464-466.

[17] P. Pinto da Silva and B. Kachar, On tight-junction structure. *Cell* **28** (1982) 441-450.

[18] S. Tsukita and M. Furuse, Occludin and claudins in tight-junction strands: leading or supporting players? *Trends Cell Biol.* **9** (1999) 268-273.

[19] W. van 't Hof and G. van Meer, Generation of lipid polarity in intestinal epithelial (Caco-2) cells: Sphingolipid synthesis in the Golgi complex and sorting before vesicular traffic to the plasma membrane. *J. Cell Biol.* **111** (1990) 977-986.

[20] P. van der Bijl, M. Lopes-Cardozo, and G. van Meer, Sorting of newly synthesized galactosphingolipids to the two surface domains of epithelial cells. *J. Cell Biol.* **132** (1996) 813-821.

[21] I.L. van Genderen and G. van Meer, Lipid sorting – measurement and interpretation. *Biochem. Soc. Trans.* **21** (1993) 235-239.

[22] K. Simons and E. Ikonen, Functional rafts in cell membranes. *Nature* **387** (1997) 569-572.

[23] M.P. Lisanti, M. Sargiacomo, L. Graeve, A.R. Saltiel, and E. Rodriguez-Boulan, Polarized apical distribution of glycosyl-phosphatidylinositol-anchored proteins in a renal epithelial cell line. *Proc. Natl. Acad. Sci. USA* **85** (1988) 9557-9561.

[24] J.E. Skibbens, M.G. Roth, and K.S. Matlin, Differential extractability of influenza virus hemagglutinin during intracellular transport in polarized epithelial cells and non-polar fibroblasts. *J. Cell Biol.* **108** (1989) 821-832.

[25] D.A. Brown and J.K. Rose, Sorting of GPI-anchored proteins to glycolipid-enriched membrane subdomains during transport to the apical cell surface. *Cell* **68** (1992) 533-544.

[26] R.G.W. Anderson, The caveolae membrane system. *Annu. Rev. Biochem.* **67** (1998) 199-225.

[27] R.E. Brown, Sphingolipid organization in biomembranes: what physical studies of model membranes reveal. *J. Cell Sci.* **111** (1998) 1-9.

[28] D.A. Brown and E. London, Structure and origin of ordered lipid domains in biological membranes. *J. Membrane Biol.* **164** (1998) 103-114.

[29] D.A. Brown and E. London, Functions of lipid rafts in biological membranes. *Annu. Rev. Cell Dev. Biol.* **14** (1998) 111-136.

[30] G.G. Forstner and J.R. Wherrett, Plasma membrane and mucosal glycosphingolipids in the rat intestine. *Biochim. Biophys. Acta* **306** (1973) 446-459.

[31] B. Siddiqui, J.S. Whitehead, and Y.S. Kim, Glycosphingolipids in human colonic adenocarcinoma. *J. Biol. Chem.* **253** (1978) 2168-2175.

[32] G. van Meer, Lipids of the Golgi membrane. *Trends Cell Biol.* **8** (1998) 29-33.

[33] H. Sprong, B. Kruithof, R. Leijendekker, J.W. Slot, G. van Meer, and P. van der Sluijs, UDP-galactose:ceramide galactosyltransferase is a class I integral membrane protein of the endoplasmic reticulum. *J. Biol. Chem.* **273** (1998) 25880-25888.

[34] H. Lannert, C. Bünning, D. Jeckel, and F.T. Wieland, Lactosylceramide is synthesized in the lumen of the Golgi apparatus. *FEBS Lett.* **342** (1994) 91-96.

[35] K.N.J. Burger, P. van der Bijl, and G. van Meer, Topology of sphingolipid galactosyltransferases in ER and Golgi: Transbilayer movement of monohexosyl sphingolipids is required for higher glycosphingolipid biosynthesis. *J. Cell Biol.* **133** (1996) 15-28.

[36] D.E. Warnock, M.S. Lutz, W.A. Blackburn, W.W. Young, Jr., and J.U. Baenziger, Transport of newly synthesized glucosylceramide to the plasma membrane by a non-Golgi pathway. *Proc. Natl. Acad. Sci. USA* **91** (1994) 2708-2712.

[37] A. van Helvoort, M.L. Giudici, M. Thielemans, and G. van Meer, Transport of sphingomyelin to the cell surface is inhibited by brefeldin A and in mitosis, where C_6-NBD-sphingomyelin is translocated across the plasma membrane by a multidrug transporter activity. *J. Cell Sci.* **110** (1997) 75-83.

[38] A. van Helvoort, A.J. Smith, H. Sprong, I. Fritzsche, A.H. Schinkel, P. Borst, and G. van Meer, MDR1 P-glycoprotein is a lipid translocase of broad specificity, while MDR3 P-glycoprotein specifically translocates phosphatidylcholine. *Cell* **87** (1996) 507-517.

[39] R.J. Raggers, A. van Helvoort, R. Evers, and G. van Meer, The human multidrug resistance protein MRP1 translocates sphingolipid analogs across the plasma membrane. *J. Cell Sci.* **112** (1999) 415-422.

[40] Y. Lavie, G. Fiucci, and M. Liscovitch, Up-regulation of caveolae and caveolar constituents in multidrug-resistant cancer cells. *J. Biol. Chem.* **273** (1998) 32380-32383.

[41] M. Seigneuret and P.F. Devaux, ATP-dependent asymmetric distribution of spin-labeled phospholipids in the erythrocyte membrane: relation to shape changes. *Proc. Natl. Acad. Sci. USA* **81** (1984) 3751-3755.

[42] X. Tang, M.S. Halleck, R.A. Schlegel, and P. Williamson, A subfamily of P-type ATPases with aminophospholipid transporting activity. *Science* **272** (1996) 1495-1497.

[43] Q. Zhou, J. Zhao, J.G. Stout, R.A. Luhm, T. Wiedmer, and P.J. Sims, Molecular cloning of human plasma membrane phospholipid scramblase. A protein mediating transbilayer movement of plasma membrane phospholipids. *J. Biol. Chem.* **272** (1997) 18240-18244.

[44] N. Andrieu, R. Salvayre, and T. Levade, Comparative study of the metabolic pools of sphingomyelin and phosphatidylcholine sensitive to tumor necrosis factor. *Eur. J. Biochem.* **236** (1996) 738-745.

[45] C.M. Linardic and Y.A. Hannun, Identification of a distinct pool of sphingomyelin involved in the sphingomyelin cycle. *J. Biol. Chem.* **269** (1994) 23530-23537.

[46] S. Tomiuk, K. Hofmann, M. Nix, M. Zumbansen, and W. Stoffel, Cloned mammalian neutral sphingomyelinase. Functions in sphingolipid signaling. *Proc. Natl. Acad. Sci. USA* **95** (1998) 3638-3643.

[47] T. Pomorski, A. Herrmann, P. Muller, G. van Meer, and K. Burger, Protein-mediated inward translocation of phospholipids occurs in both the apical and basolateral plasma membrane domains of epithelial cells. *Biochemistry* **38** (1999) 142-150.

[48] K.A. Melkonian, A.G. Ostermeyer, J.Z. Chen, M.G. Roth, and D.A. Brown, Role of lipid modifications in targeting proteins to detergent- resistant membrane rafts. Many raft proteins are acylated, while few are prenylated. *J. Biol. Chem.* **274** (1999) 3910-3917.

[49] D.B. Zilversmit and M.E. Hughes, Extensive exchange of rat liver microsomal phospholipids. *Biochim. Biophys. Acta* **469** (1977) 99-110.

[50] A. Herrmann, A. Zachowski, and P.F. Devaux, Protein-mediated phospholipid translocation in the endoplasmic reticulum with a low lipid specificity. *Biochemistry* **29** (1990) 2023-2027.

[51] X. Buton, G. Morrot, P. Fellmann, and M. Seigneuret, Ultrafast glycerophospholipid-selective transbilayer motion mediated by a protein in the endoplasmic reticulum membrane. *J. Biol. Chem.* **271** (1996) 6651-6657.

[52] J. Vidugiriene and A.K. Menon, The GPI anchor of cell-surface proteins is synthesized on the cytoplasmic face of the endoplasmic reticulum. *J. Cell Biol.* **127** (1994) 333-341.

[53] I.L. van Genderen, G. van Meer, J.W. Slot, H.J. Geuze, and W.F. Voorhout, Subcellular localization of Forssman glycolipid in epithelial MDCK cells by immuno-electronmicroscopy after freeze-substitution. *J. Cell Biol.* **115** (1991) 1009-1019.

[54] J.J.M. Smit, A.H. Schinkel, R.P.J. Oude Elferink, A.K. Groen, E. Wagenaar, L. van Deemter, C.A.A.M. Mol, R. Ottenhoff, N.M.T. van der Lugt, M.A. van Roon, M.A. van der Valk, G.J.A. Offerhaus, A.J.M. Berns, and P. Borst, Homozygous disruption of the murine *mdr2* P-glycoprotein gene leads to a complete absence of phospholipid from bile and to liver disease. *Cell* **75** (1993) 451-462.

Transmembrane domain length and protein sorting in the secretory pathway: an examination of the exit of Axl2p from the yeast ER

Timothy P. Levine and Sean Munro
MRC Laboratory of Molecular Biology
Hills Road, Cambridge CB2 2QH, UK
email: sean@mrc-lmb.cam.ac.uk

The membranes of eukaryotic cells do not have a uniform lipid composition, presumably because the different membranes have different functions. One major example of this is the plasma membrane (PM) which is enriched in both sphingolipids and sterols compared to the membranes of earlier compartments of the secretory pathway [1]. In mammals, the sphingolipids (sphingomyelin and glycolipids) are mainly on the outer leaflet of the PM, with phospholipids found on both leaflets, except for phosphatidylserine which is restricted to the inner leaflet. In yeast the sphingolipids are inositol phosphorylceramide (IPC) and its mannosylated derivatives, and, as in mammalian cells, both these lipids and the yeast sterol ergosterol are most abundant in the plasma membrane [2,3].

This raises the question of why the PM has a higher level of sterols and sphingolipids. Various structural and regulatory functions have been proposed for both classes of lipids. For instance, sphingolipids can be used as substrates by enzymes that generate second messengers such as ceramide for signaling events [4]. However, since both classes of lipids are usually present at substantially more than 10% of the total plasma membrane lipid, it seems likely that they must also have a role in determining the overall physical properties of the plasma membrane bilayer [2]. Biophysical studies show that increasing the concentration of cholesterol in phospholipid bilayers reduces their permeability to small molecules and ions. This effect is believed to be due to the rigid sterol rings ordering, or compacting, the lipid acyl chains, since at least one acyl chain per phospholipid is unsaturated with a cis-double bond which reduces tight packing. Sphingolipids usually have fully saturated chains and in some cases, such as in myelin, and in yeast, the acyl chains are particularly long. In addition, unlike the diacylglycerol core of phospholipids, the ceramide core of sphingolipids contains a hydroxyl group and a secondary amine which have the potential to participate in inter-lipid hydrogen bonds. In yeast, intestinal cells, and myelin-producing astroglia the ceramide is often found to be further hydroxylated, which presumably serves to increase yet further the tightness with which the lipids are packed together [5]. In addition, sterols interact more tightly with sphingolipids than with phospholipids. Finally, it is known that the ordering and tight packing activities of both sterols and sphingolipids results in an increase in the thickness of the bilayer.

All of above physical properties of sphingolipids and sterol add up to meaning that the plasma membrane will be less permeable than internal membranes such as the endoplasmic reticulum (ER), whilst remaining fluid to allow diffusional mobility of membrane proteins in the plane of the bilayer. This increased impermeability presumably helps the plasma membrane serve as a particularly effective barrier that can exclude small toxic molecules, and restrain within the cell useful molecules that have been synthesised or taken up, such as sugars and amino acids. In contrast, at the beginning of the secretory pathway the membrane of the ER has less of a barrier function to perform, but rather is the site of synthesis, insertion and assembly of many hydrophobic molecules including membrane proteins, GPI anchors, phospholipids, sterols and dolichols. As such it might be advantageous for the cell to have a malleable, ER bilayer which can accommodate the ready insertion of exogenous hydrophobic molecules, and as such a high level of unsaturated phospholipids and a low level of sphingolipids and sterols would seem desirable.

The above arguments may provide a rationale for the distinct lipid composition of the plasma membrane, but further questions remain. One of these is how the difference in composition between plasma membrane and ER is set up and maintained. This will not be discussed in depth here, beyond that sterols are synthesised in the ER, and sphingolipids mainly in the Golgi. This means that mechanisms must exist to selectively transport these lipids forwards in the secretory pathway, and up a concentration gradient. The mechanism of this selective forward transport is not clearly established, but for the sphingolipids at least, transport appears to occur in the vesicles of the secretory pathway [6]. This means that the Golgi apparatus must be organised to ensure that sphingolipids are preferentially sorted into the anterograde vesicles leaving for the plasma membrane, and not the retrograde vesicles returning to the ER, although the mechanism by which this is accomplished remains speculative [1,7]. Likewise cholesterol must be preferentially transported from the ER to the plasma membrane, although whether this occurs only in transport vesicles, or if there is additional transport directly across the cytoplasm is unresolved [8].

The difference in the lipid composition, and hence physical properties, of the ER and plasma membrane also raises the question of whether this change in the bilayer affects membrane proteins as they move from the site of synthesis in the ER to locations elsewhere in the cell. The different bilayers are both in liquid crystalline state, but other features of the bilayer might be expected to affect the proteins in them. In particular, an increase in sterols and sphingolipids would be expected to thicken the bilayer, and indeed there is evidence that the plasma membrane is indeed thicker than the ER. Firstly, when cells are examined by electron microscopy, the interleaflet space at the plasma membrane is greater than that of the ER for both plants and yeast [3,9]. Secondly, for mammalian cells the hydrophobic transmembrane domains of plasma membrane proteins are on average five residues longer that those of Golgi proteins [7]. Thus it seems likely that the bilayers in a eukaryotic cell do not all have the same hydrophobic thickness, a result presumably of the differences in lipid composition.

Implications of bilayer thickness changes

There is both theoretical and experimental evidence that the properties of membrane protein can be affected by how well the hydrophobic portion of the protein matches the hydrophobic portion of the bilayer [10]. Obviously the mismatch can be in either direction, with the bilayer being either thinner, or thicker, than the protein's TMD. A certain degree of mismatch in either direction can be accommodated by various means. Lipid acyl chains can become either better aligned to the perpendicular, or more crumpled; flexibility of amino acid side chains can allow charged residues to be a moved up or down with respect the TMD; and proteins can tilt to fit into thinner bilayers [10,11]. However if the mismatch become too great then proteins can aggregate, or even be expelled from the bilayer altogether [12,13]. Is there any evidence that hydrophobic mismatch is an issue in the secretory pathway, beyond a general requirement that TMDs will have evolved to fit sufficiently well in bilayers that they neither aggregate, or are ejected? It seems likely that the increase in bilayer thickness in the pathway has at least two further implications. Firstly the point in the pathway at which the bilayer gets thicker allows sorting of proteins from a thinner to a thicker bilayer. Secondly, the proteins of the plasma membrane will be inserted and assembled in a membrane, the ER, that is thinner than the one in which they will finally reside in.

The increase in bilayer thickness is likely to take place in the Golgi apparatus as this is where sphingolipids are made, and where the lipid composition starts to become more like that of the plasma membrane. We have suggested that TMD length could be used as a means of preventing Golgi resident proteins from moving beyond the Golgi apparatus [7]. Bilayer mismatch would mean that Golgi proteins with shorter TMDs would be less likely than plasma membrane proteins to enter the sterol/sphingolipid rich vesicles carrier vesicles leaving the Golgi destined for the plasma membrane. Analysis of the behaviour of proteins with synthetic TMDs of various lengths has given some support to this model, although it remains possible that other mechanisms also act on Golgi residents, especially to sort the proteins to the different cisternae within the stacked structure of the organelle [14,15].

The second area where hydrophobic mismatch may have implications for protein trafficking is in the accommodation of newly-made plasma-membrane proteins in ER membranes. Presumably ER proteins have evolved TMDs of lengths optimal for the thickness of the ER bilayer, but our observations indicate that the plasma membrane TMD will be 5-8 residues longer than this. This raises the possibility that plasma membrane proteins are at risk of aggregating in the ER. It is known from studies in model systems that some degree of mismatch can be tolerated without aggregation, and indeed it appears that protein can be more readily accommodated in a bilayer that is too thin, rather than one that is too thick [16]. Thus it is possible that the only implication of mismatch in this context is that it has put a constraint during evolution on how much thicker the plasma membrane can be than the ER. However, part of this accommodation of proteins in thin bilayers is achieved by TMD tilting, which seems likely to be of less help to larger polytopic proteins than to single TMD proteins. In addition, the degree of mismatch is likely to be particularly large in those systems where long chain sphingolipids are present, such as myelin or yeast. Indeed it is striking that for yeast several ER resident proteins have recently been identified whose function appears to be to aid the exit from the ER of particular, often polytopic, plasma membrane proteins [17-20]. This raises the intriguing possibility that "mismatch chaperones" might exist in the ER whose function would be to guide plasma membrane proteins with long TMDs through to a point in the Golgi where the bilayer thickens, thereby preventing the proteins aggregating in the thinner ER bilayer. To start to address this idea, we have examined a candidate for such a mismatch chaperone in yeast. However, we find that in this case although the TMD of the chaperoned protein is involved in the action of the chaperone, shortening the TMD does not overcome the requirement for the chaperone protein.

Results and Discussion

The *S. cerevisiae* protein Erv14p was identified as a component of ER to Golgi transport vesicles [17]. It is a small hydrophobic protein with three TMDs, and has a further relative in the yeast genome, and homologues of the protein are also present in the human EST libraries, and the *C. elegans* genome. Yeast strains lacking the *ERV14* gene are viable, but a plasma membrane protein Axl2p, required for polarised growth, was found to reach the plasma membrane much more slowly, and instead accumulate in the ER [17]. Axl2p has a single TMD and like that of other single TMD proteins of the yeast plasma membrane, it is strikingly long (26 residues, Figure 1).

```
Axl2p     ..499SHNKKAVAIACGVAIPLGVILVALICFLIFWRRRRENP..
Axl2pΔ6a  ..499SHNKKAV------AIPLGVILVALICFLIFWRRRRENP..
Axl2pΔ6b  ..499SHNKKAVAIACGV------ILVALICFLIFWRRRRENP..
```

Figure 1. TMD of Axl2p and of the deleted forms used in this study.

We thus decided to examine whether shortening the TMD of Axl2p would remove the requirement for Erv14p to get the protein out of the ER. Initially Axl2p was expressed in yeast as a fusion to GFP to allow its localisation to be readily examined in living cells of both wild type and *Δerv14* strains. Figure 2 shows that as expected the Axl2p-GFP chimera was at the plasma membrane of wild type cells, with little internal staining except for some vacuolar staining which is often seen when membrane proteins are expressed as GFP fusions in yeast, and is presumably caused by GFP remaining stable in the vacuole after the rest of the protein is turned over.

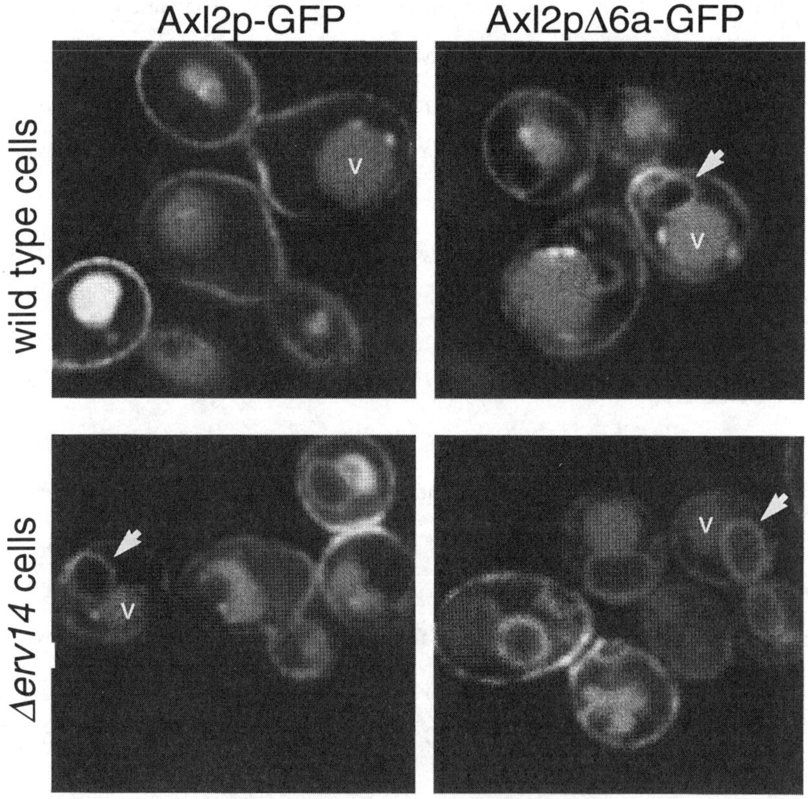

Figure 2. Fluorescence confocal micrographs of live yeast expressing Axl2p-GFP chimeras. Yeast are either SEY6210 (wild-type, [21]) or the same strain with *ERV14* deleted by homologous recombination (*Δerv14*), and are expressing either intact Axl2p-GFP or Axl2pΔ6a-GFP which has a TMD that is 6 amino acids shorter, from a constitutive *PHO5* promoter on a CEN plasmid. ER/ nuclear envelope (arrow) and vacuole (v) are indicated. Axl2pΔ6b-GFP gave indistinguishable results to Δ6a and is not shown.

By contrast to the situation in wild type cells, in *Δerv14* cells, fluorescence from the Axl2p-GFP could be clearly seen in the nuclear envelope indicating that as previously reported the protein accumulates in the ER before slowly leaving for it final destination in the plasma membrane[17].

We then made two deletions in the TMD of Axl2p, both removing six residues (See Figure 1). When these were expressed in wild type cells, we found that both proteins showed accumulation in the ER similar to that seen with the intact protein in strains lacking Erv14p (Figure 2). When placed in strains lacking Erv14p these the versions of Axl2p with deleted TMDs showed a distribution indistinguishable from that in wild type strains. These observations suggest that altering the TMD of Axl2p causes to have a slowed ER exit rate that may now be independent of Erv14p. This would be consistent with Erv14p only directly interacting with the intact TMD of Axl2p, but it indicates that requirement for Erv14p for efficient ER exit cannot be simply overcome by shortening the TMD of Axl2p. It is possible that Erv14p serves two functions - not only to chaperone Axl2p in the ER but also to recruit it into ER to Golgi transport vesicles. It may be possible to address this by using photobleaching to examine if the Axl2p-GFP fusion has aggregated in the ER of strains that lack Erv14p.

Whilst these results suggest that the TMD of Axl2p is the target of Erv14p interaction, they do not provide evidence that this purpose of this interaction is to accommodate hydrophobic mismatch. However, Erv14p is just one of several proteins that help yeast plasma membrane proteins leave the ER, and unlike Axl2p these other proteins are polytopic and so have less possibility to tilt to fit into a thinner bilayer [18-20]. Further work is clearly needed to address the question of how plasma membrane proteins cope with the hydrophobic mismatch they seem likely to experience in the ER. However, it seems certain the non-uniform lipid composition of the secretory pathway has presented eukaryotic cells with a challenge of matching TMD length and bilayer thickness. Like other physical properties of biomolecules, this will have placed constraints on evolution, but will also have provided physical phenomenon that could have been exploited by evolution to drive processes of benefit to the cell.

References

[1] G. van Meer, Lipid traffic in animal cells. *Annu. Rev. Cell Biol.* **5** (1989) 247-275.

[2] J.L. Patton and R.L. Lester, The phosphoinositol sphingolipids of *Saccharomyces cerevisiae* are highly localized in the plasma membrane. *J. Bacteriol.* **173** (1991) 3101-3108.

[3] R. Schneiter *et al.*, Electrospray ionization tandem mass spectrometry (ESI-MS/MS) analysis of the lipid molecular species composition of yeast subcellular membranes reveals acyl chain-based sorting/remodeling of distinct molecular species en route to the plasma membrane. *J. Cell Biol.* **146** (1999) 741-754.

[4] R.C. Dickson, Sphingolipid functions in *Saccharomyces cerevisiae*: comparison to mammals. *Annu. Rev. Biochem.* **67** (1998) 27-48.

[5] J.M. Boggs, Lipid intermolecular hydrogen bonding: influence on structural organization and membrane function. *Biochim. Biophys. Acta* **906** (1987) 353-404.

[6] G. van Meer, Lipids of the Golgi membrane. *Trends Cell Biol.* **8** (1998) 29-33.

[7] M.S. Bretscher and S. Munro, Cholesterol and the Golgi apparatus. *Science* **261** (1993) 1280-1281.

[8] L. Liscum and K.W. Underwood, Intracellular cholesterol transport and compartmentation. *J. Biol. Chem.* **270** (1995) 15443-15446.

[9] S.N. Grove, C.E. Bracker and D.J. Morre, Cytomembrane differentiation in the endoplasmic reticulum-Golgi apparatus-vesicle complex. *Science* **161** (1968) 171-173.

[10] J.A. Killian, Hydrophobic mismatch between proteins and lipids in membranes. *Biochim. Biophys. Acta* **1376** (1998) 401-415.

[11] T.A. Harroun, W.T. Heller, T.M. Weiss, L. Yang and H.W. Huang, Experimental evidence for hydrophobic matching and membrane-mediated interactions in lipid bilayers containing gramicidin. *Biophys. J.* **76** (1999) 937-945.

[12] R.J. Webb, J.M. East, R.P. Sharma and A.G. Lee, Hydrophobic mismatch and the incorporation of peptides into lipid bilayers: a possible mechanism for retention in the Golgi. *Biochemistry* **37** (1998) 673-679.

[13] J. Ren, S. Lew, J. Wang and E. London, Control of the transmembrane orientation and interhelical interactions within membranes by hydrophobic helix length. *Biochemistry* **38** (1999) 5905-5912.

[14] S. Munro, Localization of proteins to the Golgi apparatus. *Trends Cell Biol.* **8** (1998) 11-15.

[15] S. Munro, An investigation of the role of transmembrane domains in Golgi protein retention. *EMBO J.* **14** (1995) 4695-4704.

[16] B.A. Lewis and D.M. Engelman, Bacteriorhodopsin remains dispersed in fluid phospholipid bilayers over a wide range of bilayer thicknesses. *J. Mol. Biol.* **166** (1983) 203-210.

[17] J. Powers and C. Barlowe, Transport of axl2p depends on erv14p, an ER-vesicle protein related to the *Drosophila* cornichon gene product. *J. Cell Biol.* **142** (1998) 1209-1222.

[18] P.O. Ljungdahl, C.J. Gimeno, C.A. Styles and G.R. Fink, *SHR3*: a novel component of the secretory pathway specifically required for localization of amino acid permeases in yeast. *Cell* **71** (1992) 463-478.

[19] J.A. Trilla, A. Duran and C. Roncero, Chs7p, a new protein involved in the control of protein export from the endoplasmic reticulum that is specifically engaged in the regulation of chitin synthesis in *Saccharomyces cerevisiae*. *J. Cell Biol.* **145** (1999) 1153-1163.

[20] P.W. Sherwood and M. Carlson, Efficient export of the glucose transporter Hxt1p from the endoplasmic reticulum requires Gsf2p. *Proc. Natl. Acad. Sci. U S A* **96** (1999) 7415-7420.

[21] J.S. Robinson, D.J. Klionsky, L.M. Banta and S.D. Emr, Protein sorting in *Saccharomyces cerevisiae* - isolation of mutants defective in the delivery and processing of multiple vacuolar hydrolases. *Mol. Cell. Biol.* **8** (1988) 4936-4948.

Protein, Lipid and Membrane Traffic: Pathways and Targeting
J.A.F. Op den Kamp (Ed.)
IOS Press, 2000

Distinct Lipid Microdomains within the Apical Plasma Membrane of Polarized Epithelial Cells

A potential role of the lipid bilayer in the retention of prominin in microvilli

Katja Röper*, Denis Corbeil* and Wieland B. Huttner

Max-Planck-Institute of Molecular Cell Biology and Genetics, Pfotenhauserstrasse 110, D-01307 Dresden, and Department of Neurobiology, Interdisciplinary Center of Neuroscience, University of Heidelberg, Im Neuenheimer Feld 364, D-69120 Heidelberg, Germany

*both authors contributed equally to this work

Introduction

One interesting field of cell biology concerns the molecular mechanisms underlying the sorting of membrane proteins to, and their retention in, specific subcellular organelles. It becomes more and more evident that the plasma membrane is subdivided into distinct regions, which might constitute specific "organelle-like" functional compartments. A classical example is the plasma membrane of epithelial cells, which is characterized by two functionally and structurally distinct domains, the apical and basolateral membrane. Both membrane domains differ from each other in protein and lipid composition, and are spatially separated by the presence of tight junctions [1-3].

Asymmetric distribution of plasma membrane proteins in polarized epithelial cells is achieved by three major mechanisms, which are not mutually exclusive [4-6]. First, newly synthesized membrane components are sorted at the level of the trans-Golgi network (TGN) or the basolateral endosome into carrier vesicles which are targeted to the appropriate membrane domain [7-11]. Second, the selective proteolytic processing of randomly delivered membrane proteins at one specific cell surface leads to their enrichment in the opposite domain [12]. Third, certain integral membrane proteins are selectively retained in the appropriate plasma membrane domain, whereas they are endocytosed from the other membrane domain and become quickly degraded [13-16].

Mechanisms underlying the sorting of transmembrane proteins to the apical and basolateral plasma membrane domains have been the subject of intense studies during the last twenty years leading to the identification and characterization of several sorting signals (for a review, see ref. [72].). For instance, short cytoplasmic amino acid stretches consisting of either a tyrosine or a di-leucine motif have been found to constitute basolateral sorting signals

[17,18]. N- and possibly O-linked glycans [19,20], as well as association with specific cholesterol/glycosphingolipid-rich membrane microdomains termed rafts [21], mediated by either a GPI-moiety or a transmembrane segment, have been identified as potential apical sorting signals [22-24]. Recently, cytoplasmic amino acid-based motifs have also been proposed to act as apical sorting signals [25-27]. Finally, several studies using chimerae and/or truncated membrane proteins have shown that a single membrane protein could contain more than one type of sorting information, one acting dominant over the other [28-30].

Once targeted to the appropriate plasma membrane domain, certain integral membrane proteins are specifically retained and concentrated in subdomains of either the apical or basolateral plasma membrane [31-33]. In contrast to sorting information, very little is known regarding the mechanisms involved in this retention and concentration (see below). Such specific retention and concentration could provide a general explanation for plasma membrane compartmentalization, and this not only in epithelial cells which contain tight junctions, but also in non-epithelial cells [34]. Furthermore, it has been shown in epithelial cells that the polarized distribution of several apical cell surface glycoproteins does not depend on the presence of functional tight junctions [5,35-37], suggesting that other fence mechanisms can operate in this plasma membrane domain [38].

Our group has been studying the apical-basal polarity of neuroepithelial cells, which constitute the inner layer of the neural tube of the vertebrate embryo and give rise to all neurons and most glial cells of the central nervous system [39]. Interestingly, during the transition from the neural plate stage to the neural tube stage, neuroepithelial cells loose functional tight junctions [40] and polarity of certain plasma membrane proteins [41]. However, after neural tube closure, not all plasma membrane proteins of neuroepithelial cells become randomly distributed over the apical and basolateral plasma membranes. One of these plasma membrane proteins is prominin [42] (see below). Remarkably, prominin is exclusively associated with microvilli present at the apical domain of neuroepithelial cells which suggests that entry into, and retention in, microvilli may play an important role in the maintenance of apical-basal polarity of epithelial cells in the absence of tight junctions.

Prominin, a specific marker of plasma membrane protrusions

Prominin is a polytopic membrane protein that is present on the apical surface of several mouse embryonic epithelia including neuroepithelial cells, and on the brush border membrane of epithelial cells lining the adult kidney proximal tubules [42]. It was first identified using the monoclonal antibody (mAb) 13A4 generated against E12 mouse telencephalic neuroepithelium, and its cDNA was cloned from mouse kidney [42]. This novel 115-kD glycoprotein contains five putative transmembrane domains with an N-terminal extracellular domain and a cytoplasmic C-terminal domain (Fig. 1).

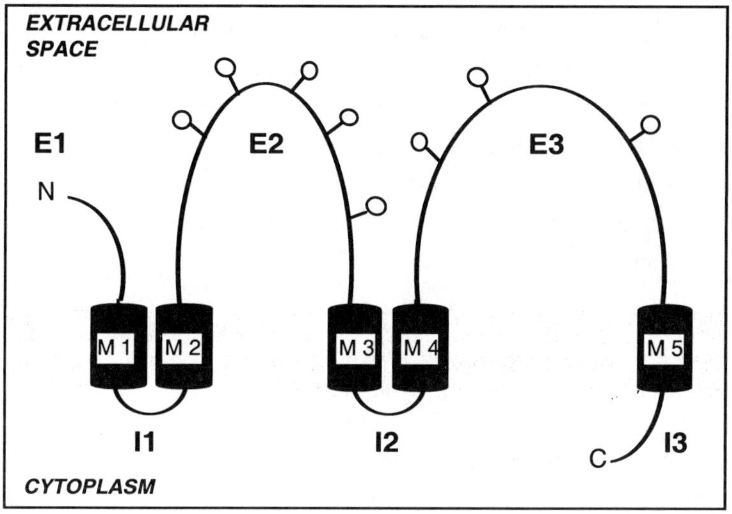

Figure 1. Topological model of prominin. M1-M5, putative transmembrane segments; E1-E3, extracellular domains; I1-I3, cytoplasmic domains. The two large extracellular loops (domains E2 and E3) contain all eight potential N-glycosylation sites (lollipops). Modified from Corbeil et al., 1999.

Its human orthologue [43,44], prominin (mouse)-like-1 (originally referred to as AC133 antigen), recently found to be expressed in epithelial cells (D.C., K.R., A. Hellwig, M. Tavian, S. Miraglia, S.M. Watt, P.J. Simmons, B. Peault, D.W. Buck, W.B.H., manuscript in preparation), was initially described in non-epithelial cells, i.e. hematopoietic progenitor cells [45,46]. Although the function of prominin is unknown, the occurrence of orthologues/paralogues in zebrafish (accession number: AF160970), *Drosophila melanogaster* (accession number: AF127935) and *C. elegans* [42-44] suggests that this polytopic membrane protein plays an important role from invertebrates to vertebrates.

One particularly interesting feature of prominin in epithelial cells is its selective occurrence in specific subdomains of the apical plasma membrane. In epithelial cells with a brush border, two structurally and functionally distinct subdomains of the apical plasmalemma can be distinguished; (i) the microvilli, which enlarge the apical surface; and (ii) an extended system of clathrin coated pits located at the base of the microvilli [47-50]. In kidney proximal tubules, prominin is restricted to the brush border membrane, where it appears to be concentrated at the tips of the microvilli, and is largely excluded from the intermicrovillar plasma membrane regions [42]. In the apical plasma membrane of neuroepithelial cells, which contains less microvilli than the kidney brush border, prominin is associated with microvilli and other small plasma membrane protrusions rather than the large planar subdomain [42]. Furthermore, this protrusion-specific localization of prominin does not depend on an epithelial state of the cell. When prominin is ectopically expressed in fibroblasts such as CHO cells, it is also selectively found in plasma membrane protrusions such as microspikes, filopodia, membrane ruffles and the leading edge of lamellipodia [42] (unpublished data).

Although the general mechanisms underlying the retention of membrane proteins in plasma membrane protrusions has not been elucidated (see below), expression and localization of prominin in epithelial and non-epithelial cells, as revealed that this mechanism is highly conserved between these types of cells and may probably reflect a cell type-specific adaptation of a process common to all eukaryotic cells.

Prominin contains dual targeting information

Recently, we could show that the enrichment of prominin in plasma membrane protrusions, i.e. microvilli at the apical surface of polarized epithelial MDCK cells, is due to both sorting to the apical membrane domain at the TGN level and subsequent retention in the microvillar subdomain [5]. This dual "targeting" does not require the cytoplasmic tail of prominin and, more interestingly, the selective retention of prominin in microvilli was maintained when prominin-transfected MDCK cells were cultured in low-calcium medium [5], i.e. in the absence tight junctions [51]. These *in vitro* data resemble the *in vivo* situation observed in the neuroepithelium (see Introduction). They suggest a potential role of the microvillar retention of certain plasma membrane proteins in maintaining their polarized distribution even without functional tight junctions, which would be achieved by preventing their lateral movement.

Retention of integral membrane proteins in the apical or basolateral domain

The retention of membrane proteins inside a specific plasmalemmal subdomain can be obtained by direct protein-protein interactions occurring in that particular domain. For example, selective anchoring of plasma membrane proteins to submembraneous cytoskeletal elements, and/or specific interaction with other membrane proteins or extracellular components, are possible alternative retention mechanisms.

Cytoskeletal interactions have been reported to explain the polarized distribution of the Na^+/K^+-ATPase [14,52], the Na^+/H^+-exchanger [53] and the amiloride-sensitive Na^+ channel [54] in epithelial cells, the latter two being captured on the apical side via association of their C-terminal proline-rich domain with the SH3-domain of α-spectrin. In contrast, the Na^+/K^+-ATPase is maintained on the (baso)lateral side via binding to an ankyrin/fodrin/uvomorulin complex in MDCK cells [14], whereas in the retinal pigment epithelium, where the polarity of some cytoskeletal components is reversed, a similar association with ankyrin restricts it to the apical side [52].

As shown more recently, several GABA-transporter subtypes contain a PDZ-interacting domain, which is necessary for their retention, and hence polarized localization, on either the basolateral membrane domain (as in the case of BGT-1 [6]) or the apical membrane domain (as in the case of GAT-3 [26]). PDZ domain-containing proteins are known to cluster interacting proteins at specific functional sites of the cell such as synapses or tight junctions in order to allow these proteins to exert their function.

Retention of prominin in microvilli: facts and a hypothesis

Microvilli are a characteristic subdomain of the plasma membrane present not only in epithelial cells [47], but also in non-epithelial cells, as for instance leukocytes. Physiological functions of microvilli are reflected by the specific membrane proteins found therein; cell adhesion molecules in the case of leukocytes; brush border enzymes and transporters in the case of resorptive epithelia. The molecular mechanisms underlying the retention of integral membrane proteins in microvilli and other membrane protrusions are poorly understood [55-58]. To our knowledge, it is not clear whether all microvillar membrane proteins interact with constituents of the subplasmalemmal actin network such as brush-border myosin 1, ezrin or other members of the ERM (ezrin-radixin-moesin) family, which are known to serve as general plasma membrane protrusion–actin microfilament linkers [59,60], although some clearly do [61,62]. For certain membrane proteins such as sucrase-isomaltase [63], the presence of "intact microvilli" appears to be essential for their cell surface expression and polarized distribution.

In order to elucidate a possible association of prominin with components of the microvillar cytoskeleton, prominin-transfected MDCK cells were solubilized in Triton X-100 in the cold and fractionated into detergent-soluble and -insoluble material. Prominin appeared to be completely soluble in Triton X-100 [5], as reported already for several brush border enzymes, such as lactase-phlorizin hydrolase [64,65]. Interestingly, this brush-border hydrolase, like prominin, has been found to be specifically associated with microvilli [66]. One interpretation of these data is that prominin does not bind to cytoskeletal components. However, an alternative possibility is that prominin is anchored to the cytoskeleton via a labile interaction which may have been disrupted under our experimental conditions. To further investigate a putative interaction of prominin with cellular components, we decided to explore other solubilization conditions that might prevent such dissociation. Interestingly, the cell surface, microvilli-associated form of prominin was found to be insoluble in cold Lubrol WX, another nonionic detergent [67,68], whereas the ER form was completely soluble (manuscript in preparation).

In principle, two modes of interaction could cause the Lubrol WX insolubility of prominin, (i) binding to cytoskeletal or extracellular matrix (ECM) constituents (referred to as vertical interaction), and/or (ii) association with other transmembrane proteins or certain lipids (referred to as lateral interaction). The latter type of interaction has been proposed to explain the microvillar presentation of integrins in leukocytes. In this particular case, it was postulated that certain tetraspan proteins and other integrin-associated molecules form multimolecular complexes with the integrins, and this complex formation may have a role in targeting/retention of integrins in microvilli [57].

Further biochemical analysis of the microvilli-associated, Lubrol WX insoluble prominin suggested a lateral interaction. Interestingly, this insoluble complex showed several characteristics of detergent resistant membrane microdomains (DRMMs) [22], including (i) floatation to lower density fractions in sucrose density gradients; (ii) loss of insolubility at 37°C; and (iii) a high content of cholesterol and glycolipids. In addition, removal of plasma membrane cholesterol by means of methyl-β-cyclodextrin (m-β-CD) treatment led to the complete solubilization of prominin in Lubrol WX at 4°C and to loss of floatation. Remarkably, m-β-CD treatment resulted in the redistribution of prominin, as observed by cell surface immunofluorescence, from an exclusively microvillar localization to a more homogeneous distribution over the apical surface. The effects of cholesterol depletion were reversed by refeeding the cells with cholesterol.

Taken together, these data suggest that a certain lipid composition and/or organization of the lipid bilayer of microvilli and other plasma membrane protrusions is necessary (and perhaps sufficient) to maintain prominin in its proper location. In analogy to the role of the length of transmembrane segments in membrane protein retention in, vs. exit from, the Golgi complex [69,70], a certain length of prominin's transmembrane segments is possibly required for its lateral diffusion into, and concentration in, plasma membrane protrusions. In this context, it is interesting to note that four of the five predicted transmembrane segments of prominin contain >25 amino acid residues and thus exceed the average length of the transmembrane segment of plasma membrane proteins [69].

Distinct lipid microdomains as building blocks of the apical plasma membrane: a model

The specific subcellular localization of prominin in the apical microvillar subdomain and its complete solubilization in Triton X-100 but not Lubrol WX strongly suggest that the lipid bilayer of this microdomain is sensitive to extraction by the former detergent but is preserved by Lubrol WX. This differential detergent extraction may reflect the existence of at least two distinct lipid microdomains at the level of the apical plasma membrane, characteristic of either the microvillar plasmalemma (Triton X-100 soluble and Lubrol WX insoluble) or the planar, intermicrovillar plasmalemma (Triton X-100 and Lubrol WX insoluble) (Fig. 2).

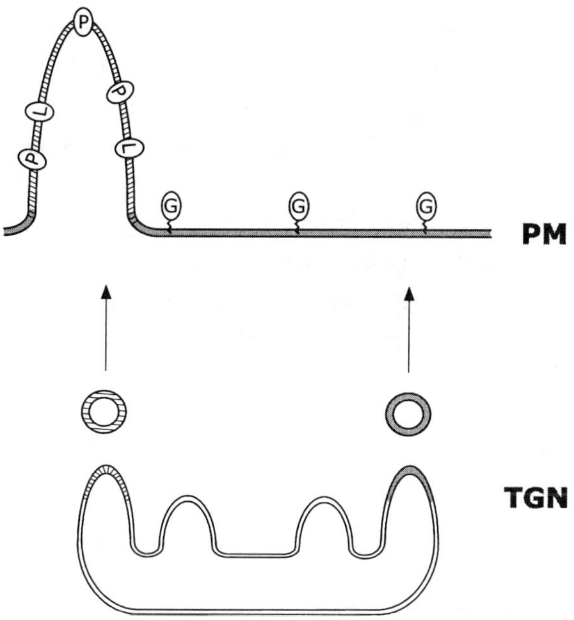

Fig.2 Formation of distinct lipid microdomains in the apical plasma membrane. The differential detergent extraction properties of prominin and other apical plasma membrane protein constituents could reflect the existence of at least two distinct lipid microdomains at the level of the apical plasma membrane. Striped area, Lubrol WX insoluble and TX-100 soluble membrane; grey area, Lubrol WX and Triton X-100 insoluble membrane). These microdomains could serve a role in retaining proteins at sites of their action, and/or, as both domains seem to be formed already at the level of the TGN, in sorting of apical plasma membrane components. PM: plasma membrane; TGN: trans-Golgi network; P: prominin; L: lactase-phlorizin hydrolase; G: GPI-linked proteins.

Consistent with this proposal, Danielsen and van Deurs (1995) have observed that a GPI-linked protein, the transferrin-like GPI-linked iron-binding protein (which was found to be insoluble in Triton X-100 like other GPI-anchored proteins) was mainly localized in patches of flat or invaginated apical membrane subdomains rather than at the surface of microvilli (see figure 3 in [50]). Thus, the "affinity" of certain plasma membrane proteins to particular plasma membrane lipids, as defined here by their solubility in different detergents (Triton X-100 vs. Lubrol WX), could target these molecules to, and maintain them in, spatially distinct microdomains. It remains to be established whether microvillar association of transmembrane proteins is generally reflected by an insolubility in Lubrol WX (and solubility in Triton X-100), as demonstrated here for prominin.

It has been proposed that the Triton X 100 insoluble raft has a function in sorting of newly synthesized plasma membrane components from the TGN to the cell surface, via the formation of lipidic platforms that incorporate the cargo proteins [71] (see also Introduction). As prominin achieves its Lubrol WX-insolubility during transport to the cell surface (manuscript in preparation), it is tempting to speculate that the Lubrol WX-insoluble/Triton X-100-soluble microdomain could also function in sorting (Fig. 2). Quantitative analysis of the lipid composition of Lubrol WX-insoluble complexes as well as a comprehensive analysis of its protein composition may reveal further details about the mechanisms involved in either sorting to the apical domain or retention in a specific apical subdomain via this type of DRMM.

Acknowledgements

K.R. was supported by the Studienstiftung des Deutschen Volkes, D.C. was the recipient of fellowships from the Medical Research Counsil of Canada and the Max-Planck-Society, W.B.H. was supported by a grant from the Deutsche Forschungsgemeinschaft (SFB 352, C1).

References

[1] Diamond, J.M. The epithelial junction: Bridge, gate and fence. *Physiologist* **20**, 10-18 (1977).

[2] Dragsten, P.R., Blumenthal, R. & Handler, J.S. Membrane asymmetry in epithelia: Is the tight junction a barrier to diffusion in the plasma membrane? *Nature* **294**, 718-722 (1981).

[3] van Meer, G. & Simons, K. The function of tight junctions in maintaining differences in lipid composition between the apical and basolateral cell surface domains of MDCK cells. *EMBO J.* **5**, 1455-1464 (1986).

[4] Nussenzveig, D.R., Matos, M.d.F.C. & Thaw, C. Human calcitonin receptor is directly targeted to and retained in the basolateral surface of MDCK cells. *Am. J. Physiol.* **275**, C1264-C1276 (1998).

[5] Corbeil, D., Röper, K., Hannah, M.J., Hellwig, A. & Huttner, W.B. Selective localization of the polytopic membrane protein prominin in microvilli of epithelial cells- a combination of apical sorting and retention in plasma membrane protrusions. *J. Cell Sci.* **112**, 1023-1033 (1999).

[6] Perego, C. *et al.* PDZ-mediated interactions retain the epithelial GABA transporter on the basolateral surface of polarized epithelial cells. *EMBO J.* **18**, 2384-2393 (1999).

[7] Rodriguez-Boulan, E. & Nelson, J. Morphogenesis of the polarized epithelial cell phenotype. *Science* **245**, 718-725 (1989).

[8] Simons, K. *et al.* Biogenesis of cell-surface polarity in epithelial cells and neurons. *Cold Spring Harbor Symp. Quant. Biol.* **57**, 611-619 (1992).

[9] Bartles, J.R., Feracci, H.M., Stieger, B. & Hubbard, A.L. Biogenesis of the rat hepatocyte plasma membrane in vivo: comparison of the pathways taken by apical and basolateral proteins using subcellular fractionation. *J. Cell Biol.* **105**, 1241-1251 (1987).

[10] Massey, D., Feracci, H., Gorvel, J-P., Rigal, A., Soulié, J.M. & Maroux, S. Evidence for the transit of aminopeptidase N through the basolateral membrane before it reaches the brush border of enterocytes. *J. Membr. Biol.* **96**, 19-25 (1987).

[11] Matter, K., Brauchbar, M., Bucher, K. & Hauri, H.-P. Sorting of endogenous plasma membrane proteins occurs from two sites in cultured human intestinal epithelial cells (Caco-2). *Cell* **60**, 429-437 (1990).

[12] Dempsey, P.J., Meise, K.S., Yoshitake, Y., Nishikawa, K. & Coffey, R.J. Apical enrichment of human EGF precursor in Madin-Darby canine kidney cells involves preferential basolateral ectodomain cleavage sensitive to a metalloprotease inhibitor. *J. Cell Biol.* **138**, 747-758 (1997).

[13] Caplan, M.J., Anderson, H.C., Palade, G.E. & Jamieson, J.D. Intracellular sorting and polarized cell surface delivery of Na^{+}/K^{+}-ATPase an endogenous component of MDCK cell basolateral plasma membranes. *Cell* **46**, 623-631 (1986).

[14] Hammerton, R.W. *et al.* Mechanism for regulating cell surface distribution of Na^{+}/K^{+}-ATPase in polarized epithelial cells. *Science* **254**, 847-850 (1991).

[15] Mays, R.W. *et al.* Hierarchy of mechanisms involved in generating Na^{+}/K^{+}-ATPase polarity in MDCK epithelial cells. *J. Cell Biol.* **130**, 1105-1115 (1995).

[16] Wozniak, M. & Limbird, L.E. The three α2-adrenergic receptor subtypes achieve basolateral localization in madin-darby canine kidney II cells via different targeting mechanisms. *J. Biol. Chem.* **271**, 5017-5024 (1996).

[17] Hunziker, W., Harter, C., Matter, K. & Mellman, I. Basolateral sorting in MDCK cells requires a distinct cytoplasmic domain determinant. *Cell* **66**, 1-20 (1991).

[18] Hunziker, W. & Fumey, C. A di-leucine motif mediates endocytosis and basolateral sorting of macrophage IgG Fc receptors in MDCK cells. *EMBO J.* **13**, 2963-2967 (1994).

[19] Scheiffele, P., Peränen, J. & Simons, K. N-glycans as apical sorting signals in epithelial cells. *Nature* **378**, 96-98 (1995).

[20] Yeaman, C. *et al.* The O-glycosylated stalk domain is required for apical sorting of neurotrophin receptors in polarized MDCK cells. *J. Cell Biol.* **139**, 929-940 (1997).

[21] Simons, K. & Ikonen, E. Functional rafts in cell membranes. *Nature* **387**, 569-572 (1997).

[22] Brown, D.A. & Rose, J.K. Sorting of GPI-Anchored proteins to glycolipid-enriched membrane subdomains during transport to the apical cell surface. *Cell* **68**, 533-544 (1992).

[23] Kundu, A., Avalos, R.T., Sanderson, C.M. & Nayak, D.P. Transmembrane domain of influenza virus neuraminidase, a type II protein, possesses an apical sorting signal in polarized MDCK cells. *J. Virol.* **70**, 6508-6515 (1996).

[24] Scheiffele, P., Roth, M.G. & Simons, K. Interaction of influenza virus haemagglutinin with sphingolipid-cholesterol membrane domains via its transmembrane domain. *EMBO J.* **16**, 5501-5508 (1997).

[25] Chuang, J.-Z. & Sung, C.-H. The cytoplasmic tail of rhodopsin acts as a novel apical sorting signal in polarized MDCK cells. *J. Cell Biol.* **142**, 1245-1256 (1998).

[26] Muth, T.R., Ahn, J. & Caplan, M.J. Identification of sorting determinants in the C-terminal cytoplasmic tails of the gamma-aminobutyric acid transporters GAT-2 and GAT-3. *J. Biol. Chem.* **273**, 25616-25627 (1998).

[27] Sun, A.Q. *et al.* Sorting of rat liver and ileal sodium-dependent bile acid transporters in polarized epithelial cells. *Am. J. Physiol.* **275**, G1045-1055 (1998).

[28] Powell, S.K., Lisanti, M.P. & Rodriguez Boulan, E.J. Thy-1 expresses two signals for apical localization in epithelial cells. *Am. J. Physiol.* **260**, C715-720 (1991).

[29] Weisz, O.A., Machamer, C.E. & Hubbard, A.L. Rat liver dipeptidylpeptidase IV contains competing apical and basolateral targeting information. *J. Biol. Chem.* **267**, 22282-22288 (1992).

[30] Jacob, R. *et al.* Hierarchy of sorting signals in chimeras of intestinal lactase-phlorizin hydrolase and the influenza virus hemagglutinin. *J. Biol. Chem.* **274**, 8061-8067 (1999).

[31] Nelson, W.J., Shore, E.M., Wang, A.Z. & Hammerton, R.W. Identification of a membrane-cytoskeletal complex containing the cell adhesion molecule uvomorulin (E-cadherin), ankyrin, and fodrin in Madin-Darby canine kidney epithelial cells. *J. Cell Biol.* **110**, 349-357 (1990).

[32] Beck, K.A. & Nelson, W.J. The spectrin-based membrane skeleton as a membrane protein-sorting machine. *Am. J. Physiol.* **270**, C1263-1270 (1996).

[33] Edidin, M. Patches and fences: probing for plasma membrane domains. *J. Cell Sci. Suppl.* **17**, 165-169 (1993).

[34] Sheets, E.D., Simson, R. & Jacobson, K. New insights into membrane dynamics from the analysis of cell surface interactions by physical methods. *Curr. Op. Cell Biol.* **7**, 707-714 (1995).

[35] Vega-Salas, D.E., Salas, P.J.I., Gunderson, D. & Rodriguez-Boulan, E. Formation of the apical pole of epithelial (MDCK) cells: Polarity of an apical protein is independent of tight junctions while segregation of a basolateral marker requires cell-cell interactions. *J. Cell Biol.* **104**, 905-916 (1987).

[36] Ojakian, G.K. & Schwimmer, R. The polarized distribution of an apical cell surface glycoprotein is maintained by interactions with the cytoskeleton of Madin-Darby canine kidney cells. *J. Cell Biol.* **107**, 2377-2387 (1988).

[37] Rodriguez, M.L., Brignoni, M. & Salas, P.J. A specifically apical sub-membrane intermediate filament cytoskeleton in non-brush-border epithelial cells. *J. Cell Sci.* **107**, 3145-3151 (1994).

[38] Salas, P.J.I., Vega-Salas, D.E., Hochman, J., Rodriquez-Boulan, E. & Edidin, M. Selective anchoring in the specific plasma membrane: a role in epithelial cell polarity. *J. Cell Biol.* **107**, 2363-2376 (1988).

[39] Huttner, W.B. & Brand, M. Asymmetric division and polarity of neuroepithelial cells. *Curr. Opin. Neurobiol.* **7**, 29-39 (1997).

[40] Aaku-Saraste, E., Hellwig, A. & Huttner, W.B. Loss of occludin and functional tight junctions, but not ZO-1, during neural tube closure - remodeling of the neuroepithelium prior to neurogenesis. *Dev. Biol.* **180**, 664-679 (1996).

[41] Aaku-Saraste, E., Oback, B., Hellwig, A. & Huttner, W.B. Neuroepithelial cells downregulate their plasma membrane polarity prior to neural tube closure and neurogenesis. *Mech. Dev.* **69**, 71-81 (1997).

[42] Weigmann, A., Corbeil, D., Hellwig, A. & Huttner, W.B. Prominin, a novel microvilli-specific polytopic membrane protein of the apical surface of epithelial cells, is targeted to plasmalemmal protrusions of non-epithelial cells. *Proc. Natl. Acad. Sci. USA* **94**, 12425-12430 (1997).

[43] Corbeil, D., Roeper, K., Weigmann, A. & Huttner, W.B. AC133 hematopoietic stem cell antigen: human homologue of mouse kidney prominin or distinct member of a novel protein family? *Blood* **91**, 2625-2626 (1998).

[44] Miraglia, S., Godfrey, W. & Buck, D. A response to AC133 hematopoietic stem cell antigen: human homologue of mouse kidney prominin or distinct member of a novel protein family. *Blood* **91**, 4390-4391 (1998).

[45] Yin, A.H. *et al.* AC133, a novel marker for human hematopoietic stem and progenitor cells. *Blood* **90**, 5002-5012 (1997).

[46] Miraglia, S. *et al.* A novel five-transmembrane hematopoietic stem cell antigen: isolation, characterization, and molecular cloning. *Blood* **90**, 5013-5021 (1997).

[47] Kerjaschki, D., Noronha-Blob, L., Sacktor, B. & Farquhar, M.G. Microdomains of distinctive glycoprotein composition in the kidney proximal tubule brush border. *J. Cell Biol.* **98**, 1505-1513 (1984).

[48] Rodman, J.S., Kerjaschki, D., Merisko, E. & Farquhar, M.G. Presence of an extensive clathrin coat on the apical plasmalemma of the rat kidney proximal tubule cell. *J. Cell Biol.* **98**, 1630-1636 (1984).

[49] Biemesderfer, D., Dekan, G., Aronson, P.S. & Farquhar, M.G. Assembly of distinctive coated pit and microvillar microdomains in the renal brush border. *Am. J. Physiol.* **262**, F55-67 (1992).

[50] Danielsen, M.E. & van Deurs, B. A transferrin-like GPI-linked iron-binding protein in detergent-insoluble noncaveolar microdomains at the apical surface of fetal intestinal epithelial cells. *J. Cell Biol.* **131**, 939-950 (1995).

[51] Gonzalez-Mariscal, L., Chávez de Ramírez, B. & Cereijido, M. Tight junction formation in cultured epithelial cells (MDCK). *J. Membrane Biol.* **86**, 113-125 (1985).

[52] Rizzolo, L.J. & Zhou, S. The distribution of Na^+/K^+-ATPase and 5A11 antigen in apical microvilli of the retinal pigment epithelium is unrelated to alpha-spectrin. *J. Cell Sci.* **108**, 3623-3633 (1995).

[53] Chow, C.-W. *et al.* Proline-rich motifs of the Na^+/H^+-exchanger 2 isoform. *J. Biol. Chem.* **274**, 10481-10488 (1999).

[54] Rotin, D. *et al.* An SH3 binding region in the epithelial Na^+ channel (arENaC) mediates its localization at the apical membrane. *EMBO J.* **13**, 4440-4450 (1994).

[55] von Andrian, U.H., Hasslen, S.R., Nelson, R.D., Erlandsen, S.L. & Butcher, E.C. A central role for microvillous receptor presentation in leukocyte adhesion under flow. *Cell* **82**, 989-999 (1995).

[56] Pavalko, F.M. *et al.* The cytoplasmic domain of L-selectin interacts with cytoskeletal proteins via alpha-actinin: receptor positioning in microvilli does not require interaction with alpha-actinin. *J. Cell Biol.* **129**, 1155-1164 (1995).

[57] Abitorabi, M.A., Pachynski, R.K., Ferrando, R.E., Tidswell, M. & Erle, D.J. Presentation of integrins on leukocyte microvilli: a role for the extracellular domain in determining membrane localization. *J. Cell Biol.* **139**, 563-571 (1997).

[58] Hamer, I. *et al.* Dual role of a dileucine motif in insulin receptor endocytosis. *J. Biol. Chem.* **272**, 21685-21691 (1997).

[59] Bretscher, A. Purification of an 80,000-dalton protein that is a component of isolated microvillus cytoskeleton, and its localization in nonmuscle cells. *J. Cell Biol.* **97**, 425-432 (1983).

[60] Mangeat, P., Roy, C. & Martin, M. ERM proteins in cell adhesion and membrane dynamics. *Trends Cell Biol.* **9**, 187-192 (1999).

[61] Coudrier, E., Reggio, H. & Louvard, D. Characterization of an integral membrane glycoprotein associated with the microfilaments of pig intestinal microvilli. *EMBO J.* **2**, 469-475 (1983).

[62] Yonemura, S. *et al.* Ezrin/Radixin/Moesin (ERM) proteins bind to a positively charged amino acid cluster in the juxta-membrane cytoplasmic domain of CD44, CD43 , and ICAM-2. *J. Cell Biol.* **140**, 885-895 (1998).

[63] Costa de Beauregard, M.-A., Pringault, E., Robine, S. & Louvard, D. Suppression of villin expression by antisense RNA impairs brush border assembly in polarized epithelial intestinal cells. *EMBO J.* **14**, 409-421 (1995).

[64] Danielsen, E.M. Involvement of detergent-insoluble complexes in the intracellular transport of intestinal brush border enzymes. *Biochemistry* **34**, 1596-1605 (1995).

[65] Zheng, X., Lu, D. & Sadler, J.E. Apical sorting of bovine enteropeptidase does not involve detergent-resistant association with sphingolipid-cholesterol rafts. *J. Biol. Chem.* **274**, 1596-1605 (1999).

[66] Jacob, R., Zimmer, K.P., Naim, H. & Naim, H.Y. The apical sorting of lactase-phlorizin hydrolase implicates sorting sequences found in the mature domain. *Eur. J. Cell Biol.* **72**, 54-60 (1997).

[67] Moosic, J.P., Sung, E., Nilson, A., Jones, P.P. & McKean, D.J. The selective solubilization of different murine splenocyte membrane fractions with lubrol WX and Triton X-100 distinguishes two forms of Ia antigens. A cell surface (alpha, beta) and an intracellular (alpha, Ii, beta). *J. Biol. Chem.* **257**, 9684-9691 (1982).

[68] Schmucker, B., Ballhausen, W.G. & Kressel, M. Subcellular localization and expression pattern of the neurofibromatosis type 2 protein merlin/schwannomin. *Eur. J. Cell Biol.* **72**, 46-53 (1997).

[69] Bretscher, M.S. & Munro, S. Cholesterol and the Golgi apparatus. *Science* **261**, 1280-1281 (1993).

[70] Munro, S. An investigation of the role of transmembrane domains in Golgi protein retention. *EMBO J.* **14**, 4695-4704 (1995).

[71] Simons, K. & Wandinger-Ness, A. Polarized sorting in epithelia. *Cell* **62**, 207-210 (1990).

[72] Yeaman, C., Grindstaff, K.K., & Nelson, W.J. New perspectives on mechanisms involved in generating epithelial cell polarity. *Physiol. Rev.* **79**, 73-98 (1999).

Spectrofluorometric Characterization of Lipid-Peptide Interactions

Susanne Gangl[1], Erich Staudegger[2], Karl Lohner[2], Sylvie Blondelle[3], Gottfried Grabner[1] and Gottfried Köhler[1]

[1] *Institut für Theoretische Chemie und Strahlenchemie, University of Vienna, Austria,*
[2] *Institut für Biophysik und Röntgenstrukturforschung, Österreichische Akademie der Wissenschaften, Austria and*
[3] *Torrey Pines Institute for Molecular Studies, San Diego, CA 92121, USA*

Abstract. Lipid-peptide interactions are analyzed by fluorescence spectroscopic techniques for a 21-residue antimicrobial peptide isolated from frog skin and model membranes. In particular specific docking of the peptide to the membrane surface and incorporation into the lipid bilayer is followed and the resulting perturbations of the acyl chain order are analyzed.

The location of this cationic peptide within model membranes composed of DMPC and DMPG was investigated by fluorescence emission using a derivative of PGLa carrying a single tryptophan residue. In the case the surface is negatively charged the tryptophan residue is well embedded within the hydrophobic part of the bilayer. Fluorescence resonance energy transfer to 12-(9-anthroyloxy) stearic acid revealed a considerable increase of the spatial separation of the two chromophors in the fluid phase thus indicating lateral phase separation induced by the peptide.

Different fluorescence probes reporting specific physico-chemical properties of the lipid bilayer are applied. The change of the shape and position of the emission spectrum of 12-(9-anthroyloxy) stearic acid in the presence of PGLa indicate restricted mobility, decreased microviscosity and a lower degree of water accessibility within the acyl chain region, both in the gel and in the liquid-crystalline phase. A model for the interactions between PGLa and membranes is proposed.

1. Introduction

Cell membranes are one of the most important constituents of the cell responsible for a great variety of biological processes including respiration, vision, photosynthesis, cell-cell recognition, nerve impulse transmission and they are a target of drug action. An important general feature is that the function of a given membrane is determined by its chemical composition, physical state and mode of organization, all of which are interdependent [1].

The current concept of biological membranes rests on a dynamic molecular assembly of their constituents characterized by the coexistence of domains with highly restricted mobility and other components having great motional freedom [2].

The mechanisms responsible for the lateral organization of membranes are rather complex and apart from lipid-lipid interactions also interactions with incorporated structures like peptides play an important role [3,4].

The first step in peptide-lipid interactions is often determined by electrostatic interactions between positively charged amino acids and negatively charged phospholipids, which can contribute to the orientation and folding of the peptide. This process may be responsible for membrane protein topology (e.g. the "*positive-inside*" rule [5]), may result in specific receptor-mediated import of proteins into organelles [6] or may lead to membrane damage as in the case of membranolytic peptides [7,8]. A proper understanding of the rules governing lipid-peptide interactions is necessary in order to get insight into the role of the lipid bilayer in mediating such diverse events. Therefore, liposomes as a relatively simple model system were chosen for our studies, because the effects of membrane perturbing peptides on the physico-chemical properties of the bilayer can be investigated systematically as a function of lipid composition, temperature, pH and salt concentration.

2. Lipid-Peptide Interactions of the Membrane-active Peptide, PGLa

Changes of the physico-chemical properties of a lipid bilayer in the presence of the membranolytic peptide PGLa are investigated.

The antimicrobial peptide peptidyl-glycylleucine-carboxyamide (PGLa; GMASK AGAIAGKIAKVALKAL-carboxyamide), isolated from the skin of the South African clawed frog, *Xenopus laevis* [9] contains 21 amino acids and was shown to be antimicrobial [10]. It is cationic and possesses an amidated C-terminus which increases its resistance to proteases [12]. The lytic activity of this peptide stems from its ability to perturb the lipid bilayer structure of membranes and is not related to binding to a specific protein receptor site [11, 13]. Like other antimicrobial peptides, PGLa displays a broad spectrum of activity lysing Gram-negative and Gram-positive bacteria, fungi and protozoa, but they rarely affect eucaryotic cells. This specificity in membranolysis is a prerequisite to avoid destruction of eukaryotic cells, a most important fact to consider in evaluating antimicrobial peptides as a source of antibiotics [7]. Therefore we investigated the effect of PGLa on liposomes composed of zwitterionic and negatively charged phospholipids (Fig. 1), the former being typical for eukaryotic and the latter for bacterial plasma membranes. In addition, a derivative of PGLa containing Trp as an intrinsic fluorophor at position 6 was used.

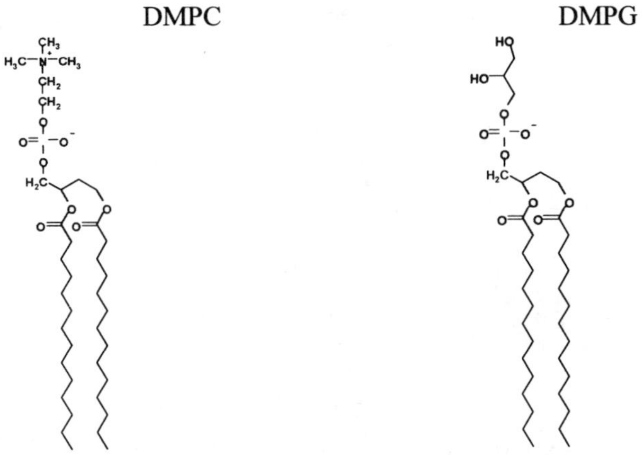

Fig. 1. Chemical structure of the zwitterionic lipid DMPC and the negatively charged lipid DMPG.

3. „Multiprobe" approach

3.1. Location of W-PGLa in the membrane

Fluorescence emission spectra of a derivative of PGLa containing one tryptophan residue (W-PGLa, see Fig. 2) were recorded in order to probe the molecular environment and the location of tryptophan (Trp) within the bilayer.

The tryptophan fluorescence emission spectra of W-PGLa in aqueous solution and in liposomes containing only DMPC (Fig. 3) show a maximum at 355 nm and a rather low fluorescence quantum yield typical for a polar, aqueous environment. The temperature dependence of the fluorescence yield parallels that of Trp in water and shows no influence on the main phase transition of a DMPC bilayer around 24 °C (Fig. 4).

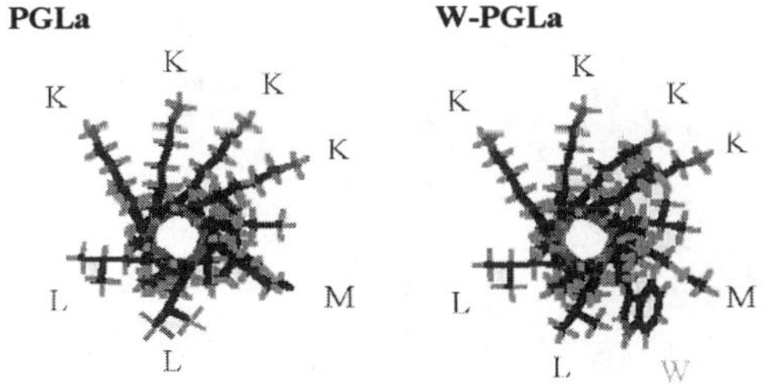

Fig.2. Helical wheel representation of PGLa and W-PGLa seen from the N-terminus.

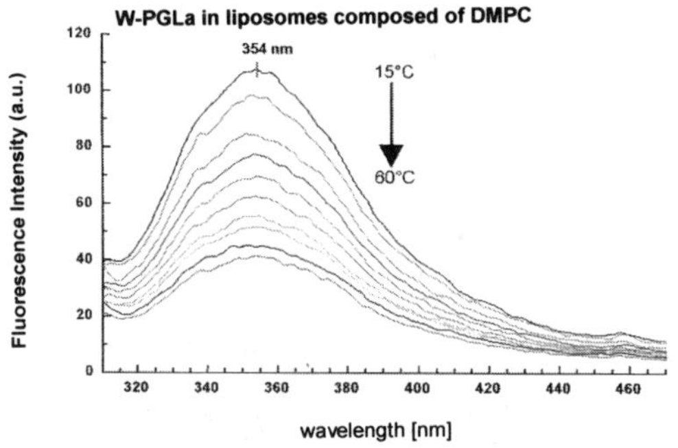

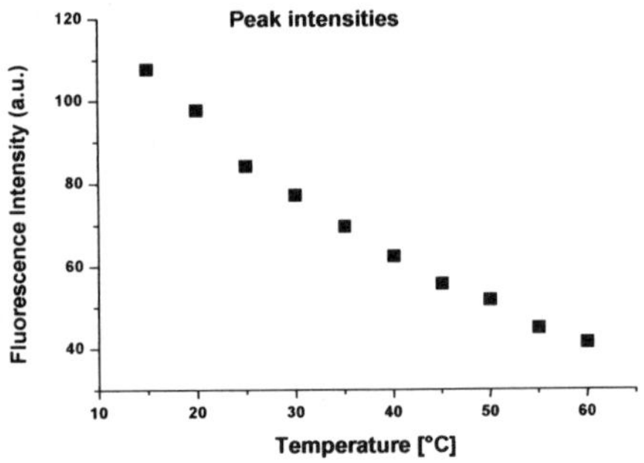

Fig.3. Fluorescence emission spectra of 5 μM W-PGLa in 0,5 mM DMPC liposomes in dependence of the temperature. The probe was excited at 280 nm.

Fig.4. Temperature dependence of the emission intensity at the maximum.

Tryptophan is thus exposed to an aqueous environment and shows no significant interaction of the peptide with choline phosphatides, the main phospholipids of erythrocytes. This is in agreement with previous findings [11] that PGLa adopts a random conformation in aqueous solution and in the presence of zwitterionic lipids.

On the other hand, the emission maximum of W-PGLa undergoes a large blue shift of 20 nm when the liposomes contain only 20 mol% of the negatively charged lipid DMPG (Fig. 5). The fluorescence intensities reflect the main phase transition of the lipid bilayer (Fig. 6). In the fluid phase the emission intensity of W-PGLa's tryptophan is quenched due to increased water accessibility [14]. Fluorescence data show the specific interaction of antibacterial PGLa (and W-PGLa) with bacterial membranes, where it was found to adopt an α-helix [11] oriented in plane with the bilayer surface [15] as depicted schematically in Fig. 7. It should be noted that the N-terminus might be partly unwinded to locate the tryptophan in the headgroup region of the lipid bilayer in accordance with the fluorescence peak of tryptophan which is found at longer wavelength as is typical for a fully hydrophobic environment.

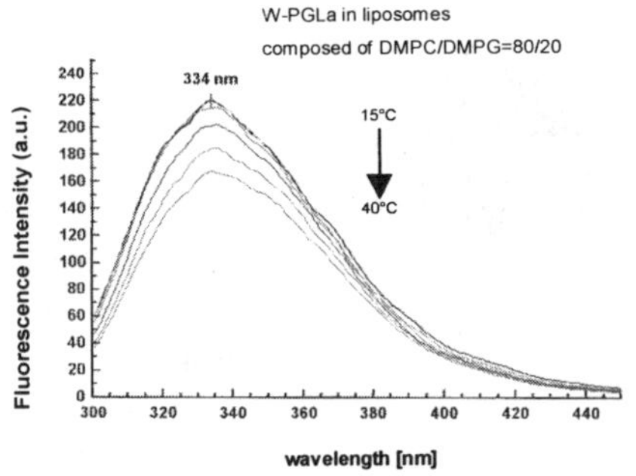

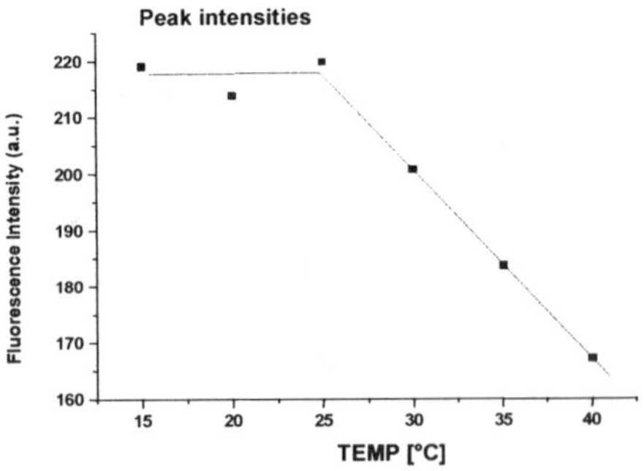

Fig.5. Fluorescence emission spectrum of W-PGLa in liposomes containing DMPC and 20 mol% DMPG.

Fig.6. Temperature dependence of the maximal emission intensity of W-PGLa.

a. **b.**

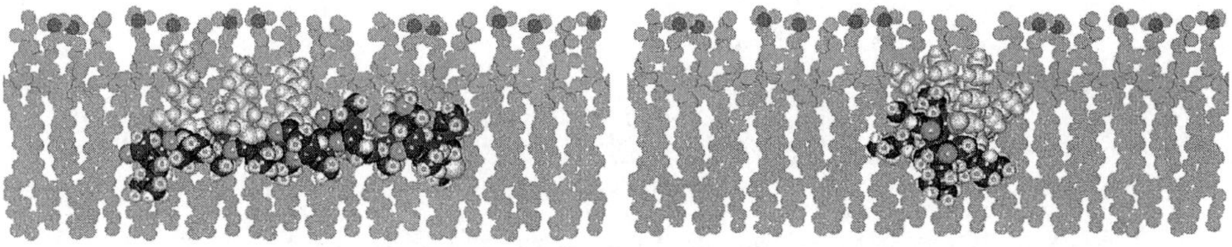

Fig.7. Schematic drawing of the structure and proposed location of PGLa within a mixed membrane with choline and glycerol headgroups. The amphipathic α-helix orients parallel to the membrane surface. The four lysine residues coloured in yellow are thought to anchor the peptide between the negatively charged DMPG headgroups and the positively charged lysine nitrogens.
a. represents a view from the side and **b.** along the helix axis.

3.2. Does PGLa induce lateral phase separation in model membranes?

In order to monitor the lateral distribution of W-PGLa within a mixed bilayer composed of DMPC and DMPG fluorescence resonance energy transfer between the tryptophan of W-PGLa and 12-(9-anthroyloxy) stearic acid (12-AS, Fig. 8) was performed. The transfer efficiency is constant at temperatures below the phase transition but decreased markedly when the liquid crystalline phase is formed. This shows that the average distance between the two chromophors increases in the liquid crystalline phase (Fig. 9). This should be due to lateral heterogeneities within the membrane: Whereas in the gel phase 12-AS is found in the vicinity of PGLa it can diffuse to further distances when the membrane becomes fluid. The basic peptide PGLa is bound to membranes largely by electrostatic forces mediated by the negatively charged lipid component in the membrane. Negatively charged lipids are thus most likely recruited in the vicinity of the peptide and this gives rise to formation of in-plane membrane domains [16]. Such domain formation due to binding of basic proteins or peptides has been observed both in natural membranes and in lipid membranes containing charged and zwitterionic lipid species [3, 4].

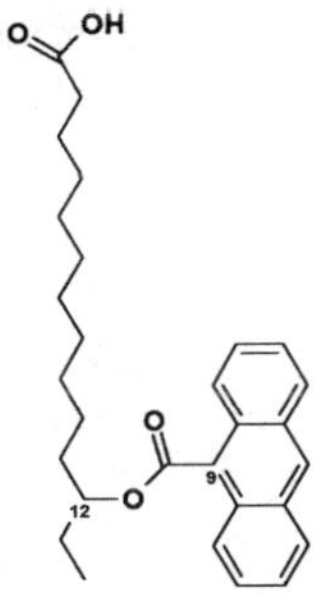

Fig.8. Chemical structure of 12-(9-anthroyloxy) stearic acid

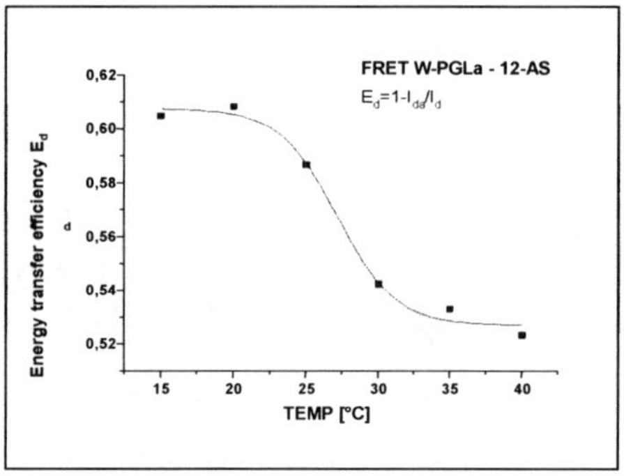

Fig.9. Fluorescence resonance energy transfer between tryptophan of W-PGLa and 12-(9-anthroyloxy) stearic acid. The sample contained 200 μM liposomes composed of DMPC and 20 mol% DMPG, W-PGLa at a peptide to lipid molar ratio of 1:40 and 12-(9-anthroyloxy) stearic acid at a peptide (donor) to probe (acceptor) molar ratio of 1:2 to 1:10.

3.3. Effect of PGLa on membrane organization: headgroup region and acyl chain order

In a „multiprobe" approach probes are selected which are (1) located at different locations in the membranes and (2) respond to different physico-chemical properties of the bilayer.

3.3.1. 12-(9-anthroyloxy) stearic acid as a sensor for microviscosity

Emission spectra of 12-(9-anthroyloxy) stearic acid at different temperatures below and above the phase transition temperature were recorded and compared to the spectra obtained after addition of the peptide PGLa and W-PGLa, respectively (Fig. 10).

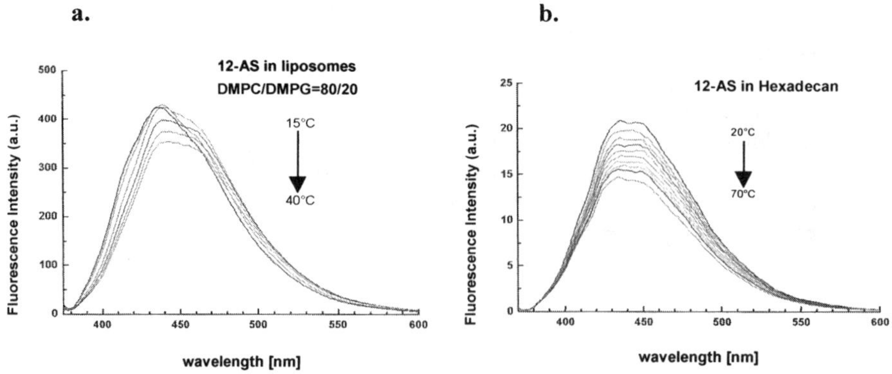

Fig.10. Fluorescence emission spectra of 12-(9-anthroyloxy) stearic acid upon excitation at 365nm
a. incorporated into liposomes composed of DMPC and 20 mol% DMPG at a probe to lipid molar ratio of 1:100, and **b.** in hexadecane as a function of temperature. In the lipid bilayer the emission intensity at 436 nm decreases relative to that at 460 nm with increasing temperature.

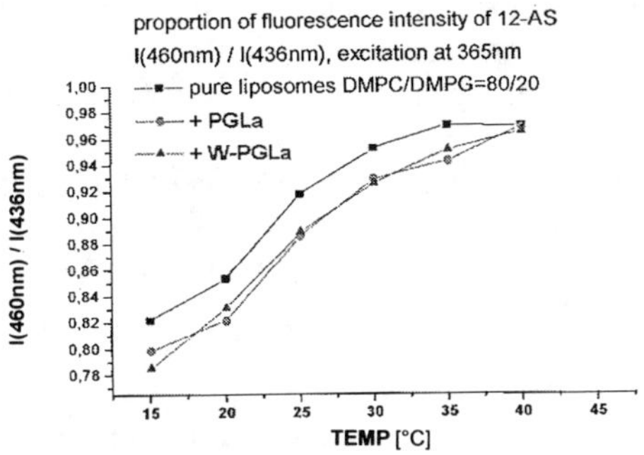

Fig.11. Proportion of the emission intensities at 436 nm and 460 nm in the pure lipid system and after addition of PGLa or W-PGLa at a peptide to lipid molar ratio of 1:40.

The shape of the fluorescence spectrum (Fig. 10 and 11) changes with viscosity and thus also with bilayer fluidity. The red shift seen at the phase transition is in agreement with this prediction [17]. The formation of kinks in the phospholipid acyl chains creates pockets large enough to accommodate water molecules and to facilitate water transport through the bilayer. Reduction in the number of gauche isomers either by decreasing the temperature or by incorporation of the peptide PGLa reduces the number of such pockets and thus the free volume available. The blue shift found on the addition of PGLa to liposome membranes is in accord with this interpretation.

Anthroic acid esters have a broad structureless emission in organic solvents compared to the characteristic vibrational structure found for anthracene. The diffuse emission of the esters is due to resonance interaction between the ester carboxyl group and the π-system of the ring [17]. In the ground state the two groups are not co-planar due to steric hindrance by the peri-hydrogens (Fig. 12).

In the excited state the carboxyl group rotates to obtain a more planar configuration which causes the observed redshift of the fluorescence.

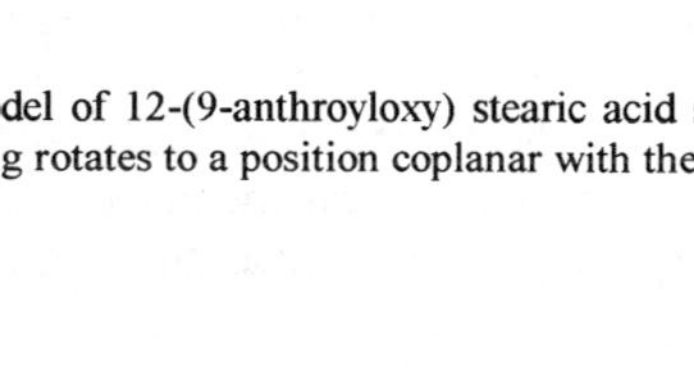

Fig.12. Three-dimensional model of 12-(9-anthroyloxy) stearic acid seen from 3 different views. In the excited state the anthracene ring rotates to a position coplanar with the carboxyl group.

If, however, such a rotation is blocked by the rigidity of the environment the fluorescence spectrum becomes the mirror image of the absorption spectrum, similar to that observed in the gel phase of the membrane. When PGLa is present 12-AS senses a phase transition at a slightly higher temperature than without added peptide (see Fig. 11). An increase in the phase transition temperature was also found by calorimetry [11]. These results indicate, therefore, that the incorporation of PGLa in the membrane forms very compact domains, most probably by clustering of negatively charged lipids. 12-AS is bound to these clusters in close proximity to the peptide.

3.3.2. Orientation of acyl chains in the presence and absence of PGLa

The energy-transfer efficiency for a given pair of donor and acceptor molecules depends mainly on the donor-acceptor separation distance R and the orientation factor κ^2. κ^2 depends on the relative orientation of the emission oscillator of the donor and the absorption oscillator of the acceptor. For random orientation a value of $\kappa^2 = 2/3$ is assumed, a value corresponding to conditions of rapid, isotropic rotation of the donor and acceptor moments, so that all mutual orientations are rapidly sampled in a time short compared to the excited state lifetime [20].

In the case donor and acceptor chromophors are embedded in the anisotropic lipid bilayer, the static limit is a more adequate model: In this case rotational relaxation is slow and donor and acceptor transition dipoles can be assumed to have a fixed relative orientation. The value of κ^2 may then range from 0 to 4, depending whether the two transition dipoles are orthogonal or parallel to each other.

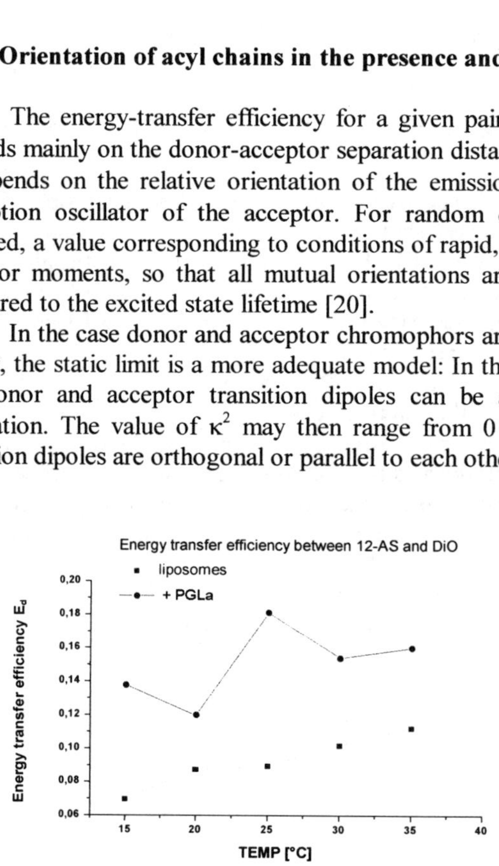

Fig.13. Energy transfer efficiency between 12-(9-anthroyloxy) stearic acid and 3,3′-dioctadecyloxa-carbocyanine as revealed by donor quenching as a function of temperature at a peptide to lipid ratio of 1:40.

As the transition dipole of $DiOC_{18}$ is nearly parallel and that of 12-AS nearly orthogonal to the membrane surface κ^2 is unfavorable for energy transfer. This explains the low efficiency found for energy transfer in this system. The increase in the transfer efficiency in the presence of PGLa indicates that the relative orientations of the transition dipole moments become more favorable, i.e. that the angle between the long axis of the anthracene moiety and the membrane surface becomes considerably smaller than 90°.

4. Summary and future directions

The location of the antimicrobial peptide PGLa and a derivative of PGLa carrying a single tryptophan residue (W-PGLa) within model membranes composed of DMPC and DMPG was investigated by environmental effects of the intrinsic tryptophan fluorescence and by fluorescence resonance energy transfer from W-PGLa tryptophan residue to 12-(9-anthroyloxy) stearic acid (12-AS). The physico-chemical properties of the molecular surrounding of the peptide were also probed by the fluorescence properties of 12-AS.

In the case of negatively charged lipids the results are consistent with a location of the amphipatic helical peptide within the hydrophobic part of the membrane fixed to the head group region by the positively charged lysine residues with the helix being nearly parallel to the membrane surface. The peptide induces the formation of lateral domains; most likely due to the clustering of negatively charged lipids. Below the phase transition temperature of the membrane 12-AS is located within these domains which are compact and do not allow rotational relaxation of the chromophor of 12-AS. They break up at a temperature slightly higher than the phase transition temperature of the bulk of the membrane. This second transition temperature is sensed by the fluorescence of 12-AS and was also found by calorimetric measurements. When the membrane becomes liquid crystalline 12-AS is released from these domains, which shows up in an increase of the average distance between the peptide and the probe.

We demonstrated the applicability of fluorescence measurements to structural studies of interactions of peptides with lipid membranes. Future work will concentrate on the quantitative evaluation of the obtained results. Time resolved studies should indicate the orientational dynamics of the chromophors and its interrelation with the orientational factor κ^2. Appropriate probes sensing charges and hydration on the surface will characterize other properties of the bilayer surface in the presence of PGLa. The application of triplet probes will reveal information on slow diffusion processes.

5. Acknowledgements

S.G. greatly acknowledges financial support by the Austrian Academy of Sciences and K.L by the Austrian Ministry of Science and Transportation.

6. References

[1] P.K.J. Kinnunen *et al.*, Lipid dynamics and peripheral interactions of proteins with membrane surfaces, *Chemistry and Physics of Lipids* **73** (1994) 181-207.

[2] J. Matko *et al.*, Lipid and protein mobility in the plasma membrane of cells: Studies by fluorescence anisotropy and Fluorescence Photobleaching Recovery (FPR), In: Z. Krasznai and L. Mátyus (Eds.), Practical guide to physical analysis of cell surface receptors, Hungary 1998, pp. 92-109.

[3] R.Welti and M. Glaser, Lipid domains in model and biological membranes, *Chemistry and Physics of Lipids* **73** (1994) 121-137.

[4] J.F. Tocanne *et al.* , Lipid domains and lipid/protein interactions in biological membranes, *Chemistry and Physics of Lipids* **73** (1994) 139-158.)

[5] W. Klompenburg, Anionic phospholipids are determinants of membrane protein topology, *EMBO Journal* **16**(14) (1997) 164261-164266.

[6] G. Lametschwandtner *et al.*, The Difference in Recognition of Terminal Tripeptides as Peroxisomal Targeting Signal 1 between Yeast and Human Is Due to Different Affinities of Their Receptor Pex5p to the Cognate Signal and to Residues Adjacent to It, *The Journal of Biological Chemistry* **50** (1998) 33635-33643.

[7] K. Lohner and R.M. Epand, Membrane interactions of hemolytic and antibacterial peptides, *Adv. Biophys.Chem.* **6** (1997) 53-56

[8] K. Matsuzaki, Magainins as paradigm for the mode of action of pore forming polypeptides, *Biochimica et Biophysica Acta* **1376** (1998) 391-400.

[9] M.G. Giovannini *et al.* , Biosynthesis and degradation of peptides derived from Xenopus laevis prohormones, *Biochem. J.* **234** (1987) 113-120.

[10] E. Soravia *et al.* (1988) Antimicrobial properties of peptides from Xenopus granular gland secretions, *FEBS Lett.* **228**, 337-340.

[11] A. Latal *et al.*, Structural aspects of the interaction of peptdyl-glycylleucine-carboxyamide, a highly potent antimicrobial peptide from frog skin, with lipids, *Eur. J. Biochem.* **248** (1997) 938-946)

[12] D. Juretic *et al.* (1989) Magainin-2 amide and analogs-antimicrobial activity, membrane depolarization and susceptibility to proteolysis, *FEBS Lett.* **249**, 219-223.

[13] N. Sitaram and R. Nagaraj (1993) Interaction of the 47-residue antibacterial peptide seminalplasmin and its 13-residue fragment which has antibacterial and hemolytic activities with model membranes, *Biochemistry* **32**, 3124-3130.

[14] C.Ho *et al.*, Hydration and Order in Lipid Bilayers, *Biochemistry* **34** (1995) 6188-6195.)

[15] B. Bechinger *et al.* , Structure and Dynamics of the Antibiotic Peptide PGLa in Membranes by Solution and Solid-State Nuclear Magnetic Resonance Spectroscopy, *Biophysical Journal* **74** (1998) 981-987.

[16] T. Heimburg *et al.*, Binding of Peripheral Proteins to Mixed Lipids Membranes: Effect of Lipid Demixing upon Binding, *Biophysical Journal* **76** (1999) 2575-2586.)

[17] K.R. Thulborn *et al.*, The use of n-(9-anthroyloxy) fatty acids to determine fluidity and polarity gradients in phospholipid bilayers, *Biochimica et Biophysica Acta* **558** (1979) 166-178.

[18] S. Scarlata *et al.*, Effect of Increased Lipid Packing on the Surface Charge of Micelles and Membranes, Biochemistry 29 (1990) 10233-10240.

[19] N.J. Zuidam and Y. Barenholz, Electrostatic parameters of cationic liposomes commonly used for gene delivery as determined by 4-heptadecyl-7-hydroxycoumarin, Biochimica et Biophysica Acta 1329 (1997) 211-222.

[20] M.R. Eftink, Fluorescence Techniques for Studying Protein Structure, In: C.H.Suelter (ed.), Methods of Biochemical Analysis **35**: Protein Structure Determination, ISBN 0-471-51326-1, John Wiley & Sons, Inc. 1991, pp.127-205.

Protein, Lipid and Membrane Traffic: Pathways and Targeting
J.A.F. Op den Kamp (Ed.)
IOS Press, 2000

Membrane protein assembly and structure

Gunnar von Heijne

Department of Biochemistry, Stockholm University, S-106 91 Stockholm, Sweden

Two major questions regarding integral membrane proteins will be dealt with in this chapter: How are membrane proteins targeted to and assembled into the membrane? How does the amino acid sequence of a membrane protein dictate its 3D structure? In particular, recent discoveries have uncovered new, hitherto unsuspected alternative targeting and assembly pathways in bacteria, and new methodological advances have made it possible to study detailed questions of how individual transmembrane segments interact with the lipid bilayer using simple molecular biology techniques.

1. Introduction

Membrane proteins are at the crossroads of many centrally important problem areas in molecular biology, structural biology, cell biology, and pharmacology, and have become the focus of intense research over the past 5-10 years. Although in many ways difficult to study, the sustained research efforts in many labs are now providing us with a wealth of new information on how membrane proteins are made, what they look like, and how they function. Suffice it to mention the number of new high-resolution structures published in the last couple of years, mainly from X-ray crystallography but also from NMR and electron crystallography labs [1], and the detailed insights into the membrane integration process available through the use of site-directed crosslinking approaches [2].

In this chapter, I will focus on two aspects of the membrane protein field: targeting and assembly of proteins into the inner membrane of *E. coli*, and the relationship between amino acid sequence and the structure of individual transmembrane helices. Since these and related questions are addressed also in other chapters in this book, I will mainly limit myself to discussing our own work in these areas.

2. Targeting and assembly pathways in *E. coli*

The so-called Sec protein translocation pathway, originally discovered by genetic analysis of protein secretion in *E. coli* [3], has long been known to be involved also in the assembly of inner membrane proteins. Recently, the concept of "Sec-independent" translocation has been re-investigated and some proteins previously believed to insert "spontaneously" into the inner membrane are now known to in fact require components of the Sec machinery. More importantly, the *E. coli* SRP (signal recognition particle), which was originally thought to target secretory proteins to the Sec machinery in analogy with its role in mammalian cells, now appears to be mainly devoted to chaperone membrane proteins to the Sec machinery - although not all SRP-dependent proteins may necessarily have to interact with Sec components.

A couple of years ago, work on thylakoid protein import uncovered a new targeting pathway - the Twin-arginine targeting (Tat) pathway - that was then found also in *E. coli* and other bacteria. The role of the Tat pathway for membrane protein assembly - if any - is still an open question.

Finally, some yeast mitochondrial inner membrane proteins are known to depend on the Oxa1p protein for proper insertion. Oxa1p homologues are present in bacteria (including *E. coli*), but again it's not yet known if they are involved in membrane protein assembly.

2.1. The classical Sec-pathway: role of the SRP

For a long time, the SRP was thought to be unique to eukaryotic cells. However, once an SRP homologue (composed of the Ffh protein and a 4.5S RNA) was discovered in *E. coli* [4-6], it became clear that the *E. coli* and ER translocation mechanisms were basically variations on a common theme. Somewhat disconcertingly perhaps, early studies only found a small number of SRP-dependent secretory proteins in *E. coli*, and the translocation of many well-studied periplasmic and outer membrane proteins turned out not to be affected when SRP function was compromised. Four papers published between 1995 and 1997 resolved this apparent dilemma: it was found that the *E. coli* SRP was important mainly for targeting highly hydrophobic inner membrane proteins to the membrane, and not for the targeting of secretory proteins with less hydrophobic signal peptides [7-10]. Crosslinking studies have since shown that hydrophobic segments interact strongly with SRP, but only when the nascent polypeptide chain is still attached to the ribosome [11].

In *E. coli*, the SRP thus appears to be a chaperone/targeting factor for inner membrane proteins, assuring their delivery to the Sec translocon. Whether this also means that SRP-targeted membrane proteins are inserted into the inner membrane co-translationally is not known, but this is an attractive idea since it would be a major problem for the cell to prevent highly hydrophobic multi-spanning membrane proteins from aggregation if they were released from the ribosome prior to integration.

2.2. *Are there really "Sec-independent" proteins?*

Early studies of inner membrane protein assembly in *E. coli* implied that there may exist two different classes of integral membrane proteins: those that depend on the Sec machinery for proper assembly and those that do not [12]. Nevertheless, since the "Sec-independence" of proteins could only be inferred from negative experimental results (lack of effect on membrane assembly when the Sec machinery was compromised), the concept of Sec-independence was usually hedged with various qualifications [13].

The question of Sec-independence has recently been re-investigated using novel Sec-strains with a much stronger phenotype than the previously available strains. A SecE depletion strain (CM124) has proved particularly useful in this regard, since the SecY protein is rapidly degraded in the absence of SecE. SecE depletion thus leads to very low levels of both SecE and SecY, and very little functional Sec translocase remains in the depleted cell [14].

Using CM124, at least one protein previously thought to be Sec-independent has now been found to in fact assemble more slowly under SecE depletion conditions [15]. So far, there is only one protein - the small phage M13 procoat protein - that appears to assemble normally in the absence of SecE. This protein can assemble spontaneously even into protein-free liposomes [16], and may thus be truly Sec-independent. Interestingly, procoat does not seem to require SRP for assembly either [15].

Whether proteins bound to SRP are committed to interact with the Sec machinery must still be regarded as an open question, although at least one protein construct has been found that cannot assemble properly in the absence of SRP but does assemble in SecE-depleted cells (Cristobal et al., JBC in press).

2.3. The Tat pathway - can it handle membrane proteins?

Work on thylakoid protein import - and a compilation of signal peptides from bacterial co-factor binding proteins [17] - led to the discovery of a new protein translocation pathway originally called the ΔpH pathway (since it seems to derive its energy from the ΔpH gradient across the thylakoid membrane) but now known as the Twin-arginine translocation (Tat) pathway [18]. The components of the Tat pathway are distinct from those of the Sec pathway, the Tat machinery can transport folded proteins across the inner membrane, and signal peptides specific for the Tat pathway are characterized by a twin arginine (-RR-) motif just upstream of the hydrophobic h-region, a slightly less hydrophobic h-region than found in typical Sec-targeting signal peptides, and a "Sec avoidance" motif composed of one or more positively charged residues between the h-region and the signal peptidase cleavage site [19].

So far, no clear case of an inner membrane protein targeted to the Tat pathway has been found. In an attempt to test whether such proteins can be handled by the Tat machinery, we made various fusions between a Tat signal peptide and the normally Sec-dependent protein leader peptidase (Lep). While a large periplasmic domain in Lep could be targeted to and translocated by the Tat machinery, Sec-targeting information present in the full-length Lep protein turned out to override the Tat-targeting information in the signal peptide, thus making it impossible to determine whether Lep could be properly assembled into the inner membrane by the Tat machinery [19]. What these results indicate, though, is that SRP probably gets the first shot when a membrane protein emerges from the ribosome, while the

Tat signal peptide receptor has to await its turn. The question of whether membrane proteins can be handled by the Tat machinery thus remains unresolved.

2.4. *Oxa1 - yap ("yet another pathway"), or are we barking up the wrong tree?*

At least some mitochondrial inner membrane proteins - those that are encoded in the mitochondrial genome and synthesized in the matrix - insert from the same side of the membrane as do bacterial inner membrane proteins. It was thus widely believed that there would be a Sec machinery in mitochondria (as had already been found in thylakoids). However, when the full genome sequence of *S. cerevisiae* was reported, it became clear that there is no mitochondrial Sec machinery [20] in yeast (although one apparently exists in the fungus *R. americana* [21]).

Instead, certain mitochondrial inner membrane proteins have been shown to require the presence of a new mitochondrial protein (Oxa1p) for proper assembly [22], and it is thought that Oxa1p is involved in the translocation of parts of these membrane proteins into the intermembrane space. Many bacteria (including *E. coli*) have Oxa1p homologues, but since no Oxa1 mutant strains exist so far, the function of this protein in bacteria has not been determined. It is nevertheless an intriguing possibility that the bacterial Oxa1p homologue is involved in the assembly of inner membrane proteins, either by itself or together with the Sec machinery.

3. Using sugars to study transmembrane helices

Structural aspects of individual transmembrane helices and the interactions between such helices and the lipid bilayer are best studied using biophysical techniques such as X-ray crystallography, NMR, or various spectroscopic methods. However, these methods are time consuming, require large amounts of protein or synthetic peptides, and cannot generally be done in the context of real biological membranes.

Over the past few years, we have explored an alternative way to study transmembrane helices in a natural membrane (ER-derived microsomes) using standard molecular biology techniques. The basis for this approach - *in vitro* transcription/translation of model membrane proteins in the presence of fog pancreas microsomes - was established almost 25 years ago [23].It was soon realized that N-linked glycosylation could only take place if an appropriate Asn-X-Thr/Ser glycosylation acceptor site was translocated into the lumen of the microsomes where the oligosaccharyl transferase (OST) enzyme is located. N-linked glycosylation thus provides a simple, reliable, and easily assayed marker (the glycan moiety adds ~2 kDa to the Mw of the protein) for the lumenal location of a glycosylation acceptor site, and has found widespread use in topology mapping of eukaryotic integral membrane proteins. As described below, we have adapted this technique to address questions about the structure and lipid interaction of transmembrane helices.

3.1. Breaking the Camel's back - tight turns in transmembrane helices

Often, transmembrane helices are separated only by very short loops and in essence form what has been called "helical hairpins". Remarkably, however, despite many years of studying membrane proteins, until recently no one had measured the turn propensities of the different amino acids in the context of a transmembrane helix. To address this question, we set up a system where a very long poly-leucine transmembrane segment - one that is long enough to in principle span the membrane twice - has been incorporated into a model membrane protein [24]. The resulting construct was translated *in vitro* in the presence of microsomes, and the lumenal or cytoplasmic location of the domain following the poly-Leu segment was assayed by the glycosylation status of a unique acceptor site located in this domain. Somewhat to our surprise, we found that a 40-residues long poly-Leu segment only spanned the microsomal membrane once.

This gave us the opportunity to measure the turn-promoting potential of the twenty naturally occurring amino acids by simply placing them one by one in the middle of the 40 residues long poly-Leu segment and checking whether the segment would still insert as a single transmembrane span or would form a helical hairpin [25]. By also varying the length of the poly-Leu stretch and placing a turn-promoting proline residue in the middle, we could define the minimum length of the hydrophobic segment compatible with the formation of a helical hairpin (our unpublished data).

Perhaps not surprisingly, all charged and highly polar residues, together with proline, were found to have high turn propensities. Residues such as Gly, Ser, Thr, Trp, and Tyr had intermediate turn propensities, and the hydrophobic residues had no turn-inducing properties in this system. The minimum poly-Leu based helical hairpin was found to be formed by two approximately 14 residues long helices separated by a 2-3 residues long turn; shorter poly-Leu segments could not form a helical hairpin even if a turn-inducing proline was inserted.

While being only a first step towards a full understanding of the "rules" for the formation of transmembrane helical hairpins, this work nevertheless shows how simple molecular biology techniques can be used to address what is fundamentally a biophysical problem.

3.2. Glycosylation mapping - a molecular ruler for transmembrane helices

The same *in vitro* system has also been used to study how different amino acids affect the position of transmembrane helix relative to the membrane-water interface. In this case, we took advantage of an earlier observation that glycosylation acceptor sites must be placed at least 10 residues downstream (or 12 residues upstream) of a transmembrane segment in order to reach the active site of the OST [26, 27]. We reasoned that any changes in the position of a transmembrane helix in the membrane - as brought about by, e.g, a point mutation - would be reflected in a corresponding change in the location of the closest modifiable glycosylation acceptor site. This turned out to be the case, and we have so far used this "glycosylation mapping" approach to study the effects of inserting a proline [28], a charged residue [29], or a tryptophan into a poly-Leu transmembrane helix (Braun and von Heijne, Biochemistry in press).

Another use of the same technique is to position any transmembrane helix relative to a reference helix, the position of which is known from some high-resolution biophysical technique such as X-ray crystallography or NMR [29].

4. Conclusions

Membrane proteins will continue to generate loads of interesting scientific problems for a long time to come (I hope!). While hard-to-do biophysical studies will be necessary in many instances, the simple architecture of the transmembrane α-helix nevertheless makes it possible to approach many basic structural questions by using much simpler techniques such as those described above. The biogenesis of membrane proteins, their subcellular sorting, and their central biological roles furthermore make them an ideal common ground for integrating work on protein structure, biochemistry and cell biology, as this summer school has so nicely illustrated. Membrane proteins are simply fun!

5. Acknowledgments

I could not have written this chapter without the financial support of the Swedish Natural and Technical Sciences Research Councils, the Swedish Cancer Foundation, the European Community, and the Göran Gustafsson Foundation. But above all, not without the inspired work performed by all my past and present collaborators, both at Stockholm University and elsewhere.

6. References

[1] Sakai, H. and Tsukihara, T., Structures of membrane proteins determined at atomic resolution, *J Biochem Tokyo* **124** (1998) 1051-1059.

[2] Matlack, K., Mothes, W. and Rapoport, T., Protein translocation: Tunnel vision, *Cell* **92** (1998) 381-390.

[3] Ito, K., The major pathway of protein translocaion across membranes, *Genes to Cells* **1** (1996) 337-346.

[4] Luirink, J., High, S., Wood, H., Giner, A., Tollervey, D. and Dobberstein, B., Signal-sequence recognition by an *Escherichia coli* ribonucleoprotein complex, *Nature* **359** (1992) 741-743.

[5] Luirink, J. and Dobberstein, B., Mammalian and *Escherichia coli* signal recognition particles, *Mol Microbiol* **11** (1994) 9-13.

[6] Luirink, J., ten Hagen-Jongman, C. M., van der Weijden, C. C., Oudega, B., High, S., Dobberstein, B. and Kusters, R., An alternative targeting pathway in *Escherichia coli*: Studies on the role of FtsY, *EMBO J* **13** (1994) 2289-2296.

[7] MacFarlane, J. and Müller, M., The functional integration of a polytopic membrane protein of Escherichia coli is dependent on the bacterial signal- recognition particle, *Eur J Biochem* **233** (1995) 766-771.

[8] de Gier, J.-W. L., Mansournia, P., Valent, Q., Phillips, G. J., Luirink, J. and von Heijne, G., Assembly of a cytoplasmic membrane protein in *Escherichia coli* is dependent on the signal recognition particle, *FEBS Lett* **399** (1996) 307-309.

[9] Seluanov, A. and Bibi, E., FtsY, the prokaryotic signal recognition particle receptor homologue, is essential for biogenesis of membrane proteins, *J Biol Chem* **272** (1997) 2053-2055.

[10] Ulbrandt, N. D., Newitt, J. A. and Bernstein, H. D., The *E. coli* signal recognition particle is required for the insertion of a subset of inner membrane proteins, *Cell* **88** (1997) 187-196.

[11] Valent, Q. A., de Gier, J. W. L., von Heijne, G., Kendall, D. A., ten Hagen-Jongman, C. M., Oudega, B. and Luirink, J., Association of nascent membrane and presecretory proteins with Signal Recognition Particle and Trigger Factor, *Molecular Microbiology* **25** (1997) 53-64.

[12] Andersson, H. and von Heijne, G., *Sec*-dependent and *sec*-independent assembly of *E. coli* inner membrane proteins - the topological rules depend on chain length, *EMBO J* **12** (1993) 683-691.

[13] von Heijne, G., *Sec*-independent protein insertion into the inner *E. coli* membrane: A phenomenon in search of an explanation, *FEBS Lett* **346** (1994) 69-72.

[14] Traxler, B. and Murphy, C., Insertion of the polytopic membrane protein MalF is dependent on the bacterial secretion machinery, *J Biol Chem* **271** (1996) 12394-12400

[15] de Gier, J., Scotti, P., Sääf, A., Valent, Q., Kuhn, A., Luirink, J. and von Heijne, G., Differential use of the signal recognition particle translocase targeting pathway for inner membrane protein assembly in *Escherichia coli*, *Proc Nat Acad Sci USA* **95** (1998) 14646-14651.

[16] Gallusser, A. and Kuhn, A., Initial steps in protein membrane insertion - bacteriophage-M13 procoat protein binds to the membrane surface by electrostatic interaction, *EMBO J* **9** (1990) 2723-2729.

[17] Berks, B. C., A common export pathway for proteins binding complex redox cofactors?, *Mol Microbiol* **22** (1996) 393-404.

[18] Dalbey, R. E. and Robinson, C., Protein translocation into and across the bacterial plasma membrane and the plant thylakoid membrane, *Trends Biochem Sci* **24** (1999) 17-22.

[19] Cristobal, S., de Gier, J., Nielsen, H. and von heijne, G., Competition between Sec and Tat dependent protein translocation in *Escherichia coli*, *EMBO J* **18** (1999)

[20] Glick, B. S. and von Heijne, G., *Saccharomyces cerevisiae* mitochondria lack a bacterial-type Sec machinery, *Prot Sci* **5** (1996) 2651-2652.

[21] Lang, B. F., Burger, G., Okelly, C. J., Cedergren, R., Golding, G. B., Lemieux, C., Sankoff, D., Turmel, M. and Gray, M. W., An ancestral mitochondrial DNA resembling a eubacterial genome in miniature, *Nature* **387** (1997) 493-497.

[22] Hell, K., Herrmann, J. M., Pratje, E., Neupert, W. and Stuart, R. A., Oxa1p, an essential component of the N-tail protein export machinery in mitochondria, *Proc Natl Acad Sci USA* **95** (1998) 2250-2255.

[23] Blobel, G. and Dobberstein, B., Transfer to proteins across membranes. II. Reconstitution of functional rough microsomes from heterologous components, *J Cell Biol* **67** (1975) 852-62.

[24] Nilsson, I. and von Heijne, G., Breaking the camel's back: Proline-induced turns in a model transmembrane helix, *J Mol Biol* **284** (1998) 1185-1189.

[25] Monné, M., Hermansson, M. and von Heijne, G., A turn propensity scale for transmembrane helices, *J Mol Biol* **288** (1999) 141-145.

[26] Nilsson, I. and von Heijne, G., Determination of the distance between the oligosaccharyltransferase active site and the endoplasmic reticulum membrane, *J Biol Chem* **268** (1993) 5798-5801.

[27] Popov, M., Tam, L. Y., Li, J. and Reithmeier, R. A. F., Mapping the ends of transmembrane segments in a polytopic membrane protein - Scanning N-glycosylation mutagenesis of extracytosolic loops in the anion exchanger, Band 3, *J Biol Chem* **272** (1997) 18325-18332.

[28] Nilsson, I., Sääf, A., Whitley, P., Gafvelin, G., Waller, C. and von Heijne, G., Proline-induced disruption of a transmembrane α-helix in its natural environment, *J Mol Biol* **284** (1998) 1165-1175.

[29] Monné, M., Nilsson, I., Johansson, M., Elmhed, N. and von Heijne, G., Positively and negatively charged residues have different effects on the position in the membrane of a model transmembrane helix, *J Mol Biol* **284** (1998) 1177-1183.

Protein, Lipid and Membrane Traffic: Pathways and Targeting
J.A.F. Op den Kamp (Ed.)
IOS Press, 2000

HELIX INTERACTIONS IN MEMBRANE PROTEINS

Donald M. Engelman*
Lillian F. Fisher
Kevin R. MacKenzie
Jean-Luc Popot
William P. Russ
G. Koji Sonoda

Yale University
260 Whitney Avenue, New Haven, CT 06520, USA

**Professeur de Recherche Blaise Pascal, Region Ile de France, France*

1. Introduction

Understanding the chemistry of membrane protein structures is a major goal in cell biology, physiology, and biophysics. For those membrane proteins consisting of transbilayer α-helices, an understanding of helix-helix interactions can lead to an improved ability to interpret amino acid sequence information in terms of structure and therefore to provide a basis for understanding essential membrane functions such as energy transduction, information transfer, ion channel properties, and transport. While many membrane proteins have now been cloned and sequenced, only a few structures are known at atomic resolution; hence, interpretation of membrane protein sequence information has been the concern of many investigators. Ideally, one would hope to obtain high resolution information from crystallographic studies or, perhaps, somewhat lower resolution information from two-dimensional crystals seen by electron cryo-microscopy. However, progress in these areas has been slow, and the crystal structures to date are largely related to photosynthetic events (it is interesting to ask why this is - perhaps photosynthetic proteins are exceptionally rigid and therefore more easily crystallized). In fact, groups as large as a Max Planck Institute have been established to study membrane proteins from a crystallographic standpoint. After many years, a few important structures have resulted, but they represent only a small fraction of the number of proteins implied by genomic sequence data. Therefore, we are exploring a different approach. We have shown in our previous work that it is feasible to use a combination of other methods, principally those of computational chemistry, mutagenesis, and optical spectroscopy, to establish models of chemical structures. These can be tested in structural detail by NMR methods, and free energies of association can be measured by SDS-PAGE, analytical ultracentrifugation, small angle X-ray scattering and, perhaps, by assays in *E. coli* membranes . Our recent studies of bacteriorhodopsin, glycophorin, and phospholamban are highly encouraging, and prompt us to suggest that chemical structures may be built by understanding motifs defined by non-crystallographic approaches. The major background points are summarized below and in the preliminary studies.

2. *Helical Secondary Structure in Membrane Proteins*

The determination of the primary structure of a large number of integral membrane proteins over the past decade or so has shown that they often contain one or more stretches of 20 to 25 largely hydrophobic amino acid residues that could form transmembrane α-helices [1]. Furthermore, the determination of a number of helical membrane protein structures has shown a large number of transmembrane α-helices, which constitute the membrane-spanning portions of these proteins. Consequently, α-helices are thought likely to be a common theme as the membrane-spanning portions of integral membrane proteins (bacteriorhodopsin, [2]; the photosynthetic reaction centers of *R. viridis* and *R. sphaeroides* ([3]; [4]; [5]); the plant light-harvesting complex II ([6]; [7]); system I of photosynthesis (Krauss *et al.* 1996); LHII from Rps. acidoplida (McDermott *et al.* 1995); cytochrome oxidase [8], [9]; and the K+ channel [10]. Clear exceptions to this generalization do exist, however, such as the β structures of outer membrane porins of bacteria [11], [12], [13]. Most of the intramembranous portion of the nicotinic acetylcholine receptor may also have β-structure [14], despite the appearance in its sequence of four hydrophobic stretches, each previously proposed to form a transmembrane α-helix. Nonetheless, it is generally considered, with considerable experimental precedent, that a sufficiently long stretch of hydrophobic residues in an integral membrane protein is most likely to traverse the lipid bilayer as an α-helix.

3. *The Two-Stage Model for Integral Membrane-Protein Folding*

We have proposed [15] a conceptual separation of the thermodynamics of folding of predominantly helical integral membrane proteins into two stages (Figure 1). The first stage involves the establishment of independently stable transmembrane α-helices, which are considered as autonomous folding domains. In the second stage these helices associate with one another, with no major rearrangement of secondary structure, to generate the tertiary and quaternary structure of the protein. The second of these stages can be applied equally well to the interactions between helices within a single polypeptide chain and to those interactions between the transmembrane α-helices of separate proteins (e.g., [16]; [2]). A number of rather successful algorithms have been developed that serve to predict from their sequences which regions of integral membrane proteins will form transmembrane α-helices (e.g.[17]; [18]; [19]; [20]). In the scheme outlined in Figure 1 below, it may be argued that the ΔG of formation of a transmembrane α-helix is similar to the free energy for the partitioning of that α-helix into a lipid bilayer. It is in the assessment of this quantity that the predictive algorithms for membrane protein structure are of use, and available data concerning the structure of integral membrane proteins suggests that they are rather successful [17]. The driving force for this partitioning is the hydrophobic effect; depending upon curvature effects [21], this would result in a ΔG of -40 to -80 kcal mol^{-1} for insertion of the α-helix into the lipid bilayer (taking the strength of the hydrophobic effect to be 25 cal mol^{-1} $Å^{-2}$ [22] to 47 cal mol^{-1} $Å^{-2}$ [21]. These values for ΔG are overestimates, since they ignore the fact that part of the α-helix surface is polar; they also do not take into account the unfavorable entropic terms associated with the restricted mobility of a polypeptide embedded in a lipid bilayer; and effects upon the lipid bilayer itself are ignored. A more reasonable value might be -5 to -20 kcal mol^{-1} [23]. These considerations suggest that transmembrane α-helices can be very stable structures - sufficiently stable to exist as independent domains and accommodate the unfavorable energy due to polar groups often seen in transmembrane helices.

4. Driving forces for Transmembrane α-Helix Association

For consideration of the possible biological roles of interactions between transmembrane α-helices, it is instructive to separate energetics from specificity. For α-helices in a lipid bilayer to associate, the entropy that favors their dispersal must be overcome. The magnitude of this effect was estimated by Popot & Engelman [15] to be of the order of 1-10 kcal mol^{-1} for a pair of helices. As discussed above, the hydrophobic effect is the major driving force for the establishment of independently stable transbilayer α-helices in stage I of the above model (Figure 1), and is not expected to influence their mutual interactions within the lipid bilayer [24]. Therefore, in the second stage of the model depicted in Figure 1, the association of α-helices within lipid bilayers is most likely to be driven by

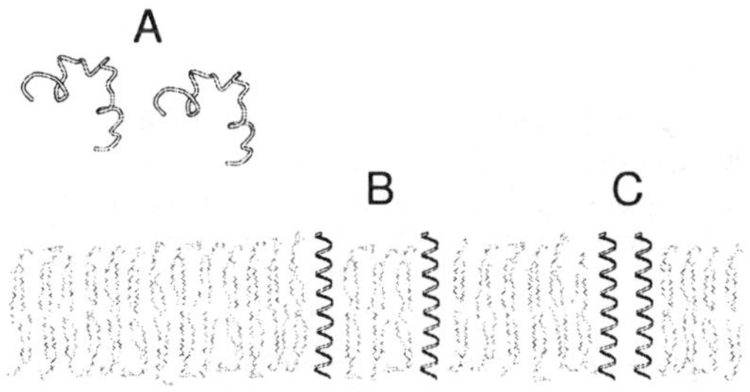

Figure 1. A schematic representation of the two-stage model for the folding of integral membrane proteins [15]. In stage I, stretches of residues with hydrophobic side-chains (A) form independently stable transmembrane α-helices (B). In stage II, these helices interact with one another, with no major rearrangement in secondary structure, to generate the tertiary and quaternary structures of the protein (C).

"solvophobic" effects (see below), packing effects, and/or electrostatic interactions. ΔG_{net} for the oligomerization of transmembrane α-helices may be written:

$$\Delta G_{net} = \Delta G_{H\text{-}H} + n\Delta G_{L\text{-}L} - 2n\Delta G_{H\text{-}L},$$

where $\Delta G_{H\text{-}H}$, $\Delta G_{L\text{-}L}$, and $\Delta G_{H\text{-}L}$ refer to free energies for helix-helix, lipid-lipid, and lipid-helix interactions respectively. For $2n$ lipid molecules that are returned to the bulk lipid upon helix association, it is assumed that n lipid-lipid interactions are gained, and $2n$ lipid-helix interactions are lost. Additional contributions include interactions with prosthetic groups, such as those seen in the structures of bacteriorhodopsin, the photosynthetic reaction centers and the pea light-harvesting complex II, and the bacterial light harvesting complex II (references above). Furthermore, the extramembraneous covalent linkages between helices will reduce the unfavorable entropy term in the free energy of helix-helix interaction. A number of transmembrane α-helices are known to include residues with polar side-chains. Hydrogen bonding between such side-chains on adjacent helices, or the formation of ion-pairs, could provide sufficient energy for the association of α-helices [24], [25]. However, although two symmetrically disposed potential (Glu-Arg) salt-bridges are seen between helices A and B in the atomic model of LHC-II, no such interactions are seen in the structures determined for photosynthetic reaction centers or for bacteriorhodopsin, and on average, less than one inter-helical hydrogen bond per α-helix is inferred. Rather, the polar residues in the transmembrane domains of these proteins appear to have functional roles. In the interaction of the single putative transmembrane α-helix of CD3δ with that of TCR-α in the assembly of the T-cell antigen receptor interaction between an aspartic acid residue in one helix (CD3δ) and two basic residues on the other (TCR-α) [26], [27] may occur, although the precise location of these residues within the membrane is unknown. However, other single transmembrane α-helices containing no highly polar side-chains, most notably that of glycophorin A [16]; [28], are known to form stable oligomers, indicating that strong polar interactions between transmembrane α-helices are not essential for their association.

5. *Origins of Specificity in Transmembrane α-Helix Association*

Many of the interactions discussed above contribute to the energetics of transmembrane helix association in a relatively non-specific manner. For example, the extramembraneous covalent loops linking transmembrane helices in a polytopic membrane protein will bring them into close proximity, but will not necessarily drive a specific helix interaction.

Extramembraneous linking regions with well-defined structure may result in a particular arrangement of transmembrane α-helices, but the specificity in such a case is actually defined outside the bilayer. Transmembrane α-helix association that is driven by the formation of an inter-helical salt bridge (or strong hydrogen bond) also need not be specific: one or two such interactions will not necessarily define the remainder of the interface between transmembrane α-helices. Similarly, packing effects, such as those modeled by Wang and Pullman [29] for polyalanine helices, are argued to result from the fact that the association of α-helices in general is energetically more favorable than is their association with lipid molecules, and are not necessarily specific in nature. Given that a single salt bridge or hydrogen bond appears to have sufficient energy to stabilize helix associations, the explanation for their apparent absence as major themes may well be that a relatively non-specific associations that could result in massive aggregation of helices in the plane of a membrane.

For transmembrane α-helix association of a specific nature, which is required for biology and is defined within the bilayer, two possibilities exist. In the case of membrane proteins that contain one or more prosthetic groups, simultaneous ligation of these groups with more than one helix can define a specific arrangement of transmembrane α-helices, as is seen in the structures of the photosynthetic reaction centers, bacteriorhodopsin and the plant light-harvesting complex. In the absence of such prosthetic groups, specificity in helix-helix interactions must arise from a detailed stereochemical fit between sets of dynamically accessible states of particular helices. Specific close packing of complementary faces of α-helices would result in the maximization of van der Waals interactions between the helices compared with the case for the packing of these faces with lipid molecules, or their packing with non-complementary faces of other transmembrane α-helices. A wide repertoire of specific interactions of this nature is possible, and may be important in a number of roles for transmembrane helix interactions in biology. Such specific interactions would be expected to play a part in defining the structures of polytopic integral membrane proteins in which a large portion of the protein is membrane-embedded. They may also be important in defining the correct assembly of complexes of integral membrane proteins.

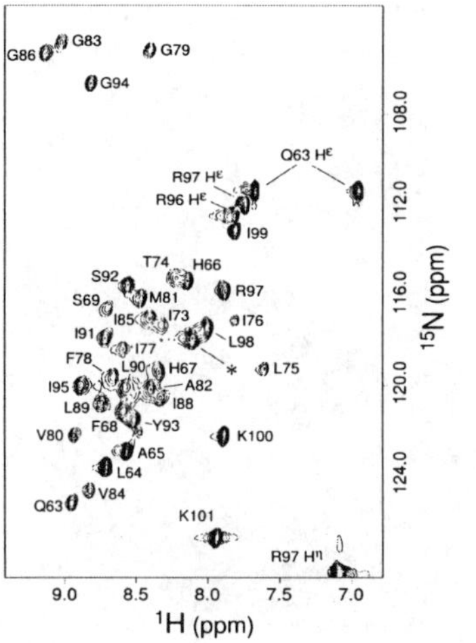

Figure 2. 2D ^{1}H-^{15}N HSQC spectrum of a 2.5 mM sample of a peptide corresponding to the TM domain of GpA dissolved in 5% DPC. Data was acquired at 40°C and 500 MHz.

We have determined the three-dimensional structure of the dimeric transmembrane domain of human glycophorin A using solution heteronuclear NMR spectroscopy of a 40-residue peptide in a micelle of dodecylphosphocholine (DPC) [30]. The micelles and associated peptides were freely tumbling in an aqueous environment, and although the effective molecular weight is large, the fact that only a subset of the complex contains peptide material results in a sufficient simplification of the number of peaks so that the use of solution NMR methods is possible. Using peptides prepared from chimeric constructs containing the transmembrane peptide of glycophorin derived from *E. coli* grown in an ^{15}N, ^{13}C medium, we obtained gradient-selected two-dimensional HSQC spectra in which most amide proton-amide nitrogen correlations were resolved (Figure 2).

Assignment of the peaks was successfully carried out using HNCA experiments to observe the carbon chemical shifts, and the entire transmembrane region was assigned. Nuclear Overhauser effect distance restraints and J coupling-derived dihedral restraints were used to generate a family of twenty structures using the program X-PLOR (A.T. Brunger, Yale University). The structures were very similar to each other (0.4 angstroms RMSD on the backbone), and reveal a closely packed interface (see Figure 3). The rotamers of the side chains in the interface are those expected as the preferred rotamers for a helix in the monomeric state. Thus, the positions of most interfacial atoms within a monomer are determined by helix formation, and dimerization follows from the association of two largely preformed surfaces. Under these circumstances the correlation (which cannot be as great in the fluid environment of lipid or detergent molecules) of van der Waals interactions accounts, we believe, for the stable association of the helices in this case. Thus, chemical insights may be derived from examination of the structure (see below).

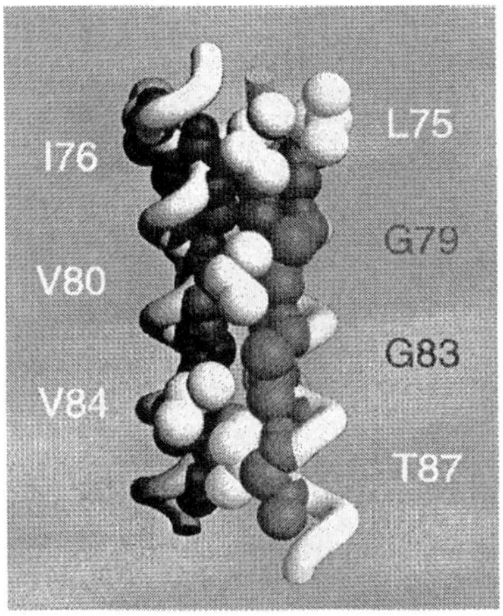

Figure 3. Side view of the GpA dimer. Atoms of interfacial; backbone atoms are dark and side-chain atoms are light.

It is notable that the structure is very similar (0.8 angstroms RMSD) to the structure derived from mutagenesis and modeling [31]. The residues that form the observed interface are those identified as a dimerization motif in mutagenesis studies using SDS gel associations [28, 32], and contribute essentially all the atoms that participate in inter-monomer van der Waals interactions. The excellent agreement between the conclusions of the indirect structural methods previously applied to those derived from the structure encourage the idea that indirect approaches can succeed in studies of membrane protein organization.

6. Structure-Based Prediction Of The Stability Of Transmembrane Helix-Helix Interactions: The Sequence Dependence Of Glycophorin A Dimerization

Given a structure for the glycophorin transmembrane helix dimer and a large body of mutant phenotypes based on SDS gel electrophoresis, we have attempted to use structure-based empirical parameters to interpret the observed phenotypes. The mutagenesis data consists of the relative dimerization propensities of 105 hydrophobic single point mutants of the glycophorin A transmembrane domain. Using the observed structure of the glycophorin dimer, the mutant changes in favorable van der Waals interactions, restrictions of side chain rotamer distributions, and the creation of steric clashes were assessed using simplified parameters. A fit to the database gave coefficients for the parameters, and a reasonable correlation was obtained (R^2=0.74). Other variables, such as the volume of side-chains or their hydrophobicity had no correlation with the disruptive effect of the mutants. The details of this treatment are found in [33].

Limiting the coefficients of the fit to a single decimal place, the calculated stability (CALC) can be expressed:

$$\text{CALC} = 2.7 - 1.6\ \text{CLSH} + 0.5\ \text{DSROT} + 0.2\ \text{VDW}$$

where CLSH is the parameter for van der Waals clashes, DSROT is the parameter for rotamer restrictions, and VDW is the parameter for favorable van der Waals interactions. It is notable that the application of the model to a set of independent SDS gel data from the laboratory of Gunnar von Heijne shows an excellent correlation (see Figure 4). Further, a good fit is found to the current data derived from analytical ultracentrifugation.

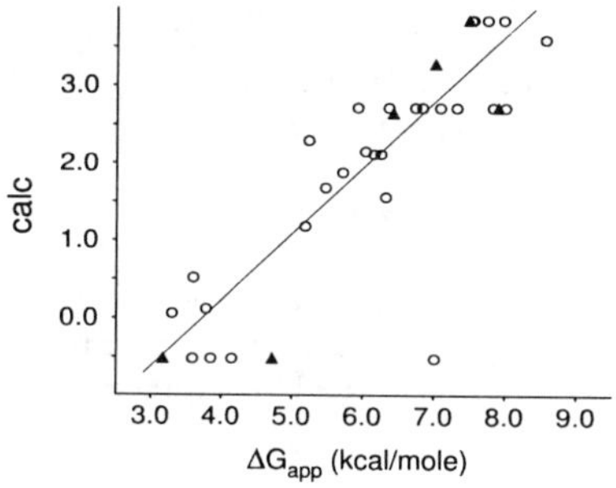

Figure 4. Correlation of calculated dimerization propensity (calc) and apparent free energy of dissociation (□ G_{app}) for wildtype and 33 multiple mutants of GpAtm.

This modeling provides a basis for understanding the sequence specificity of glycophorin transmembrane helix dimerization. Three structure-based empirical parameters representing a set of simple concepts are able to reproduce the qualitative dimerization propensities of more than 100 hydrophobic single point mutants as well as thirty insertion mutants of the glycophorin A transmembrane domain. The model demonstrates that the determinants of stability and specificity of helix interactions in detergent environments may be at least qualitatively understood by using extensions of the ideas outlined in the two-stage model for membrane protein folding [15]. This finding further reinforces the view that arriving at an understanding of the structures and stabilities of alpha-helical integral membrane proteins may be feasible using simple methods and ideas.

7. The Use Of Analytical Ultracentrifugation To Determine Mutation-Induced Free Energy Changes In Helix Association

We have developed techniques that exploit analytical ultracentrifugation to measure the energy of dimerization in a non-denaturing detergent solution, and have observed the changes in energy arising from mutants in the glycophorin dimer sequence. These results are described in a publication [34].

Equilibrium ultracentrifugation exploits the balance of centripetal forces against diffusion to establish a thermodynamic equilibrium for a molecule or mixture of molecules.

The equilibrium distribution of a single species, described by an exponential, measures the molecular weight. If two species are present, as in a monomer-dimer equilibrium, two exponentials can be fit to the resulting distribution of material to describe a set of equilibrium conditions at different concentrations represented by the radial distributions of the components. In her work, Fleming found a detergent (C_8E_5) that has two important properties: its micellar buoyant density is approximately one and it allows equilibration of transmembrane helices solubilized in the micelles on the time scale of the (C_8E_5) ultracentrifuge experiment. Thus, by examining a set of concentrations and centripetal fields, and by obtaining a global fit to all of these equilibrium distributions, data can be obtained that give information on the association of transmembrane helices, including the equilibrium constant. Changes in the equilibrium constant can then be interpreted in terms of free energy. A set of measurements has been put in place, initially involving the wild type glycophorin sequence and two mutants [34].

If the simple calculated association parameter described above is applied to the cases measured so far, there exists a good correlation, reinforcing the idea that a simple chemical theory may be useful in describing helix associations.

8. *A Genetic Screen For The Study Of Helix Interactions In Membranes*

Our assay is based on the ToxR transcriptional activation domain from *Vibria cholerae*, which activates transcription from the cholera toxin (ctx) promoter in response to dimerization [35]. A construct has been created for expression in *E. coli*, as shown in Figure 5.

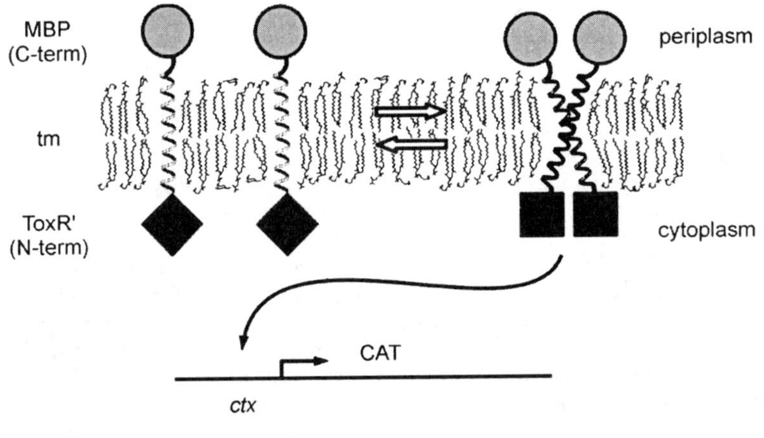

Figure 5. ToxR assay for TM oligomerization. TM-mediated dimerization of the N-terminal ToxR Cytoplasmic domains (diamonds) results in activation of CAT at the ctx promoter. The periplasmic MBP domain (circles) anchors the chimera in the membrane.

A protein is expressed in *E. coli* that consists of the maltose binding protein (a periplasmic protein) the transmembrane domain of interest, and the ToxR transcriptional activation domain. This construct is found to be inserted across the inner membrane of *E. coli*, and oligomerization places the ToxR domains in close proximity, activating the ctx promoter, which we have placed in front of the reporter gene, chloramphenicol acetyltransferase (CAT). Thus, interaction of the transmembrane domains confers chloramphenicol resistance on the *E. coli* in which the chimera is expressed. Quantitation of the level of CAT activation indicates the extent of dimerization. Our initial experiments involve transmembrane domains corresponding to residues L75-T87 of the wild type glycophorin TM domain and a set of mutants. A similar assay, but without a selectable reporter or control of chimera expression level, has been developed by [36].

The use of CAT as a reporter gene allows the extent of transmembrane domain-induced dimerization to be assayed both functionally (increased resistance to chloramphenicol) and enzymatically (direct measurement of chloramphenicol acetylation). Incorporation of the reporter construct into the expression plasmid provides several copies of the reporter per cell. This increase in reporter gene copy number is expected to provide greater sensitivity to low level dimerization and enhanced upper range of measurement.

The transmembrane domain of GpA was used to test the utility of the ToxR system, using the wildtype and mutant sequences. Chloramphenicol resistance was measured using disk diffusion assays (Figure 6A), in which cells are tested for their ability to grow close to a

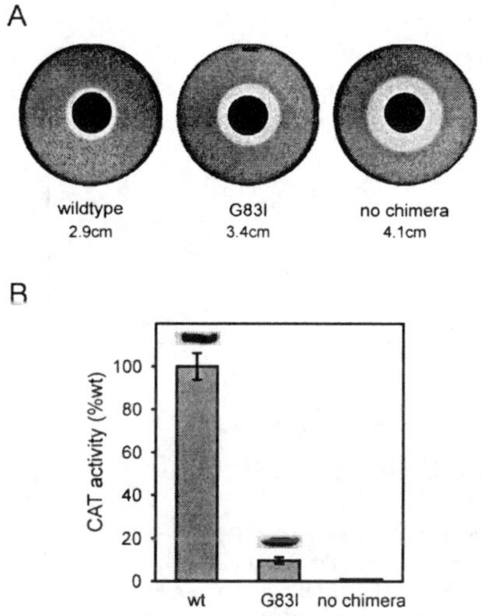

Figure 6. ToxR assay of GpA TM dimerization. **A.** Disk diffusion assays of *E. coli* expressing GpA wildtype, G83I, or no chimera. **B.** CAT activity (relative to wildtype) of cell extracts.

chloramphenicol-containing disk. The size of the resulting zone of inhibition around the disk indicates the degree of antibiotic resistance. A small zone of inhibition is seen for the wildtype GpA transmembrane domain, whereas the dissociating mutant G83I increases the size of the zone, and a still larger zone is seen if no chimera is present. This assay corresponds well to direct measurements of chloramphenicol activity in cell extracts (see Figure 6B).

Controls were run to verify correct membrane insertion and equal expression levels of ToxR chimerae. Membrane insertion and topology were measured in two ways. Protease sensitivity assays directly probe the accessibility of the MBP domain to proteolysis in spaeroplasted cells. MBP complementation assays test the ability of periplasmic MBP domains to complement an MBP deficient phenotype. Chimera expression levels were measured using western blot analysis.

A measurement of the helix-helix association constants in membranes complements our ultracentrifugation measurements in detergent micelles. By measuring the equilibrium established in membranes, the relationship between expressed chimera concentration and CAT activity should yield the association constant for the dimerization reaction. To test the dependence of CAT activity on chimera expression level, the ToxR chimera was expressed at a range of concentrations by placing it under the control of the lac promoter, and inducing with varying amounts of IPTG. As seen in Figure 7, the degree of dimerization (as reported by CAT activity) varied with chimera expression level (measured by quantitative western blot analysis). The dimerization of a disruptive mutant, G83I, was depressed relative to the wildtype at all concentrations assayed. These initial experiments suggest that it may be feasible to measure an equilibrium constant in the *E. coli* inner membrane.

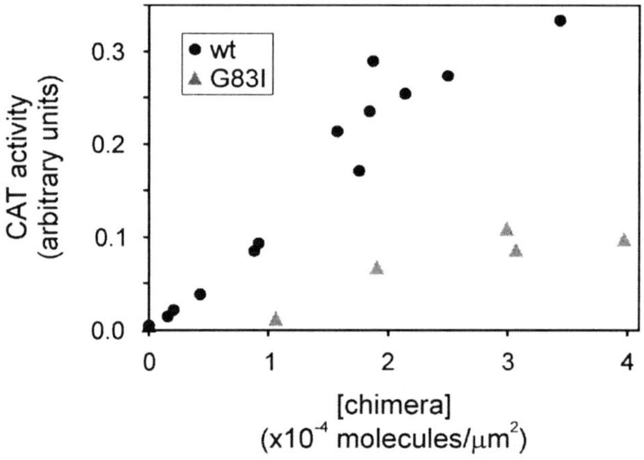

Figure 7. Concentration dependent dimerization of ToxR chimerae. Wildtype or G83I mutant chimerae at various induction levels were assayed for CAT Activity. [Chimera] is given as (moles protein) / (membrane volume).

The disk diffusion assays demonstrate that transmembrane sequences can be selected for their ability to oligomerize. We have applied this selection to a library of ToxR chimerae containing randomized substitution of G, A, V, L, I, S, T, P or R with equal probability at positions corresponding to the interface of a right-handed crossing of α-helices, based on the structure of the glycophorin dimer. Residues occupying non-interfacial positions were fixed as leucine. Figure 8 shows that the selection method is highly effective, since increasingly fewer colonies survive at higher CAM concentrations. The observed logarithmic decay in the number of surviving colonies provokes the speculation that a free energy scale may be revealed in the distribution of the transmembrane sequences in the population.

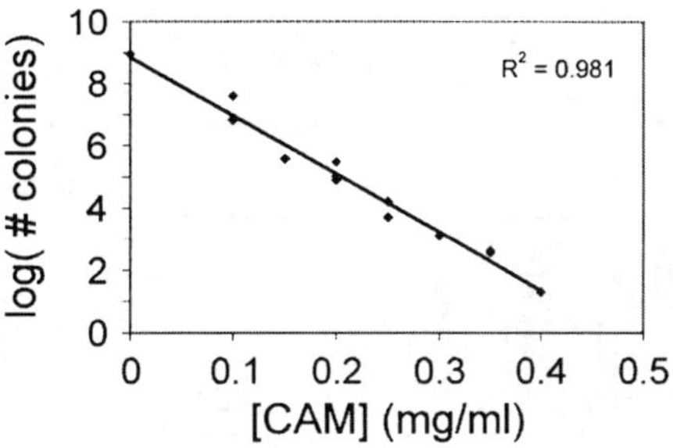

Figure 8. Selection of oligomeric TM domains from a randomized library. The number of colonies surviving at varying chloramphenicol concentrations is reported.

Sequence specificity of the GpA tm domain was assayed using the ToxR system. Figure 9 compares of the CAT activities of GpA tm mutants relative to wildtype. In each case, insertion and expression level were characterized; no effects of point mutations on levels of expression have been observed to date. The association of each mutant in the SN chimera in SDS micelles is presented (where available), and the ToxR assay largely parallels the SDS data. Interestingly, in SDS the Y mutations are invariably disruptive, probably due to artifactual interactions with detergent. In the ToxR assay, Y residues have minimal effect away from the interface, suggesting that this system will enable the study of TM domains containing polar sidechains. The ToxR assay should thus be able to accurately report on TM associations for a wide range of sequences, which will be critical for determining the importance of inter-helical hydrogen bonding to helix-helix interactions.

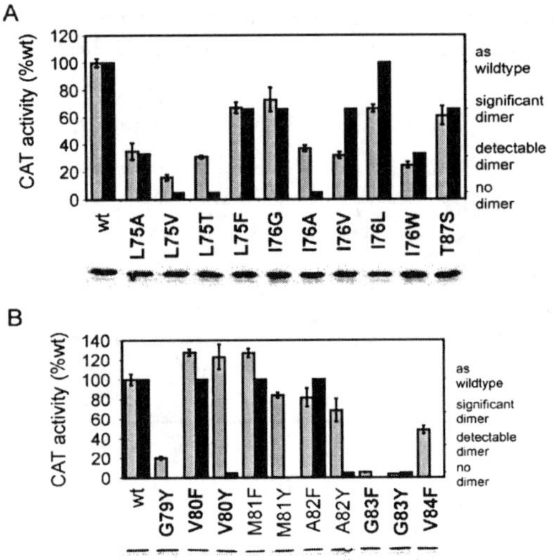

Figure 9. Mutations of GpA TM modulate dimerization in E. *coli* membranes. Light bars represent 2-3 independent CAT Assays; dark bars are SDS PAGE data of Lemmon *et al.* (1992). Anti-MBP western blots of each mutant are shown below.
A. Mutations show differential sensitivity to substitution.
B. Hydrophobic (F) or polar (Y) changes in one turn of α-helix.

9. Linearity Is Seen In Plots Of The Logarithm Of Sequence Abundance Versus Chloramphenicol Concentration

In applying the randomized library search to the right-handed motif for helix interactions, a linear relationship was discovered between the logarithm of the number of colonies and the chloramphenicol concentration at which those colonies were formed (Figure 8). Since each colony is expected to represent a unique sequence (to a good approximation), it is possible that the measurement is related to the free energy of association for each sequence. If such a relationship can be established, it would provide a measure of relative association energies between elements that can be inserted into the *E. coli* inner membrane, and would have value in a wide range of studies.

10. Conclusion: There Is Cause For Optimism In Attempts To Understand The Folding Of Membrane Proteins

Given that many membrane proteins have helix interactions as a dominant feature of their architecture and that the helices tend to be independently stable, understanding the interactions between helices could, in principle, lead to views that would allow prediction of membrane protein structures from sequence information. We believe that there is cause for optimism in this area, based on our findings using the glycophorin system. The fact that a modeling exercise coupled with molecular dynamics simulations successfully predicted the structure close to the one obtained using heteronuclear NMR is a most encouraging finding. Further, the finding that the relative perturbations of the dimerization by mutations remains on a similar scale despite shifts in its magnitude when different environments are studied is further cause for optimism. Nonetheless, the problem remains complex, even though it may prove much simpler than the problem of folding soluble proteins in an aqueous environment.

ACKNOWLEDGEMENTS

We thank the U. S. National Institutes of Health, the U. S. National Science Foundation, and the National Foundation for Cancer Research for support. DME thanks the region Ile de France and the Ecole Normale Supérieure, Paris, for their sponsorship of the Blaise Pascal Chair. We also thank A. Senes and other members of the laboratory who offered comments on the manuscript.

REFERENCES

1. Popot, J.-L., *et al.*, *The "microassembly" of integral membrane proteins : applications and implications.* Prog. Clinical Biol. Res., 1990. **343**: p. 237-262.
2. Henderson, R., *et al.*, *Model for the structure of bacteriorhodopsin based on high resolution electron cryo-microscopy.* J. Mol. Biol., 1990. **213**: p. 899-929.
3. Deisenhofer, J., *et al.*, *Structure of the protein subunits in the photosynthetic reaction center of Rhodopseudomonas viridis at 3 Å resolution.* Nature, 1985. **318**: p. 618-624.
4. Yeates, T.O., *et al.*, *Structure of the reaction center from Rhodobacter sphaeroides R-26: Membrane-protein interactions.* Proc. Natl. Acad. Sci. USA, 1987. **84**: p. 6438-6442.
5. Allen, J.P., *et al.*, *Structure of the reaction center from Rhodobacter sphaeroides R-26: the protein subunits.* Proc. Natl. Acad. Sci. USA, 1987. **84**: p. 6162-6166.
6. Kühlbrandt, W. and D.N. Wang, *Three-dimensional structure of plant light-harvesting complex determined by electron crystallography.* Nature, 1991. **350**: p. 130-134.
7. Kühlbrandt, W., D.N. Wang, and Y. Fujiyoshi, *Atomic model of plant light-harvesting complex.* Nature, 1994. **367**: p. 614-621.
8. Iwata, S., *et al.*, *Structure at 2.8 Å resolution of cytochrome c oxidase from Paracoccus denitrificans.* Nature, 1995. **376**: p. 660-669.
9. Ostermeier, C., *et al.*, *Structure at 2.7 Å resolution of the Paracoccus denitrificans two-subunit cytochrome c oxidase complexed with an antibody F_V fragment.* Proc. Natl. Acad. Sci. USA, 1997. **94**: p. 10547-10553.
10. Doyle, D.A., *et al.*, *The structure of the potassium channel : molecular basis of K^+ conduction and selectivity.* Science, 1998. **280**: p. 69-77.
11. Weiss, M.S., *et al.*, *The three-dimensional structure of porin from Rhodobacter capsulatus at 3 Å resolution.* FEBS Lett., 1990. **267**: p. 268-272.
12. Weiss, M.S., *et al.*, *The structure of porin from Rhodobacter capsulatus at 1.8 Å resolution.* FEBS Lett., 1991. **280**: p. 379-382.
13. Cowan, S.W., *et al.*, *Crystal structures explain functional properties of two E. coli porins.* Nature, 1992. **358**: p. 727-733.
14. Unwin, N., *The nicotinic acetylcholine receptor at 9 Å resolution.* J. Mol. Biol., 1993. **229**: p. 1101-1124.
15. Popot, J.-L. and D.M. Engelman, *Membrane protein folding and oligomerization : the two-stage model.* Biochemistry, 1990. **29**: p. 4031-4037.
16. Bormann, B.-J., W.J. Knowles, and V.T. Marchesi, *Synthetic polypeptides mimic the assembly of transmembrane glycoproteins.* J. Biol. Chem., 1989. **264**: p. 4033-4037.
17. Engelman, D.M., T.A. Steitz, and A. Goldman, *Identifying nonpolar transbilayer helices in amino acid sequences of membrane proteins.* Annu. Rev. Biophys. Biophys. Chem., 1986. **15**: p. 321-353.
18. Kyte, J. and R.F. Doolittle, *A simple method for displaying the hydropathic character of a protein.* J. Mol. Biol., 1982. **157**: p. 105-132.
19. Eisenberg, D., *et al.*, *Analysis of membrane and surface protein sequences with the hydrophobic moment plot.* J. Mol. Biol., 1984. **179**: p. 125-142.
20. Persson, B. and P. Argos, *Prediction of transmembrane segments in proteins utilising multiple sequence alignments.* J. Mol. Biol., 1994. **237**: p. 182-192.

21. Sharp, K.A., *et al.*, *Extracting hydrophobic free energies from experimental data: relationship to protein folding and theoretical models.* Biochemistry, 1991. **30**: p. 9686-9697.
22. Chothia, C., *The nature of the accessible and buried surfaces in solution.* J. Mol. Biol., 1976. **105**: p. 1-14.
23. Hunt, J.F., *et al.*, *Spontaneous, pH-dependent membrane insertion of a transbilayer alpha- helix.* Biochemistry, 1997. **36**(49): p. 15177-92.
24. Engelman, D.M., A. Goldman, and T.A. Steitz, *The identification of helical segments in the polypeptide chain of bacteriorhodopsin.* Methods Enzymol., 1982. **88**: p. 81-88.
25. Honig, B.H. and W.L. Hubbell, *Stability of "salt bridges" in membrane proteins.* Proc. Natl. Acad. Sci. USA, 1984. **81**: p. 5412-5416.
26. Manolios, N., J.S. Bonifacino, and R.D. Klausner, *Transmembrane helical interactions and the assembly of the T cell receptor complex.* Science, 1990. **249**: p. 274-277.
27. Cosson, P., *et al.*, *Membrane protein association by potential intermembrane charge pairs.* Nature, 1991. **351**: p. 414-416.
28. Lemmon, M.A., *et al.*, *Glycophorin A dimerization is driven by specific interactions between transmembrane α-helices.* J. Biol. Chem., 1992. **267**: p. 7683-7689.
29. Wang, J.a.P., A., *Do helices in membranes prefer to form bundles or remain dispersed in the lipid phase?* Biochim. Biophys. Acta, 1991. **1070**: p. 493-496.
30. MacKenzie, K.R., J.H. Prestegard, and D.M. Engelman, *A transmembrane helix dimer : structure and implications.* Science, 1997. **276**: p. 131-133.
31. Adams, P.D., D.M. Engelman, and A.T. Brunger, *Improved prediction for the structure of the dimeric transmembrane domain of glycophorin A obtained through global searching [published erratum appears in Proteins 1997 Jan;27(1):132].* Proteins, 1996. **26**(3): p. 257-61.
32. Lemmon, M.A., *et al.*, *Sequence specificity in the dimerization of transmembrane α-helices.* Biochemistry, 1992. **31**: p. 12719-12725.
33. MacKenzie, K.R. and D.M. Engelman, *Structure-based prediction of the stability of transmembrane helix- helix interactions: the sequence dependence of glycophorin A dimerization.* Proc Natl Acad Sci U S A, 1998. **95**(7): p. 3583-90.
34. Fleming, K.G., A.L. Ackerman, and D.M. Engelman, *The effect of point mutations on the free energy of transmembrane alpha- helix dimerization.* J Mol Biol, 1997. **272**(2): p. 266-75.
35. Ottemann, K.M. and J.J. Mekalanos, *Analysis of Vibrio cholerae ToxR function by construction of novel fusion proteins.* Mol Microbiol, 1995. **15**(4): p. 719-31.
36. Langosch, D., *et al.*, *Dimerisation of the glycophorin A transmembrane segment in membranes probed with the ToxR transcription activator.* J Mol Biol, 1996. **263**(4): p. 525-30.

Chaperones and Osmolytes in Protein Folding

Susan L. Lindquist

The University of Chicago

5841 South Maryland Avenue, MC 1028

Chicago, Illinois 60637, USA

Introduction

Changes in protein conformation and assembly govern most processes in cell biology and defects in protein folding provide some of cell biology's most common and perplexing problems. They cause devastating human illnesses, limit the ability of organisms to survive environmental stress and, most surprisingly, underlie a newly discovered mechanism of heredity. Many protein-folding transitions are regulated by chaperone proteins and cellular osmolytes. Here, I briefly review the general functions of chaperone proteins, describe some recent work we have done on the molecular function of one chaperone (Hsp104), discuss the role of osmolytes in protein folding, and describe how self-perpetuating changes in protein folding can serve as a genetic mechanism for inheritance.

Chaperone proteins

For proteins to function properly, it is essential that they fold into just the right structure. Some proteins also need to change their structures at just the right time, in just the right place. This presents many problems. There are a multitude of conformations that proteins can assume and finding the correct fold, especially for multidomain proteins, takes time. Moreover, the protein concentration inside cells is extraordinarily high - it is estimated that the average soluble protein bumps into another protein ~ one million times per second! Sticky amino acids and hydrophobic patches on proteins that are not fully folded can very easily associate with the wrong partners.

To cope with this problem all cells produce a special group of proteins dedicated to helping other proteins assume their proper conformations. These are called molecular chaperones by analogy with human chaperones: they interact transiently with highly reactive immature proteins to prevent them from making improper associations, rescue them from improper associations if they do occur, and leave mature proteins alone to perform their normal functions (Ellis and Hartl, 1999).

Many chaperone proteins are heat shock proteins or HSPs. They were named for their rapid induction in response to heat and other stresses, long before their functions were known (Parsell and Lindquist, 1993). It is now clear that Hsps are induced because high temperatures greatly intensify the protein-folding crisis. As proteins start to unfold, aggregation can reach disastrous proportions in a hurry. Therefore, all organisms react to moderate temperature increases (and other protein folding stresses) by inducing HSPS. In fact, the induction of HSPs is the most dramatic and highly conserved genetic response known (Figure 1).

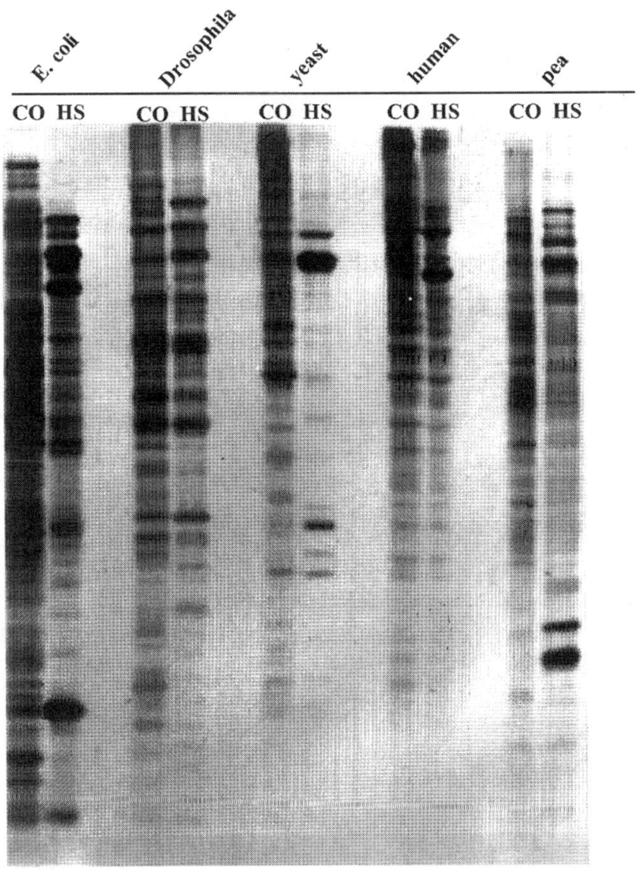

Figure 1.
Proteins synthesized at normal temperatures (C) and during heat shock at temperatures 5-10 degrees above normal growth temperatures (HS) were visualized in diverse organisms by the incorporation of radio-labelled isotopes, followed by gel electrophoresis and fluor

No doubt organisms have had to cope with changes in their environment since the dawn of life and the emergency nature of the response has caused the evolution of a remarkable variety of regulatory mechanisms to ensure that the proteins will be made as quickly as possible. I will concern myself only with the function of these proteins, but it is worth noting that studies of HSP regulation have provided some of the first examples of many key regulatory mechanisms. It is a large literature, well worth consulting (Brostrom and Brostrom, 1998; Lis and Wu, 1993; Welihinda *et al.*, 1999; Yost *et al.*, 1990; Yura T, 1999).

The HSPs are a special class of chaperones: they are "generalists" with broad substrate specificities. It is important to remember that there are many other types of chaperones that do not function in stress tolerance. Three beautiful examples are nucleoplasmin (Baker *et al.*, 1993; Eder and Fersht, 1995; Laskey *et al.*, 1993; Thanassi *et al.*, 1998), which chaperones histones for chromatin assembly, PapD (Thanassi *et al.*, 1998), which chaperones the proteins that assemble into bacterial pili, and the very specific intramolecular chaperones of various pro-proteins, such as pro-subtilisin (Baker *et al.*, 1993; Eder and Fersht, 1995). As a class, however, the HSPs are made in the largest quantities and were identified long ago. They are, therefore, among the best understood molecular chaperones. Moreover, in many cases, the same protein, or a close relative, functions in the maturation of proteins at normal temperatures. Thus, the study of Hsps has provided some of our most important insights on protein folding in a cellular context.

Table 1 lists the major Hsps and summarizes what we know about their activities. But first, a brief aside on nomenclature: Hsps are generally named according to the approximate molecular weight of the best studied, or first studied, members. Thus, Hsp70 is roughly 70 kDa, Hsp60, is about 60 kDa, etc. This can be rather confusing, however, since related proteins with similar molecular functions can have very different molecular weights. Another source of confusion is that the same protein often has many names. This is because the proteins have such basic functions in cell biology they have been "discovered" independently in many different systems, using different genetic or biochemical screens, examining many different processes. I have listed some of the most common

Table 1. Summary of major Hsps and their mode of action. See references cited in text for further details.

chaperone		biochemical function	substrate recognition	co-chaperone	other names; other family members
Hsp100	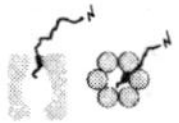	ATP-dependent disaggregation and unfolding, regulates protein conformation-based inheritance	unclear	Hsp70 and Hsp40	Hsp104, Hsp101, ClpB; ClpA, ClpC, CplX
Hsp90		maturation of transcription factors, signal-transducing kinases and other metastable proteins	proteins with secondary structure, incomplete tertiaty structure	many!, including Cdc37, cyclophilins, FKBPs, Hsp70, p23	Grp94, HtpG; Hsp82
Hsp70		ATP-dependent stabilization of unfolded proteins, release allows refolding to native state	hydrophobic extended polypeptide chains (~7aa) that are normally buried in folded proteins bind to Hsp70's C-terminal domain	Hsp40, GrpE, Bag1	Ssa1-4, DnaK; grp78, hsc70, BiP, Kar2, Ssb, Ssc, etc.
Hsp60	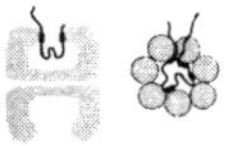	ATP-dependent facilitation of folding to native state, may help remove kinetic traps	molten globules: multivalent binding of exposed hydrophobic patches on substrate to flexible "fingers" on apical domain of chaperone	Hsp10	GroEL, Hsp65, Cpn60, Rubisco binding protein; TRiC/TTC
small Hsps		stabilizes unfolded proteins, prevents aggregation during HS	?	?	Hsp18, Hsp22, Hsp23, hsp26, Hsp27
Calnexin Calreticulin		folding of glucosylated proteins in the ER	primarily recognizes carbohydrate chains that are characteristic of immature protein, but some protein: protein recognition occurs as well.	UDP-glucose glycoprotein glucosyl transferase	

aliases for some of these proteins in Table 1. Yet another source of confusion is that proteins related to Hsps are synthesized at normal temperatures; a few of these are also listed in Table 1.

Although all of the major HSPs have related functions -- they ameliorate problems caused by protein misfolding and aggregation -- each HSP chaperone family has a unique mechanism of action. Most bind to various types of folding intermediates and prevent them from aggregating (Hsp70, Hsp60, Hsp40, and the small HSPs); others promote the reactivation of proteins that have already aggregated (Hsp100). Among the former, HSPs with different mechanisms of substrate recognition and release complement each other's activities to ensure that a wide variety of protein folding intermediates can be accommodated. Many laboratories have contributed to this work, too many to cite in this brief discussion. There are several excellent reviews I urge you to consult to obtain a more detailed picture of chaperone biology and biochemistry and to locate the primary sources [Parsell, 1993 #2; Beissinger, 1998 #33; Bukau, 1998 #30; Caplan, 1999 #32; Ellis, 1999 #29; Fink, 1999 #31; Mayer, 1999 #34]. Here I mention only two examples of substrate recognition and biochemical function, from the two chaperones whose functions are currently the best understood.

Hsp60 (GroEL) is a large barrel-shaped chaperone protein made up of 14 subunits, arranged as two stacked, seven-membered rings (Bukau and Horwich, 1998; Ellis and Hartl, 1999). It binds substrates by virtue of seven flexible hydrophobic "fingers" located at the top of the barrel. Substrates are recognized as whole molten globule, which have a collapsed structure but multiple exposed hydrophobic patches. The barrel expands and protein substrates are released into it when ATP and a partner chaperone, Hsp10 (GroES) bind. Inside, substrates can undergo folding, or partial folding, before they are spit back out into circulation. (For some substrates, binding to

Hsp60 followed by release into the barrel may help to nudge them out of kinetic folding traps.) ATP hydrolysis is required to reset the chaperone for another round of substrate binding.

Hsp70 binds substrates in a very different way (Bukau and Horwich, 1998; Ellis and Hartl, 1999). The chaperone's carbox- terminal wraps around a short extended region of the polypeptide chain (~7 amino acids long), with a strong preference for segments with hydrophobic side chains. These regions would normally be buried in the hydrophobic core of a folded protein. Although the binding mechanism is distinct from that of Hsp60, it shares with Hsp60 a capacity to bind a wide variety of proteins yet discriminate against proteins with a native structure. Hsp70's interaction with substrates is profoundly influenced by Hsp40. This protein is a chaperone in its own right, although its mechanism of substrate recognition is still not clear. It therefore provides a combinatorial substrate-recognition mechanism for Hsp70. It also helps Hsp70 bind substrates more tightly, by stimulating ATP hydrolysis. This closes Hsp70's substrate binding pocket. Up to this point, Hsp70's reaction mechanism is common to both eukaryotic and prokaryotic proteins, although the J proteins of eukaryotic cells are much greater in number and much more diverse in function. In prokaryotic cells, Hsp70 then intereacts with another co-chaperone, GrpE, which stimulates ADP release and exchange for ATP. This opens the substrate pocket, promoting release and resetting the stage for another round of substrate binding. The released substrate can refold or, if it has trouble refolding, rebind to Hsp70. In eukaryotic cells, Hsp70 does not appear to employ a GrpE-like exchange factor. Instead a variety of other co-chaperones, as yet poorly understood, appear to regulate ATP exchange.

HSP104

As discussed above, in all organisms moderate stresses induce tolerance to more extreme stresses. In S. cerevisiae, a 30 min. pretreatment at 37°C increases survival 10,000-fold at 50°C. HSPs are induced during such pretreatments but, surprisingly, most HSP mutations have no effect on survival at extreme temperatures, although they do reduce growth at high temperatures (Parsell and Lindquist, 1993). Hsp104 mutations are a striking exception. They don't impair growth, even at high temperatures, but reduce survival > 500 fold after exposure to extremes (Sanchez and Lindquist, 1990). The likely importance of Hsp104 function in the wild is vividly illustrated by the synergistic effects of different stresses (Lindquist *et al.*, 1995). Two mild stresses, innocuous on their own, are deadly in combination, unless cells express Hsp104 (Fig 2).

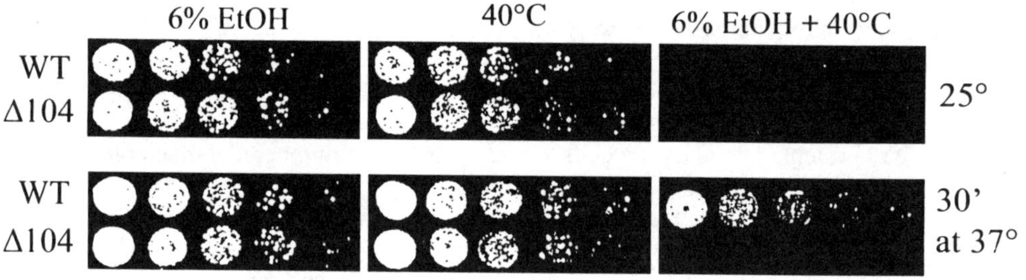

Fig 2.
Wild-type and Hsp104-deficient yeast cultures were grown at 25°C (top rows) or given a mild conditioning heat pretreatment at 37°C for 30 min (bottom rows). Cells were maintained at the same conditions (data not shown) or exposed to 6% ethanol or 40°C for two hours, or to both stresses. Cells were serially diluted (5-fold each step) and plated on rich medium to determine the number of live cells remaining in the culture. Growth after treatment with 6% ethanol only or 40°C only was the same as with no stress. However both stresses killed all of the cells in the culture unless they had been given a conditioning pre-treatment and were able to make Hsp104.

Unlike other HSPs, Hsp104 has the unique ability to promote the reactivation of aggregated proteins. Visualized by electron microscopy, Hsp104 mutant and wild-type (WT) cells have similar levels of protein aggregation after sublethal heat treatments. However, when cells are returned to 25°C, aggregates rapidly disappear from WT but not hsp104 cells. Similarly, ts luciferase luciferase activity is lost at equal rates in both cell types, and the protein aggregates. However, during recovery at 25°C luciferase is rapidly resolubilized and reactivated in WT cells, but not in the mutants (Parsell *et al.*, 1994).

The notion that a protein could actually disaggregate other proteins was so contrary to expectation, however, it required independent verification in vitro with purified components. Recently, using reporter enzymes, Hsp104 was found to have a general capacity to promote the reactivation of denatured, previously aggregated proteins (Glover and Lindquist, 1998). To accomplish this, Hsp104 requires the help of two other chaperones, Hsp70 and Hsp40. These two proteins, therefore, have dual roles in stress tolerance: they reduce the rate at which proteins aggregate in the first place, and they promote the refolding of previously aggregated proteins released by Hsp104.

Our current model for Hsp104 function is based on the observations that a) Hsp104 assembles into hexameric particles in the presence of ATP (Parsell *et al.*, 1994), b) has a basal ATPase activity that is stimulated by peptides and substrates (Schirmer and Lindquist ; Schirmer and Lindquist, 1997; Schirmer *et al.*, 1998), and c) undergoes conformational changes in response to ATP (Hattendorf *et al.*,). Hsp104's hexameric structure provides multiple sites for polypeptide substrate binding. Upon ATP binding and hydrolysis conformational changes in Hsp104 perforce change the positions of bound substrates relative to each other. Conformationally-trapped substrates can thereby be released from aggregates, but they are in a non-native state and require other chaperones, such as Hsp70 and Hsp40, to usher them towards reactivation.

This molecular function of Hsp104 perfectly reflects the phenotypes of Hsp104 mutants. Hsp104 is vital only under extreme conditions, when the rate of protein denaturation outstrips the capacity of other mechanisms to prevent aggregation. Genetic interactions between Hsp70 and Hsp104 illustrate the principle: mutations that reduce Hsp70 impair protein folding and cause aggregation even at 30°C (Lindquist *et al.*, 1995). In these cells, Hsp104 mutations drastically reduce growth. Conversely, over-expression of Hsp70 reduces the rate of aggregation at high temperatures and mitigates the effects of Hsp104 mutations on survival (Sanchez *et al.*, 1993). Proteins closely related to Hsp104 are found in most bacteria, fungi, plants, and protozoa. In these organisms it appears to have a similar, vital role in stress tolerance. It is not, however, found in metazoan organisms. Whether there are other proteins that have disaggregation functions in these organisms is not yet clear.

Trehalose

HSPs are by no means the only mechanism cells employ to help them recover from stress-induced crises in protein folding. Another is the production of osmolytes. Trehalose, and other osmolytes like betaine, mannitol, and proline, are produced by a wide variety of organisms in response to stress. Trehalose is found in the greatest variety of species1, and has been most studied in the anhydrobionts. These unusual organisms, which include 'resurrection plants', brine shrimp, nematodes, and baker's yeast (Saccharomyces cerevisiae), are able to withstand almost complete desiccation. They endure long periods in seeming suspended animation, returning from dormancy to full activity once water again becomes available. Under such conditions osmolytes are produced in massive quantities (up to 10% dry weight) (Singer and Lindquist, 1998).

Trehalose is also induced by high temperatures. In vitro experiments have demonstrated that trehalose, like other osmolytes, stabilizes proteins at high temperatures in vitro. Its role in stress tolerance, however, had been controversial for many years (Singer and Lindquist, 1998). And when genetic analysis was initiated in yeast, results were initially quite baffling. Cells carrying mutations in the enzyme that makes trehalose are much more sensitive to stress than wild-type cells. However, cells carrying mutations in the enzyme that degrades trehalose were also very sensitive to stress. This perplexing picture was resolved when three effects of trehalose were demonstrated: 1) trehalose stabilizes proteins against stress-induced aggregation in vivo, 2) trehalose reduces the aggregation of proteins that have already been denatured, and 3) trehalose interferes with chaperone-mediated protein refolding (Singer and Lindquist, 1998). These observations clarify the role of trehalose in thermotolerance and explained several previous contradictions. In order for cells to survive high temperatures, it is very important for them to induce trehalose expression. However, for them to recover normal growth, it is important for them to get rid of it.

HSP104 and protein based inheritance

Protein conformational changes also lie at the heart of a new form of genetic inheritance: an altered phenotypic trait can be based on the propagation of proteins with altered conformations rather than nucleic acids with altered sequences (Wickner, 1994). There are two well characterized elements of this type in yeast, although there may well be many more. We have worked on one of these, because its inheritance is controlled by the same chaperone protein that plays such an important role in thermotolerance, Hsp104. This element, [*PSI*+], changes the fidelity of protein synthesis and exhibits dominant, cytoplasmicall inheritance (Cox *et al.*, 1988). Strangely, this trait could not be associated with any extra-chromosomal nucleic acid. However, an unusual relationship between [*PSI*+] and a nuclear-encoded protein, Sup35, was described (Chernoff *et al.*, 1993; Ter-Avanesyan *et al.*, 1993). Sup35 is a translation-termination factor with an unusual amino-acid composition. The C-terminus has termination-factor activity; the N-terminus (N) is extremely rich in gln, asn, gly and tyr and contains oligopeptide repeats; the middle region (M) is highly charged. Remarkably, transient expression of N is sufficient to form new, heritable [*PSI*+] elements and deletion of N makes cells immune to [*PSI*+]. These observations led Reed Wickner to suggest that [*PSI*+] may be inherited by a protein-only, prion-like mechanism (Wickner, 1994). Specifically, Sup35 could exist in two different conformations, one functional in translation, the other non-functional but self-perpetuating.

Our work on [*PSI*+] began in collaboration with Yury Chernoff and Susan Liebman (Chernoff *et al.*, 1995). We established that Hsp104 over-expression cures cells of [*PSI*+]. Most remarkably, transient Hsp104 over-expression was sufficient. Because the only known function of Hsp104 is to alter the conformational state of other proteins, this provided a strong genetic argument that [*PSI*+] is due to the inheritance of a protein with an altered conformation.

Recently, we have produced cell biological and biochemical evidence in support of the protein-only model of [*PSI*+] inheritance and have begun to assess the role of Hsp104. Sup35 is in a different physical state in [*PSI*+] and [*psi*-] cells. Whenever [*PSI*+] is present, most Sup35 is in aggregates; in [*psi*-] strains, most Sup35 is soluble (Patino *et al.*, 1996). Visualization of [*PSI*+] propagation in living cells was accomplished by fusing either whole Sup35 or NM to Green Fluorescent Protein (GFP). Newly synthesized protein aggregated as soon as it could be visualized in [*PSI*+] cells, but remained soluble in [*psi*-] cells (Patino, Science). Thus, the conformational state of newly synthesized Sup35 depends upon the conformational state of pre-existing protein, a major tenet of the protein-only hypothesis. GFP also allows the visualization of new prion formation. When the fusion protein is transiently over-expressed in [*psi*-] cell, Sup35 begins to coalesce and new [*PSI*+] elements are acquired (Patino *et al.*, 1996).

Fragments of Sup35 containing the N-terminal domain form highly ordered amyloid fibers in vitro (Glover *et al.*, 1997). Moreover, these fibers are self-seeding: when protein is diluted into aqueous buffer fibers form very slowly, but if a small quantity of pre-existing fibers is added, they form much more rapidly (Glover *et al.*, 1997). Although we bypassing the requirement for Hsp104 in these experiments (presumably because we are starting with denatured protein) this process provides a plausible molecular model for the inheritance of [*PSI*+]: ordered aggregates of Sup35 are rare, but once formed they are self-perpetuating, reducing the concentration of soluble Sup35 and causing heritable nonsense suppression. Several things link self-seeded conformational change to [*PSI*+] inheritance. Lysates from [*PSI*+], but not [*psi*-] cells seed amyloid formation in vitro with purified Sup35 (Glover *et al.*, 1997). Moreover, mutations that inhibit [*PSI*+] formation in vivo inhibit amyloid formation in vitro and mutations that enhance [*PSI*+] formation in vivo enhance amyloid formation in vitro.

References

Baker, D., Shiau, A.K. and Agard, D.A. (1993) The role of pro regions in protein folding. *Curr Opin Cell Biol*, **5(6):** 966-970.

Brostrom, C.O. and Brostrom, M.A. (1998;) Regulation of translational initiation during cellular responses to stress. *Prog Nucleic Acid Res Mol Biol*, **58:** 79-125.

Bukau, B. and Horwich, A.L. (1998) The Hsp70 and Hsp60 chaperone machines. *Cell*, **92(3):** 351-366.

Chernoff, Y.O., Derkach, I.L. and Inge-Vechtomov, S.G. (1993) Multicopy SUP35 gene induces de-novo appearance of psi-like factors in the yeast Saccharomyces cerevisiae. *Curr Genet*, **24(3):** 268-270.

Cox, B.S., Tuite, M.F. and McLaughlin, C.S. (1988) The psi factor of yeast: a problem in inheritance. *Yeast*, **4(3):** 159-178.

Eder, J. and Fersht, A.R. (1995) Pro-sequence-assisted protein folding. *Mol Microbiol*, **16(4):** 609-614.

Ellis, R.J. and Hartl, F.U. (1999) Principles of protein folding in the cellular environment. *Curr Opin Struct Biol*, **9(1):** 102-110.

Glover, J.R., Kowal, A.S., Schirmer, E.C., Patino, M.M., Liu, J.J. and Lindquist, S. (1997) Self-seeded fibers formed by Sup35, the protein determinant of [PSI+], a heritable prion-like factor of S. cerevisiae. *Cell*, **89(5):** 811-819.

Glover, J.R. and Lindquist, S. (1998) Hsp104, Hsp70, and Hsp40: a novel chaperone system that rescues previously aggregated proteins. *Cell*, **94(1):** 73-82.

Hattendorf, D., Cashikar, A. and Lindquist, S. *Unpublished Observations.*

Laskey, R.A., Mills, A.D., Philpott, A., Leno, G.H., Dilworth, S.M. and Dingwall, C. (1993) The role of nucleoplasmin in chromatin assembly and disassembly. *Philos Trans R Soc Lond B Biol Sci.*, **339(1289):** 263-269;.

Lindquist, S., Patino, M.M., Chernoff, Y.O., Kowal, A.S., Singer, M.A., Liebman, S.W., Lee, K.H. and Blake, T. (1995) The role of Hsp104 in stress tolerance and [PSI+] propagation in Saccharomyces cerevisiae. *Cold Spring Harb Symp Quant Biol.*, **60:** 451-460.

Lis, J.W. and Wu, C. (1993) Protein traffic on the heat shock promoter: parking, stalling, and trucking along. *Cell*, **74(1):** 1-4.

Parsell, D.A., Kowal, A.S., Singer, M.A. and Lindquist, S. (1994) Protein disaggregation mediated by heat-shock protein Hsp104. *Nature*, **372(6505)**, 475-478.

Parsell, D.A. and Lindquist, S. (1993) The function of heat-shock proteins in stress tolerance: degradation and reactivation of damaged proteins. *Annu Rev Genet*, **27** :437-496.

Patino, M.M., Liu, J.J., Glover, J.R. and Lindquist, S. (1996) Support for the prion hypothesis for inheritance of a phenotypic trait in yeast. *Science*, **273(5275):** 622-626.

Sanchez, Y. and Lindquist, S.L. (1990) HSP104 required for induced thermotolerance. *Science*, **248(4959)**, 1112-1115.

Sanchez, Y., Parsell, D.A., Taulien, J., Vogel, J.L., Craig, E.A. and Lindquist, S. (1993) Genetic evidence for a functional relationship between Hsp104 and Hsp70. *J Bacteriol*, **175(20):** 6484-6491.

Schirmer, E. and Lindquist, S. *Unpublished Observations.*

Schirmer, E.C. and Lindquist, S. (1997) Interactions of the chaperone Hsp104 with yeast Sup35 and mammalian PrP. *Proc Natl Acad Sci U S A*, **94(25):** 13932-13937.

Schirmer, E.C., Queitsch, C., Kowal, A.S., Parsell, D.A. and Lindquist, S. (1998) The ATPase activity of Hsp104, effects of environmental conditions and mutations. *J Biol Chem*, **273(25)**, 15546-15552.

Singer, M.A. and Lindquist, S. (1998) Thermotolerance in Saccharomyces cerevisiae: the Yin and Yang of trehalose. *Trends Biotechnol*, **16(11):** 460-468.

Ter-Avanesyan, M.D., Kushnirov, V.V., Dagkesamanskaya, A.R., Didichenko, S., Chernoff, Y.O., Inge-Vechtomov, S.G. and Smirnov, V.N. (1993) Deletion analysis of the SUP35 gene of the yeast Saccharomyces cerevisiae reveals two non-overlapping functional regions in the encoded protein. *Mol Microbiol*, **7(5):** 683-692.

Thanassi, D.G., Saulino, E.T. and Hultgren, S.J. (1998) The chaperone/usher pathway: a major terminal branch of the general secretory pathway. *Curr Opin Microbiol*, **1(2):** 223-231.

Welihinda, A.A., Tirasophon, W. and Kaufman, R.J. (1999;) The cellular response to protein misfolding in the endoplasmic reticulum. *Gene Expr*, **7(4-6):** 293-300.

Wickner, W.T. (1994) How ATP drives proteins across membranes. *Science*, **266(5188):** 1197-1198.

Yost, H.J., Petersen, R.B. and Lindquist, S. (1990) RNA metabolism: strategies for regulation in the heat shock response. *Trends Genet*, **6(7):** 223-227.

Yura T, N.K. (1999) Regulation of the heat-shock response. *Curr Opin Microbiol*, **2(2):** 153-158

J.A.F. Op den Kamp (Ed.)
IOS Press, 2000

Crystallization of Membrane Proteins in Lipidic Cubic Phases

Ehud M. LANDAU
Department of Molecular Microbiology, Biozentrum, University of Basel, 4056 Basel, Switzerland

Abstract. Crystallization of membrane proteins is a notoriously difficult process, because of the coexistence of hydrophobic and hydrophilic surfaces in these proteins. Conventionally, crystallization experiments were conducted in the presence of detergents, whose colloidal properties are not well understood. We have initiated a novel approach using lipidic cubic phases, which are membranous materials that can accommodate a remarkably large range of proteins, additives and precipitants, and which proved to be efficient matrices for the crystallization of membrane proteins.

1. Introduction

Biological membranes, consisting of protein, lipid and solute molecules, surround cells and subcellular compartments, thereby distinguishing the inside from the outside. More than merely providing protection as selective permeability barriers, membranes host a large number of highly specialized membrane proteins, which participate in all the interactions between the cell and its surroundings: they perform solute transport, charge separation, conversion of energy and signal transduction as well as form junctions between cells and carry out cell-cell recognition. Over a third of the 6000 or so proteins of baker's yeast are predicted to be integral membrane proteins [1], and this fraction is thought to be even higher in cells of higher organisms. Yet, in contrast to soluble proteins, of which well over 4000 structures have been published to resolutions better than 2.5 Å, fewer than 20 high resolution structures of membrane proteins have been determined thus far (see:http://blanco.biomol.uci.edu/Membrane_Proteins_xtal.html for an updated list of membrane protein structures).

Our understanding of the structural principles of membrane proteins is thus still at its infancy. The reasons for this state of affairs are threefold: (a) Isolation and purification of membrane proteins out of their membranes in their active state and native conformations poses great difficulties. (b) Production of membrane proteins in sufficient amounts by recombinant DNA technology using heterologous expression systems is very demanding [2] (c) Establishing high resolution structures of membrane proteins is a formidable challenge. The most widely used methodology for the elucidation of high resolution structures of biological macromolecules is X-ray crystallography, which relies critically on the reproducible availability of well-ordered three-dimensional (3D) crystals. To date, this requirement has met with success only in exceptional cases, and clearly represents the bottle neck in structural biology of membrane proteins.

The first reported 3D crystallizations of membrane proteins - those of bacteriorhodopsin (bR) [3] and porin [4] - were obtained from aqueous detergent solutions. These early successes resulted in great excitement, and most of the subsequent work in the area was carried out with various detergents, the colloidal properties of which are rather complex and poorly understood. In the following years it became clear that novel crystallization approaches need to be attempted. Recently two novel concepts were introduced, both yielding crystals which diffract to high resolution: In the first, an Fv fragment of a conformation-specific monoclonal antibody, which binds to a discontinuous epitope at the periplasmic side of bacterial cytochrome *c* oxidase, was used in the co-crystallization with the enzyme. It was shown that the Fv fragment constitutes the only part involved in protein-protein contacts in one plane [5]. The second approach introduces lipidic cubic phases for the crystallization of membrane bacteriorhodopsin and other membrane proteins [6]. This membrane-mimetic system, consisting of lipid, aqueous solution and protein in appropriate proportions, forms a transparent 3D curved bilayer matrix, which is pervaded by two non-connected aqueous channels. Membrane proteins, once inserted into this complex array, diffuse laterally to nucleate, and eventually to yield well-ordered crystals [7,8]. This paper provides a brief overview of membrane protein crystallization, followed by a detailed description of the special properties of lipidic cubic phases and their utilization for structural studies of membrane proteins.

2. Crystallization of Membrane Proteins

Michel [9] introduced the notion that membrane proteins can form two types of crystals (Figure 1). Type I crystals are composed of stacks of two dimensional crystals. Lipid molecules reside in the plane of the membrane and mediate crystal contacts by hydrophobic interactions in that plane, and by polar interactions perpendicular to the plane of the membrane. The entities which crystallize in type II crystals are detergent-protein mixed micelles, in which the hydrophobic surfaces of the proteins are coated with detergent belts, thus rendering them hydrophilic. In these crystals, the lattice contacts are polar. Most of the membrane protein structures elucidated thus far belong to type II, two high resolution structures of type I have also been reported: porin [10] and bacteriorhodopsin [11-13].

The major challenge in detergent-based crystallization is to replace the native lipid bilayer with detergent, preserving the protein's integrity during crystal. This task is complicated because of the mutually exclusive solubility properties of the hydrophobic and hydrophilic surfaces of membrane proteins. Prior to crystallization, membranes have to be isolated and the proteins solubilized and purified in an appropriate detergent solution. Often, the detergent has to be exchanged for the purpose of crystallization, and in most cases a concentration step is needed before crystallization is set up. Detailed descriptions and procedures of these steps have been published [14,15]. The crystallization

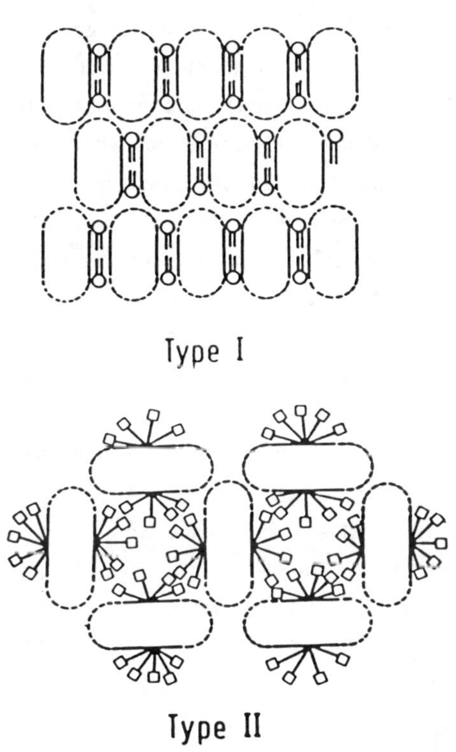

Figure 1. Schematic representation of the two basic packing types of membrane proteins. Type I is composed of stacked two-dimensional crystalline membranes which are ordered in the direction perpendicular to the membrane. Type II is composed of detergent/protein mixed micelles. Dashed lines indicate the hydrophilic surfaces, continuous lines the hydrophobic surfaces of the proteins. Lipid molecules are depicted as circles with two tails, detergent molecules as squares with one tail. With permission from [9].

strategy thereafter is identical to that of soluble proteins, involving supersaturation, nucleation and crystal growth. In order to yield type I crystals, detergent molecules have to be removed as has been demonstrated for bR [16] and lipids have to be added in some controlled fashion. Type II crystals are crystallized similarly to soluble proteins - the mixed micellar solutions are salted out when using precipitating agents, and often undergo phase separation: one phase, the detergent-rich one, contains most of the micelles, whereas the other contains detergent at a concentration close to the CMC. Crystals thus grown contain a substantial number of detergent molecules, which correspond to concentrations much higher than the starting concentrations.

The first crystals of porin were grown in "oily drops" [4] from a solution containing 1% n-octyl-tetraoxyethylene, which has a cloud point at 46°C. During the vapor diffusion experiment, water diffuses out of the protein/detergent solution, and the cloud point decreases. Additional decrease is brought about by the gradual increase in PEG and salt concentration, till the phase separates. The detergent rich phase, "oily droplets" as they were described, contain most of the micelles, and obviously also of the detergent-protein mixed micelles, which are now massively concentrated and eventually crystallize.

Many of the attempts to crystallize membrane proteins used variations of this methodology, employing combinations of detergents, additives and precipitants, while trying to avoid phase separation and formation of oily drops. In some cases the choice of detergent type and chain length is very narrow, as has been shown for the light-harvesting complex LH-II, which crystallizes with nonylglucoside but not with the homologous octylglucoside [14] or as has been described for the reaction center of *Rhodopseudomonas viridis* . In other instances, most notably with *E. coli* OmpF, a wide range of detergents could be used for crystallization [17], the only specific requirement being the use of non-ionic, non-steroid head group. In parallel, during the past years various ingenious approaches to increase the stability of membrane proteins and enhance their tendency to crystallize have been attempted. One idea was to engineer larger polar surface domains. This was successfully accomplished in the case of lactose permease by attaching a soluble cytochrome b_{562} between transmembrane helices: inserting the cDNA of soluble cytochrome b_{562} into its DNA [18]. and expressing the chimaeric protein in *E. coli* yielded a fully functional protein. Stroud et al. developed an approach to keep membrane proteins in a soluble form by covering the hydrophobic parts of the proteins with specially designed 24-residue amphipathic peptides, which pack around the protein in a well-ordered, α-helical arrangement [19]. For mostly α-helical proteins, they could demonstrate that 85 % of bR and 60 % of rhodopsin remained in their native forms for two days in solution with these "peptitergents", whereas PhoE porin, which has a β-sheet structure, precipitated completely. The plant light harvesting complex LH-II, unlike most membrane proteins, was crystallized in the presence of lipids - digalactosyl diacyl glycerol [20]. These three-dimensional crystals were not highly ordered to allow structural analysis at atomic resolution, and the structure was subsequently solved with electron crystallography using two-dimensional crystals [21]. Finally, there have been a few reports of membrane attached proteins whose soluble part is the functionally interesting one. In these cases, the membrane anchor can be cleaved off enzymatically in order to allow crystallization [22], or the soluble parts can be produced by recombinant DNA technology [23].

Unlike soluble proteins, which are often crystallized using factorial and sparse matrix strategies, in most cases of membrane protein crystallization the full factorial search strategy was employed. Gouaux and coworkers have recently succeeded in devising a systematic approach to the crystallization of α-hemolysin [24]. They have developed a sparse matrix strategy which complements the full factorial approach, selecting crystallization conditions which are close to the consolution boundary. After determining the phase behavior of octyl tetraoxyethylene in the presence of 19 different salts and 3 different PEGs at 4 and 20° C, they set up 60 different conditions at both temperatures, all close to the consolution boundary, and obtained microcrystals in 11 of these after 2 weeks. Using full factorial designs, they have subsequently succeeded in obtaining 2 crystal forms amenable to high resolution structure analysis.

Based on the observation that the most prominent crystal lattice contacts are made by the polar extramembranous domains, whereas those made by hydrophobic interactions seem to be the exception, Michel at al. [5]. have chosen to co-crystallize membrane proteins with antibody fragments that are known to specifically and strongly bind to it. The properties of monoclonal antibody fragments - they exhibit high affinity and specificity, are readily available and are homogeneous - make them ideal candidates for this task. F_V fragments can be produced efficiently by *Escherichia coli* [25] and purified by affinity chromatography using specific peptide tags [26]. Moreover, crystals of F_V fragments are often readily available and diffract well. The fragment used here was previously characterized [27,28]. in terms of cloning, cDNA expression and amino acid sequence, and has been crystallized and its structure determined to 1.28 Å. The Fv fragment is conformation specific, recognizing the native cytochrome *c* oxidase from *Paracoccus denitrificans.* The crystallization set up, consisting of a 1:1 complex of the cytochrome *c* oxidase and Fv fragment in dodecyl-β-D-maltoside, yielded red-brown tetragonal crystals of sizes up to 2 mm length and 0.6 mm diameter after four weeks in a vapor diffusion experiment with PEG monomethyl ether 2000 as precipitant. No crystals of cytochrome *c* oxidase were observed under analogous conditions without the antibody fragment.

Analysis of the packing in the complex, based on the X-ray structure to 2.8 Å resolution [29], reveals that all polar interactions are mediated by the Fv fragments There are no direct interactions between cytochrome *c* oxidase molecules in the *a,b* plane. Each Fv fragment is in contact with two cytochrome c oxidase molecules and with another Fv fragment: one lattice contact is with its antigen, the periplasmic domain of subunit II of the oxidase molecule, the other with the polar surface of subunit III of a second cytochrome *c* oxidase molecule, and the contact with a second Fv fragment is established by a two-fold crystallographic rotation axis. It is also evident that binding of the antibody fragment increases the polar surface area, which leads to the successful crystallization.

3. Lipidic cubic phases

Lipidic mesophases are liquid crystalline materials consisting of water and lipid. The cubic phases are soft condensed materials which are the most structured and rigid mesophases. Their material properties are characterized by very high viscosity, optical transparency and by non-birefringence. Cubic phases belong to one of two classes: the first is based on discrete micelles, the other on a bicontinuous membrane structure. The latter constitute a continuous lipid membrane whose hydrophilic headgroup region is curved towards the water compartments (Figure 2).

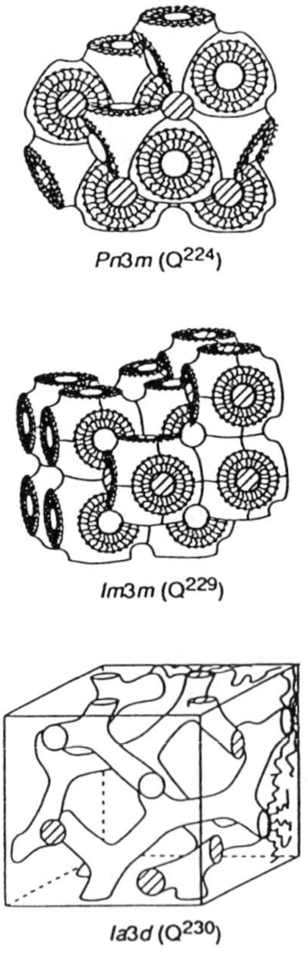

Figure 2. Schematic representation of the structures of three bicontinuous cubic phases, denoted with their respective crystallographic space groups. The non-connected water channels are shaded and cross-hatched. With permission from [30,31].

Lipidic mesophases with a cubic symmetry were discovered 40 years ago and have recently become the focus of intensive investigations by several complementary physical methods: small angle X-ray diffraction, nuclear magnetic resonance, freeze-fracture electron microscopy and fluorescence recovery after photobleaching. Three different inverted bicontinuous cubic phases with space groups of Ia3d, Pn3m and Im3m have been identified. The chemical composition of such phases consists of an extremely large amount of lipid, usually between 50 to 80 % (w/w), and 50 to 20 % (w/w) water. The interface between these two components exhibits a very large membrane surface which is surrounded on both sides by water.

In contrast to planar bilayers, the membrane of a bicontinuous lipidic cubic phase is curved and consists of a three-dimensional periodic array of saddles, constituting a continuous space-filling membrane. The accepted model is that the bilayer midplane of bicontinuous lipidic cubic phases constitutes *infinite periodic minimal surfaces* (IPMS). Such surfaces exhibit negative or zero Gaussian curvature and mean curvature of zero at any point. This holds because the two principal curvatures at any point are equal in absolute value but opposite in sign. Contrary to detergent micelles, the surface of which curves away from the water, each leaflet of membranes that are based on IPMS curve towards the water. All such cubic phases ideally consist of a single continuous lipid bilayer that divides space into two congruent hydrophilic compartments. The unit cells have cubic symmetry. In monoolein/water systems the cubic unit cell sizes depend on hydration and temperature and range from approximately 97 Å to 185 Å for the Ia3d and approximately 69 Å to 113 Å for the Pn3m cubic.

The complex packing of a curved bilayer surrounded by two aqueous channel systems has intriguing consequences. The materials thus formed are absolutely isotropic, transparent and non-birefringent by virtue of the cubic symmetry. Furthermore, due to the connectivity of each, the lipidic and aqueous systems, long range diffusion is possible in both. A hydrophilic solute molecule that is membrane impermeable is confined to one aqueous channel system and may thus diffuse into half of the available aqueous space. Lipids are likewise confined to one of the respective membrane leaflets. The diffusion coefficients of lipids in these materials are of the same order of magnitude as that of lipids in planar bilayers [32,33] and in biological membranes. Small angle X-ray diffraction studies revealed that millimeter-sized cubic phases usually do not form single crystals, but consist of domains that give rise to powder X-ray diffraction patterns. A further consequence of these soft materials is that they will adopt the macroscopic shape of the container in which they are formed. They can also be manipulated at will with tools, e.g. can be filled inside a capillary or squeezed between parallel plates. Finally, embedded objects, such as crystals are immobilized inside these materials, which is useful for a variety of applications, such as spectroscopy, crystallography and transport.

4. Crystallization in lipidic cubic phases

Most membrane proteins residing in cytoplasmic membranes are very labile, and often tend to aggregate, denature and possibly undergo degradation by proteases once solubilized out of their native membranes. In order to overcome these difficulties, and to limit as much as possible exposure to detergent solution conditions, we postulated that membrane proteins should be kept in a solid-like membranous environment throughout the crystallization process [7]. We proposed that such a system, which would promote nucleation and growth of 3D crystals, should fulfill the following requirements:

(a) Its chemical and viscoelastic properties should be mimic those of biological membranes,
(b) The system should be capable of incorporating large amounts of proteins, detergents, precipitants and additives without perturbation of the matrix.
(c) Incorporated membrane proteins should retain their activity and their native conformation.
(d) It should provide a structured yet flexible matrix for crystal nucleation and growth.

Naturally, such a system should preferentially comprise lipids, since these are the molecules that endow the bilayer membrane its characteristic properties. We proposed that lipidic cubic phases, which are the most viscous of all lipidic phases, might fulfill these requirements [7]. The phase diagrams of a commonly used cubic phase, the bicontinuous-type monoolein (MO), is depicted in figure 3, demonstrating the rather extended ranges in composition and temperature at which these materials are stable. Obviously, not all of the known cubic phases display such large existence regions.

Our investigations have demonstrated that soluble and membrane proteins could be incorporated into these matrices without impairing the macroscopic properties of the cubic phases. The soluble enzyme α-chymotrypsin was shown to retain its native conformation and exhibits enzymatic activity upon incorporation into cubic phases [35]. Rheological characterization of these systems revealed that the viscoelastic properties of cubic phases are not altered upon incorporation. Extending these studies to membrane proteins, it could be shown that bR, a light driven proton pump, and melittin, the main component of the honey bee venom, could be solubilized in a lysolecithin-based cubic phases [36]. In both cases, the proteins retained their native conformations, which were temperature independent in the range of existence of the cubic phase. In addition, conditions for stabilizing bR in its native conformation in this membrane mimetic matrix were established. A larger membrane protein, the photosynthetic reaction center from *Chloroflexus aurantiacus,* could also be immobilized in its native secondary structure in a cubic phase [37]. The photochemical activity of the reaction center was retained, and immobilized reaction centers were significantly stabilized as compared with those solubilized in detergent solutions.

Different cubic phases were doped with wide ranges of concentrations of additives, and their tolerance was monitored as a function of time in a range commensurate with protein crystallization [38]. The remarkable finding was that they do tolerate inclusion of a variety of such added compounds, presumably due to the plasticity of their curved lipidic bilayers. A corollary of the hypothesis that membrane proteins should nucleate in the lipidic compartments of cubic phases was tested. This implies that crystallization of small soluble molecules and hydrophilic proteins should occur in the aqueous compartments of both bicontinuous and closed micellar cubic phases, into which such solutes partition, as in both types the aqueous channels pervade the lipidic compartments. Indeed, it was established that nucleation and growth of sodium chloride, glycine and lysozyme crystals are independent of the type of cubic phase and the packing arrangements of the lipidic compartments [38].

Following these results, crystallization of bR in cubic phases was attempted. BR in purple membranes was purified, dissociated and delipidated in octylglucoside [7]. Crystallization was accomplished in the batch method from two homologous bicontinuous cubic phases, composed of monoolein (MO) / water or monopalmitolein (MP) / water. In both cases, the cubic phases appeared

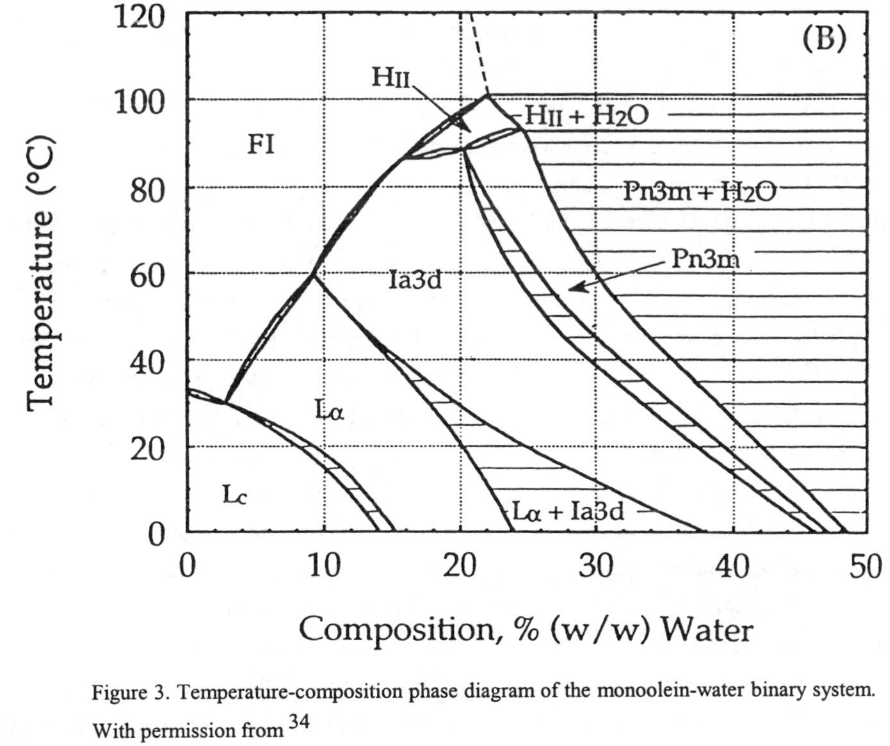

Figure 3. Temperature-composition phase diagram of the monoolein-water binary system. With permission from [34]

uniformly purple colored initially, indicating a homogeneous distribution of the protein. After several days, plates of hexagonal morphology appeared inside the MO matrix, whereas the MP system yielded rhombic crystals. In both cases, crystals grew to mature size within 14 days. During crystal formation and growth, the color intensity of the host phase faded gradually and finally became colorless, while a concomitant gradual increase in the intensity of the colored bR crystals was observed. Since the crystals are immobilized in the lipidic matrix, their growth could be monitored by inspection under the microscope. The ratios between their long and short axes remained constant throughout the process, indicating a uniform growth rate. The transparency and viscoelastic appearance typical to the cubic phases were not impaired throughout, revealing their remarkable plasticity towards the growing crystals.

Comparing these results with those obtained from soluble species crystallized in lipidic matrices revealed marked differences: Lysozyme could be crystallized from both micellar and bicontinuous cubic phases, yielding crystals that diffract to 2Å. This finding demonstrates that the properties of the lipidic compartments do not affect crystallization of soluble proteins, which are determined solely by the conditions of the aqueous compartments. Conversely, the nucleation and growth of membrane proteins is strongly dependent on the compositions and structures of the lipidic compartments. The curved bilayer seems to be a necessary structural element for the crystallization, as has been demonstrated by the appearance of bR crystals in bicontinuous systems (MO and MP) as opposed to their absence in the micellar one (PLPC). Even more intriguing is the effect of slight structural changes of the lipidic components on the crystallization within these complex materials: MO and MP, which are homologous monoglycerides of C_{18}- and C_{16}-side chains, respectively, give rise to cubic phases with markedly different radii of curvature, that of MO being much smaller than that of MP. This difference is reflected in the habits of the bR crystals. These results strongly support the notion that bR is integrated in the bilayer system, that crystal nucleation is strongly dependent on the molecular structure and packing of the bilayer system, and that growth occurs through a process of diffusion along the curved bilayers. In addition, the plasticity of lipidic cubic phases allows the growth of large crystals within the structured lattice of the host materials, while providing an expanding lipidic environment to the growing crystals.

The next step was to test the feasibility of obtaining well-ordered membrane protein crystals excluding the use of detergents. We reasoned that by direct incorporation of the purple membrane of *Halobacterium salinarum* into lipidic cubic phases we might be able to omit the solubilization in detergent micelles. Cubic phases were prepared by mixing monoolein with an aqueous suspension of the purple membrane to yield a 60% (w/w) monoolein lipidic matrix [39] and crystallization was induced by the addition of solid Na/K phosphate to the preformed cubic phase. Under similar experimental conditions, *i.e.* similar crystal size and beam intensity, the diffraction quality of the crystals thus obtained and their mosaicity were comparable, and the crystal habit and space group were identical to those of bR crystals reported earlier [7,13]which were grown in the presence of octylglucoside. We have thus demonstrated that this membrane protein can be crystallized directly from its native membrane, omitting any detergent-dependent procedures, and anticipate this crystallization method to be applicable likewise to other proteins that are enriched in a biological membrane.

Finally. For certain applications the availability of free standing, single crystals might be necessary. As the mechanical removal of the adhesive lipid matrix is impractical without damaging such delicate microcrystals, we adopted a biochemical solution to release the crystals from their host matrix [40] We have applied an enzymatic ester hydrolysis of monoolein using a lipase, which gently converts the lipidic mesophase to its fluid reaction products that phase-separate. Crystals devoid of adhering cubic phase can then be manipulated mechanically. This is especially important for X-ray studies, in which single crystals need to be mounted on cryo-loops.

5. Conclusions and outlook

Since the first reports on the crystallization of porin [4] and bacteriorhodopsin [3] in 1980, a number of membrane proteins have been crystallized and their structures determined at atomic resolution. Yet the bottleneck in structural biology of membrane proteins remains the routine availability of well-diffracting crystals. So far, for each membrane protein a unique set of crystallization conditions had to be determined empirically, reflecting the lack of a comprehensive theoretical understanding of these complex colloidal systems, practical difficulties in overexpression and stability problems of membrane proteins. Very recently, however, novel approaches to the problem have yielded crystals which diffract X-rays to high resolution, demonstrating the potentials that lie ahead. At the same time, dramatic developments in synchrotron technology have made it possible to obtain high resolution data from very small microcrystals [41] at very fast time scales [42,43]. There is now good reason to expect a more rational understanding of the field, resulting in an increase in the number of high resolution structures and a deeper understanding of the structural features that are fundamental to the specialized function of membrane proteins.

Acknowledgements

I am grateful to H. Belrhali, M.L. Chiu, K. Edman, J. Hajdu, A. Hardmeyer, R. Neutze, P. Nollert, E. Pebay-Peyroula, J.P.Rosenbusch and A. Royant for their advice, help and encouragement. This work was supported by grants from the Swiss National Science Foundation, EU-BIOMED and EU-BIOTECH.

References

1. J. Walker and M. Saraste. Membrane protein structure. Editorial overview. *Curr. Opin. Struct. Biol.* **6,** (1996) 457-459.

2. G. F. X. Schertler. Overproduction of membrane proteins. *Curr. Opin. Struct. Biol.* **2,** (1992) 534-544.

3. H. Michel and D. Oesterhelt. Three-dimensional crystals of membrane proteins: bacteriorhodopsin. *Proc. Natl. Acad. Sci. USA* **77,** (1980) 1283-1285.

4. R. M. Garavito and J. P. Rosenbusch. Three-dimensional crystals of an integral membrane protein: an initial x-ray analysis. *J. Cell Biol.* **86,** (1980) 327-329.

5. C. Ostermeier, S. Iwata, B. Ludwig and H. Michel. Fv fragment-mediated crystallization of the membrane protein bacterial cytochrome c oxidase. *Nat. Struct. Biol.* **2,** (1995) 842-846.

6. M. L. Chiu, *et al.* Crystallization in cubo: general applicability to membrane proteins. *submitted* , (1999) .

7. E. M. Landau and J. P. Rosenbusch. Lipidic cubic phases: A novel concept for the crystallization of membrane proteins crystallization. *Proc. Natl. Acad. Sci. USA* **93,** (1996) 14532-14535.

8. G. Rummel, *et al.* Lipidic cubic phases: new matrices for the three-dimensional crystallization of membrane proteins. *J. Struct. Biol.* **121,** (1998) 82-91.

9. H. Michel. Crystallization of membrane proteins. *Trends Biochm. Sci.* **8,** (1983) 56-59.

10. R. A. Pauptit, *et al.* Trigonal crystals of porin from *Escherichia coli. J. Mol. Biol.* **218,** (1991) 505-507.

11. E. Pebay-Peyroula, G. Rummel, J. P. Rosenbusch and E. M. Landau. X-ray structure of bacteriorhodopsin at 2.5 Angstroms from microcrystals grown in lipidic cubic phases. *Science* **277,** (1997) 1676-1681.

12. H. Luecke, H. T. Richter and J. K. Lanyi. Proton transfer pathways in bacteriorhodopsin at 2.3 angstrom resolution. *Science* **280,** (1998) 1934-1937.

13. H. Belrhali, *et al.* Protein, lipid and water organization in bacteriorhodopsin crystals: a molecular view of the purple membrane at 1.9 A resolution. *Structure Fold. Des.* **7,** (1999) 909-917.

14. W. Kühlbrandt. Three-dimensional crystallization of membrane proteins. *Q. Rev. Biophys.* **21,** (1988) 429-477.

15. Reiss-Husson. in *crystallization of nucleic acids and proteins. A practical approach* (eds Ducruix, A. & Giegé, R.) 175-193 (IRL Press, New York, 1992).

16. R. Henderson and D. Shotton. Crystallization of purple membrane in three dimensions. *J. Mol. Biol.* **139,** (1980) 99-109.

17. J.-L. Eiselé and J. P. Rosenbusch. Crystallization of porin using short chain phospholipids. *J. Mol. Biol.* **206,** (1989) 209-212.

18. G. Privé, R. H. Kaback and D. Eisenberg. Fusion proteins as tools for crystallization: the lactose permease from *Escherichia coli. Biol. Crystallogr.* **D50,** (1994) 375-379.

19. C. E. Schafmeister, L. J. W. Miercke and R. M. Stroud. Structure at 2.5 Å of a designed peptide that maintains solubility of membrane proteins. *Science* **262,** (1993) 734-737.

20. S. Nussberger, K. Dörr, W. D.N. and W. Kühlbrandt. Lipid-protein interactions in crystals of plant light-harvesting complex. *J. Mol. Biol.* **234,** (1993) 347-356.

21. W. Kühlbrandt, D. N. Wang and Y. Fujiyoshi. Atomic model of plant light-harvesting complex by electron crystallography. *Nature* **367,** (1994) 614-621.

22. P. J. Bjorkman, *et al.* Structure of the human class I histocompatibility antigen, HLA-A2. *Nature* **329,** (1987) 506-512.

23. S. R. Hubbard, L. Wei, L. Ellis and W. A. Hendrickson. Crystal structure of the tyrosine kinase domain of the human insulin receptor. *Natute* **372,** (1994) 746-754.

24. L. Song and J. E. Gouaux. Membrane protein crystallization: application of a new sparse matrix to the a-hemolysin heptamer. *Methods Enzymol.* **in press,** (1997) .

25. A. Skerra and A. Plückthun. Assembly of a functional immunoglobulin Fv fragment in *Escherichia coli. Science* **240,** (1988) 1038-1041.

26. T. G. M. Schmidt and A. Skerra. The random peptide library-assisted engineering of a C-terminal affinity peptide, useful for the detection and purification of functional Ig Fv fragment. *Protein Engineering* **6,** (1993) 109-122.

27. G. Kleymann, O. C., L. B., A. Skerra and H. Michel. Engineered Fv fragments as a tool for the one-step purification of integral multisubunit membrane proteins. *Bio/Technology* **13,** (1995) 155-160.

28. C. Ostermeier, L.-O. Essen and H. Michel. Crystals of an antibody Fv fragment against an integral membrane protein diffract to 1.28 Å resolution. *Proteins* **21,** (1995) 74-77.

29. S. Iwata, C. Ostermeier, B. Ludwig and H. Michel. Structure at 2.8Å resolution of cytochrome *c* oxidase from *Paracoccus denitrificans. Nature* **376,** (1995) 660-668.

30. M. W. Tate, E. F. Eikenberry, D. C. Turner, E. Shyamsunder and S. M. Gruner. *Chem. Phys. Lipids* **57,** (1991) 147-164.

31. J. M. Seddon. *Biochim. Biophys. Acta* **1031,** (1990) 1-69.

32. S. Cribier, *et al.* Cubic phases of lipid-containing systems. A translational diffusion study by fluorescence recovery after photobleaching. *J. Mol. Biol.* **229,** (1993) 517-525.

33. G. Lindblom. Nuclear magnetic resonance on lipids and surfactans. *Curr. Op. Colloid Interface Struct.* **1,** (1996) 287-295.

34. J. Briggs, H. Chung and M. Caffrey. The temperature-composition phase diagram and mesophase structure characterization of the monoolein/water system. *J. Physique II* **6,** (1996) 723-752.

35. M. Portmann, E. M. Landau and P. L. Luisi. Spectroscopic and rheological studies of enzymes in rigid matrices: The case of α–chymotrypsin in a lysolecithin/water cubic phase. *J. Phys. Chem.* **95,** (1991) 8437-8440.

36. E. M. Landau and P. L. Luisi. Lipidic cubic phases as transparent rigid matrices for the direct spectroscopic study of immobilized membrane proteins. *J. Am. Chem. Soc.* **115,** (1993) 2102-2106.

37. A. Hochkoeppler, *et al.* Photochemistry of a photosynthetic reaction center immobilized in lipidic cubic phases. *Biotechnol. Bioeng.* **46,** (1995) 93-98.

38. E. M. Landau, G. Rummel, S. W. Cowan-Jacob and J. P. Rosenbusch. Crystallization of a polar protein and small molecules from the aqueous compartment of lipidic cubic phases. *J. Phys. Chem. B* **101,** (1997) 1935-1937.

39. P. Nollert, A. Royant, E. Pebay-Peyroula and E. M. Landau. Detergent-free membrane protein crystallization. *FEBS Lett.* **457,** (1999) 205-208.

40. P. Nollert and E. M. Landau. Enzymic release of crystals from lipidic cubic phases. *Biochem. Soc. Trans.* **26,** (1998) 709-713.

41. H. Belrhali, A. Bram, S. Cusack and C. Riekel. in *ESRF Newsletter* 15-16 (1997).

42. V. Srajer, *et al.* Photolysis of the carbon monoxide complex of myoglobin: nanosecond time-resolved crystallography. *Science* **274,** (1996) 1726-1729.

43. K. Edman, *et al.* High-resolution X-ray structure of an early intermediate in the bacteriorhodopsin photocycle. *Nature* **401,** (1999) 822-826.

MITOCHONDRIAL TARGETING SIGNALS

Connie M. Lee, Walter Neupert and Rosemary A. Stuart
Institut für Physiologische Chemie der Universität München
80336 München, Germany

Mitochondria, like several other cellular organelles, confront the dilemma of importing the majority of its proteins from the cytoplasm. In order to reach mitochondria, proteins that are nuclear-encoded contain specific sequences that target them to mitochondria. The best characterized mitochondrial targeting sequences have been from matrix proteins and consist of several positively-charged amino acid residues at the N-terminus of the protein, which are cleaved upon import into mitochondria. Although much less characterized, internal targeting sequence have also been identified in several proteins of the outer and inner membranes. These internal targeting sequences often consist, in part, of hydrophobic transmembrane anchor signals. In addition, a cleavable C-terminal targeting signal has recently been identified on a protein destined for the mitochondrial matrix. The diversity in targeting signals indicates that proteins may be targeted in a variety of ways to mitochondria based upon their final topology, destination and function within mitochondria.

1. Introduction

Mitochondria are complex cellular organelles that contain two highly specialized membranes, which divide mitochondria into four different subcompartments: matrix, inner membrane, intermembrane space, and outer membrane. The biogenesis of mitochondria requires products of both the nuclear and mitochondrial genomes, with the majority of the proteins being encoded in the nucleus. Most proteins, therefore, must be imported into mitochondria from the cytoplasm. In order to accomplish this task, translocases exist in both the outer (TOM machinery) and inner (TIM machinery) mitochondrial membranes [reviewed in 1-3].

Many proteins destined for import into mitochondria, especially those proteins of the mitochondrial matrix, bear cleavable targeting information at their N-termini. Because matrix targeting signals (MTSs) are located at the N-termini, these proteins are thought to traverse the import machineries in a linear and an N- to C-terminal (N→C) fashion. Unlike matrix targeted proteins, however, several proteins of the inner and outer membranes have been found to contain internal targeting information. In addition, we have recently identified a matrix protein that bears a novel targeting signal at the very C-terminus of the protein [4]. The identification of proteins with internal and C-terminal targeting signals suggests that not all proteins may enter mitochondria in a linear or in an N→C manner. Here we briefly review the mitochondrial import machinery and the different types of signals that target proteins to mitochondria. We also discuss why individual mitochondrial proteins, depending on their final location within mitochondria, may use different methods of targeting to mitochondria.

2. Mitochondrial import machinery

2.1 Preprotein Translocase of the Outer Membrane (TOM)

Mitochondria possess distinct import translocases in both their outer and inner membranes. These machineries cooperate in the translocation of preproteins into mitochondria, as precursors arrested during translocation have been found to span both the outer and inner membrane [5]. Each translocase, however, can also act independently. Isolated outer membrane vesicles have the ability to translocate outer membrane and intermembrane space proteins [6,7]. Outer membrane vesicles also can initiate import of matrix-targeted preproteins so that the presequences reach the inner face of the outer membrane [8]; however, completion of translocation requires components of TIM. Similarly, preproteins destined for the inner membrane or matrix can be imported *in vitro* into isolated mitoplasts in the absence of any outer membrane machinery [9].

Studies with *Neurospora crassa* and *Saccharomyces cerevisiae* have identified several proteins of the TOM machinery [reviewed in 2,10]. The components of the TOM complex can be grouped into two classes: those that function as import receptors and those that constitute a putative general import pore. Tom70, Tom37, Tom20, and Tom22 proteins, which all bear cytosolic domains, are proposed to be involved in the recognition of preproteins on the surface of the outer membrane. Matrix-targeting preproteins with N-terminal targeting signals are recognized mainly by a Tom20-Tom22 subcomplex [11,12], while Tom70 binds preferentially to preproteins with internal targeting information [12,13]. The exact function of Tom37 remains unknown; however, it has been suggested to form a subcomplex with Tom70 [14] and was therefore assumed to function in the recognition of preproteins with internal targeting signals. An additional protein believed to be involved in the TOM complex, Tom71 [15,16], has a close structural similarity to Tom70, although its exact function is not known. Other components of TOM are more firmly embedded into the outer membrane and include Tom40 [17-19], Tom7 [20], Tom6 [21,22], and Tom5 [23]. Tom40, which spans the membrane several times [18,19], is considered to be the main component of the translocation pore. Of all the TOM proteins identified thus far, only Tom40 and Tom22 have been found to be essential for viability in yeast [18,24].

2.2 Preprotein translocases of the inner membrane

2.2.1 TIM17-23 translocase

At least two different import translocases exist in the mitochondrial inner membrane that differ in their substrate specificity. Import of presequence targeted proteins into the mitochondrial matrix generally requires a membrane potential ($\Delta\psi$) across the mitochondrial inner membrane and the TIM17-23 translocase machinery [25-29]. Tim23 and Tim17 are both hydrophobic proteins that are homologous to each other in their membrane-spanning regions, and are thought to constitute the core of the import channel. Tim23 forms dimers that are promoted by the membrane potential $\Delta\psi$, while binding of a matrix targeting sequence to Tim23 triggers dimer dissociation [29]. Also associated with the Tim17-23 translocase is the hydrophilic protein, Tim44. Tim44 can be found in close contact with preproteins during import [30,31]; however, it is not membrane-embedded but rather is peripherally associated on the matrix side of the inner membrane. Tim44 has also recently been found to form dimers and to be in equimolar amounts with Tim23 and Tim17 [32]. The mitochondrial matrix chaperone mt-Hsp70, in cooperation with the membrane-associated Tim44, mediates the ATP-dependent translocation of preproteins across the inner membrane into the matrix [31,33-37].

2.2.2 TIM22-54 translocase

Import of several inner membrane proteins, such as the ADP/ATP carrier (AAC) and components of the TIM machinery itself is mediated by the TIM22-54 translocase [38-40]. Tim22 is a hydrophobic mitochondrial inner membrane protein with homology to Tim17 and Tim23. Tim54, together with Tim22, appear to form the import channel. Several smaller proteins have also been identified as playing a role in the TIM22-54 translocase. Tim9, Tim10, and Tim12 are structurally related proteins of ca. 10 kDa which bear zinc-finger-like motifs [39,41]. Tim9 and Tim10 form a soluble 70 kDa complex in the intermembrane space which binds to an incoming preprotein and transfers it from the outer membrane to the inner membrane [41,42]. Tim9, Tim10, and Tim12, together with Tim22 and Tim54, form a membrane bound ca. 300 kDa complex that then mediates the insertion of proteins into the inner membrane [41,42].

3. Mitochondrial Targeting Signals

3.1 N-terminal targeting signals of matrix proteins

As mentioned above, preproteins that are destined for the mitochondrial matrix typically bear cleavable targeting signals at the N-termini. These N-terminal targeting signals are necessary and sufficient to target the proteins to mitochondria. The fusion of MTSs to non-mitochondrial proteins, such as dihydrofolate reductase (DHFR), can direct normally cytosolic proteins into mitochondria [43]. MTSs are commonly 20-60 amino acids in length and are cleaved upon import into the mitochondrial matrix by the mitochondrial processing peptidase (MPP) [44,45]. There are, however, proteins in which the N-terminal MTS remains and is present in the imported functional protein (i.e., rhodanese, 3-oxo-acyl-CoA thiolase). N-terminal presequences normally contain several positively-charged amino acids while being notably devoid of negative charges. Presequences are also predicted to form amphipathic α-helices [46], which are thought to be important for the interaction with the import machinery. In addition, MTSs frequently contain periodic hydroxylated amino acid residues, which typically align along one face of the α-helix. The consensus cleavage site for MPP is R X (F/I/L/Y) (S/A) [47], with the cleavage event usually occurring two residues downstream from the conserved arginine residue.

Since the targeting signal of proteins bearing the classical MTS is located at the N-terminus, these proteins are generally thought to enter mitochondria in a linear and in an N- to C-terminal (N→C) direction. This notion is suggested by the ability to arrest preproteins by having a folded domain present in the sequence. The folded domain then is close to the TOM complex whereas the segment spanning both outer and inner membranes is in a largely extended state. Such N-terminal presequences on matrix-targeted proteins serve not only to target the preprotein to mitochondria by their interactions with receptors of the TOM machinery but also facilitate the early steps in protein translocation across the inner membrane in a $\Delta\psi$-dependent manner [48]. The preproteins are then imported across the inner membrane via the TIM17-23 machinery and driven by ATP-hydrolysis in cooperation with mt-Hsp70. The import of a representative matrix protein, such as pSu9(1-69)-DHFR, is strongly reduced in mitochondria harboring a modified form of Tim23, Tim23fs [38]. Similarly, import of pSu9(1-69)-DHFR is also reduced in mitochondria from the *ssc1-3* strain bearing a mutated form of mt-Hsp70, as well as in mitochondria which have been depleted of their matrix ATP [49].

3.2 Internal targeting signals of outer membrane proteins

In contrast to the extensive data regarding the N-terminal MTSs, relatively little is known about the signals that direct proteins to the other mitochondrial subcompartments. Proteins of the outer and inner membranes and intermembrane space contain, in the majority of cases, internal targeting information that is not proteolytically processed upon import. Frequently these internal signals encompass hydrophobic membrane anchor segments. Several outer membrane proteins have been characterized to contain, at least in part, such membrane anchor signals. The N-proximal signal anchor of the Tom70p (amino acid residues 11-29) has been determined to contain the necessary information to target and insert Tom70 into the outer membrane in an N_{in}-C_{out} orientation [50,51]. The positively-charged N-terminal 10 amino acid residues of Tom70p may have a supporting role, but are not a requirement for import [50,51]. Similarly, the transmembrane domain of the human Bcl-2 has been shown to function as a mitochondrial signal anchor sequence that targets and inserts the protein into the outer membrane [52]. Unlike the Tom70 transmembrane anchor, however, the Bcl-2 signal anchor is near the C-terminus and the protein is inserted into the outer membrane in a C_{in}-N_{out} orientation.

In addition to its transmembrane domain, essential targeting information of Tom22p also resides in the cytosolic domain of the protein [53]. The N-terminal 84 residues of Tom22p are exposed to the cytosol while the C-terminal 49 residues reside in the intermembrane space [54]. Through deletion mutants, it was determined that the cytosolic residues 45-75 are essential for import [53]. Furthermore, this import sequence is quite similar to classical MTSs in that it is enriched in serine residues and carries a net positive charge that was determined to be important for its function [53].

In general, the import of proteins containing internal targeting information has not been as thoroughly characterized as those proteins bearing the classical MTS. The mechanisms used to recognize and insert outer membrane proteins containing membrane anchor signals remain unclear except for a few observations. The import of Tom70 and Bcl-2 appears to require cytosolic ATP, mitochondrial surface-exposed proteins, and physiological temperatures [51,52]. Import is, however, independent of a $\Delta\psi$ across the inner membrane [51,52]. Receptors of the TOM complex are not always involved, and thus can perhaps be bypassed in certain cases.

3.3 Internal targeting signals of inner membrane proteins

Several inner membrane proteins are also targeted, at least in part, by their transmembrane domains. The inner membrane protein Bcs1p has a short N-terminal tail exposed to the intermembrane space, while the majority of the protein is located in the matrix [55]. In Bcs1p, a segment of positively-charged amino acids located directly C-terminal to a single transmembrane domain constitute an internal targeting signal [55]. The region following the transmembrane domain resembles very closely the classical N-terminal targeting signal in regards to its net positive charge and amphipathicity. Import of Bcs1p is dependent upon a $\Delta\psi$ across the inner membrane and requires the TIM17-23 machinery with the cooperation of ATP-driven mt-Hsp70 [55].

Bcs1p appears to be translocated across the inner mitochondrial membrane in a hairpin loop structure [55]. The loop structure is formed between the transmembrane domain and the stretch of positively-charged amino acids directly following it. Both regions, the transmembrane domain and the positively-charged portion, were found to be necessary for import into mitochondria [55]. Translocation of a looped targeting signal across the inner membrane implies that the first N-terminal part of the protein (prior to the transmembrane domain) would be entering mitochondria in a 'backwards' $C \rightarrow N$ direction.

Alternatively, the majority of the protein (from the positively-charged segment to the C-terminal end) would be imported in a 'normal' N→C fashion. When the first 65 amino acid residues of Bcs1p were deleted (including the transmembrane domain) so that the protein started with the positively charged segment, Bcs1p then became misrouted to the matrix and was even processed by MPP [55]. Because of the neighboring transmembrane domain normally proceeding the positively-charged stretch, however, the protein becomes correctly sorted to the mitochondrial inner membrane. Thus, the cooperation between the transmembrane and the stretch of positively-charged amino acids provides the specificity for mitochondrial targeting and sorting of Bcs1p to the inner membrane.

Another protein of the inner membrane, cytochrome c_1, has been characterized to contain two separate and independent targeting signals [56]. Cytochrome c_1, a subunit of the cytochrome bc_1 complex, is anchored to the inner membrane in an N_{out}-C_{in} orientation via a single transmembrane segment near its C-terminus [57]. Cytochrome c_1 contains an N-terminal cleavable bipartite presequence that is processed by MPP in the matrix and the Imp2p protease in the intermembrane space [58,59]. In addition, an internal targeting signal is located close to the C-terminus, consisting of the transmembrane domain followed by a stretch of positively charged amino acids. This internal signal is similar to the internal signal in the Bsc1p in that the transmembrane domain and the amphipathic helix are proposed to form a hairpin loop during import [56]. This internal signal operates from the intermembrane space side of the inner membrane to sort the C-terminal region of cytochrome c_l to the matrix and to position the transmembrane segment into the inner membrane [56]. Other proteins of the inner membrane which are potential candidates to contain both N-terminal and internal targeting signals are Yta10p and Yta12p [60-63] and Shy1p [64]; however, their internal mitochondrial targeting signals have not been thoroughly characterized.

A special type of targeting signals is present in inner membrane proteins that are imported via the TIM22-54 translocase, such as the mitochondrial carrier proteins. The family of carrier proteins in yeast is extensive and has more than 30 members. These proteins contain six membrane spanning segments, exposing three hydrophilic loops into the matrix. They have apparently evolved by triplication of modules consisting of two membrane spanning segments and a loop. Each of these loops in the various members of the family contains a matrix-exposed sequence motif that is conserved and called a 'carrier' signature [65,66] which, like Tim23 and Tim17, are imported by TIM22-54. A similar motif is also present in Tim22p [38]. The pattern of charged and hydrophobic amino acids residues in this motif is complementary to a pattern of charged and hydrophobic residues present in the putative zinc-fingers of Tim10 and of Tim12, but in an antiparallel orientation [39]. It has been proposed that this may constitute a structural basis for the recognition of precursors by Tim10 and Tim12 [39]; however, this remains to be further characterized.

3.4 C-terminal targeting signals

The DNA helicase Hmi1p from *S. cerevisiae* is required for mitochondrial DNA inheritance [67]. It is located in the mitochondrial matrix as a soluble protein [4,67]. Analysis of the N-terminal sequence of the helicase revealed no obvious MTS; however, a potential targeting sequence was observed at the very C-terminus of the protein. Import of artificial fusion proteins has recently demonstrated that a MTS placed at the C-terminus of a protein can direct efficient import both *in vitro* and *in vivo* [49]. A stretch of ~40 amino acid residues at the C-terminus of the helicase represents a targeting sequence responsible for directing it into the mitochondrial matrix [4]. Notably, the targeting signal was observed to be processed by MPP. The targeting signal of the helicase resembles classical N-terminal signals in terms of its net positive charge and amphipathic character. Deletion of the signal resulted in loss of import into isolated mitochondria, while its addition to the C-terminus (or N-terminus) of the DHFR passenger protein ensued import into the mitochondrial matrix [4]. The helicase is the only mitochondrial matrix protein identified thus far with a cleavable targeting signal at its C-terminus. The identification of this novel C-terminal targeting signal in the helicase indicates that it may be a representative of a new class of mitochondrial matrix proteins that bear cleavable C-terminal targeting signals.

Import of the helicase Hmi1p with a C-terminal targeting signal occurs in a manner similar to that of proteins bearing the classical N-terminal MTSs. Import was dependent on a $\Delta\psi$ across the inner mitochondrial membrane and utilized the TIM17-23 translocase of the inner membrane. Completion of translocation of the helicase into the matrix was dependent on matrix ATP and mt-Hsp70 [4]. The only obvious difference between the N-targeted proteins and the C-targeted helicase would be in the directionality of their import process. Proteins bearing the classical MTSs at their N-terminus are imported in a N→C direction while the helicase is imported in the opposite direction, C→N.

The DNA helicase Hmi1p has high degree of similarity to the *S. aureus* PCRA helicase [68] and the *E. coli* Rep helicase [69]. Compared to these proteins, Hmi1p has an extended C-terminus (ca. 60 amino acids) of unrelated sequence. One other mitochondrial DNA helicase has been characterized in *S. cerevisiae*, Pif1p, which contains a predicted N-terminal presequence [70]. Why might the two mitochondrial helicases in *S. cerevisiae* be targeted to mitochondria in different fashions? Both helicases contain conserved ATP-binding motifs. In the case of the Pif1p, the ATP-binding site is located internally in the protein. In contrast, with the Hmi1p it is present at the extreme N-terminal region of the protein. The presence of an N-terminal presequence in close proximity to the ATP-binding domain may somehow interfere with or modulate folding of this domain before or after translocation across the mitochondrial membranes.

The C→N directionality of import in the helicase brings up an interesting point regarding the topic of cotranslational versus posttranslational import into mitochondria. Because the classical MTSs are localized at the N-terminus of proteins, it has been proposed that many nascent mitochondrial polypeptide chains may be cotranslationally inserted into mitochondria [71,72]. Although preproteins can be imported in a posttranslational manner *in vitro*, it cannot be excluded that proteins may be imported cotranslationally *in vivo*. The pathway of C→N import, however, implies a post-translational mechanism of translocation occurs in the case of the C-terminally targeted helicase. Additionally, the fact that several internal targeting sequences have been discovered in the C-terminal portions of mitochondrial proteins (i.e., Bcl-2 and cytochrome c_1) also implies that they would be imported post-translationally.

4. Concluding remarks

In the past several years, it has become evident that several types of targeting signals exist in addition to the classical N-terminal targeting signals of matrix proteins. Many proteins destined for the outer and inner membranes have been found to contain internal targeting information, often consisting in part of transmembrane domains. Additionally, not all mitochondrial matrix proteins bear N-terminal targeting signals, as is the case with Hmi1 helicase that is targeted to the matrix via a cleavable C-terminal signal. The discovery of new targeting signals, such as the C-terminal mitochondrial signal, will aid in the subcellular localization of open reading frames, identified through genome sequencing projects. Furthermore, it is likely that several additional types of targeting signals have yet to be discovered within the hundreds of different mitochondrial proteins, adding further complexity and diversity to the biogenesis of mitochondria.

5. Acknowledgments

This work was supported by grants from the Deutsche Forschungsgemeinschaft (Sonderforschungsbereich 184, Teilprojeck B2), the Münchener Medizinische Wochenschrift (to R. A. S.) and the Alexander von Humboldt-Stiftung (to C. M. L.).

6. References

[1] G. Schatz and B. Dobberstein, Common Principles of Protein Translocation Across Membranes, *Science* **271** (1996) 1519-1526.
[2] W. Neupert, Protein Import into Mitochondria, *Annu. Rev. Biochem.* **66** (1997) 863-917.
[3] N. Pfanner *et al.*, Mitochondrial Preprotein Translocase, *Annu. Rev. Cell. Dev. Biol.* **13** (1997) 25-51.
[4] C. Lee *et al.*, The DNA Helicase, Hmi1p, is Transported into Mitochondria by a C-Terminal Cleavable Targeting Signal, *J. Biol. Chem* (1999) in press.
[5] M. Schleyer and W. Neupert, Transport Of Proteins Into Mitochondria: Translocational Intermediates Spanning Contact Sites Between Outer And Inner Membranes, *Cell* **43** (1985) 339-350.
[6] A. Mayer *et al.*, Translocation and Insertion of Precursor Proteins into Isolated Outer Membranes of Mitochondria, *J. Cell Biol.* **121** (1993) 1233-1243.
[7] A. Mayer *et al.*, Translocation of Apocytochrome C Across the Outer Membrane of Mitochondria, *J. Biol Chem.* **270** (1995) 12390-12397.
[8] A. Mayer *et al.*, Mitochondrial Protein Import: Reversible Binding of the Presequence at the Trans Side of the Outer Membrane Drives Partial Translocation and Unfolding, *Cell* **80** (1995) 127-37.
[9] S. Hwang *et al.*, Protein Import into the Yeast Mitochondrial Matrix. A New Translocation Intermediate Between the Two Mitochondrial Membranes, *J. Biol. Chem.* **266** (1991) 21083-21089.
[10] N. Pfanner and M. Meijer, The Tom and Tim Machinery, *Curr. Biol.* **7** (1997) R100-103.
[11] A. Mayer *et al.*, MOM22 is a Receptor for Mitochondrial Targeting Sequences and Cooperates with MOM19, *EMBO J.* **14** (1995) 4204-4211.
[12] J. Brix *et al.*, Differential Recognition of Preproteins by the Purified Cytosolic Domains of the Mitochondrial Import Receptors Tom20, Tom22, and Tom70, *J. Biol. Chem.* **272** (1997) 20730-20735.
[13] J. Schlossmann *et al.*, Specific Recognition of Mitochondrial Preproteins by the Cytosolic Domain of the Import Receptor MOM72, *J. Biol. Chem.* **269** (1994) 11893-11901.
[14] S. Gratzer *et al.*, Mas37p, a Novel Receptor Subunit for Protein Import into Mitochondria, *J. Cell Biol.* **129** (1995) 25-34.
[15] J. Schlossmann *et al.*, Tom71, a Novel Homologue of the Mitochondrial Preprotein Receptor Tom70, *J. Biol. Chem.* **271** (1996) 17890-17895.
[16] U. Bömer *et al.*, Identification of a Third Yeast Mitochondrial Tom Protein with Tetratrico Peptide Repeats, *FEBS Lett.* **382** (1996) 153-158.
[17] D. Vestweber *et al.*, A 42K Outer-Membrane Protein is a Component of the Yeast Mitochondrial Protein Import Site, *Nature* **341** (1989) 205-209.
[18] K. Baker *et al.*, A Yeast Mitochondrial Outer Membrane Protein Essential for Protein Import and Cell Viability, *Nature* **348** (1990) 605-609.
[19] M. Kiebler *et al.*, Identification of a Mitochondrial Receptor Complex Required for Recognition and Membrane Insertion of Precursor Proteins, *Nature* **348** (1990) 610-616.
[20] A. Hönlinger *et al.*, Tom7 Modulates the Dynamics of the Mitochondrial Outer Membrane Translocase and Plays a Pathway-Related Role in Protein Import, *EMBO J.* **15** (1996) 2125-2137.
[21] C. Kassenbrock *et al.*, Genetic and Biochemical Characterization of ISP6, a Small Mitochondrial Outer Membrane Protein Associated with the Protein Translocation Complex, *EMBO J.* **12** (1993) 3023-3034.
[22] A. Alconada *et al.*, The Mitochondrial Receptor Complex: The Small Subunit Mom8b/Isp6 Supports Association of Receptors with the General Insertion Pore and Transfer of Preproteins, *Mol. Cell. Biol.* **15** (1995) 6196-6205.
[23] K. Dietmeier *et al.*, Tom5 Functionally Links Mitochondrial Preprotein Receptors to the General Import Pore, *Nature* **388** (1997) 195-200.
[24] T. Lithgow *et al.*, The Mitochondrial Outer Membrane Protein Mas22p Is Essential For Protein Import and Viability of Yeast, *Proc Natl Acad Sci* **91** (1994) 11973-11977.
[25] P. Dekker *et al.*, Identification of MIM23, a Putative Component of the Protein Import Machinery of the Mitochondrial Inner Membrane, *FEBS Lett.* **330** (1993) 66-70.
[26] J. Emtage and R. Jensen, MAS6 Encodes an Essential Inner Membrane Component of the Yeast Mitochondrial Import Pathway, *J. Cell Biol.* **122** (1993) 1003-1012.

[27] A. Maarse *et al.*, Identification of the Essential Yeast Protein MIM17, an Integral Mitochondrial Inner Membrane Protein Involved in Protein Import, *FEBS Lett.* **349** (1994) 215-221.

[28] J. Berthold *et al.*, The MIM Complex Mediates Preprotein Translocation Across the Mitochondrial Inner Membrane and Couples It to the mt-Hsp70/ATP Driving System, *Cell* **81** (1995) 1085-1093.

[29] M. Bauer *et al.*, Role of Tim23 as Voltage Sensor and Presequence Receptor in Protein Import into Mitochondria, *Cell* **87** (1996) 33-41.

[30] J. Blom *et al.*, The Essential Yeast Protein MIM44 (encoded by MPI1) is Involved in an Early Step of Preprotein Translocation Across the Mitochondrial Inner Membrane, *Mol. Cell. Biol.* **13** (1993) 7364-7371.

[31] M. Horst *et al.*, Protein Import into Yeast Mitochondria: Inner Membrane Import Site Protein ISP45 is the MPI1 Gene Product, *EMBO J.* **12** (1993) 3035-3041.

[32] F. Moro *et al.*, The TIM17-23 Preprotein Tranlocase of Mitochondria: Composition and Function in Protein Transport into the Matrix, *EMBO J.* (1999) in press.

[33] A. Maarse *et al.*, MPI1, An Essential Gene Encoding a Mitochondrial Membrane Protein, is Possibly Involved in Protein Import into Yeast Mitochondria, *EMBO J.* **11** (1992) 3619-3628.

[34] J. Rassow *et al.*, Mitochondrial Protein Import: Biochemical and Genetic Evidence for Interaction of Matrix Hsp70 and the Inner Membrane Protein MIM44, *J. Cell Biol.* **127** (1994) 1547-1556.

[35] H. Schneider *et al.*, Mitochondrial Hsp70/MIM44 Complex Facilitates Protein Import, *Nature* **371** (1994) 768-774.

[36] H. Schneider *et al.*, The Nucleotide Exchange Factor MGE Exerts a Key Function in the ATP-Dependent Cycle of Mt-Hsp70-Tim44 Interaction Driving Mitochondrial Protein Import, *EMBO J.* **15** (1996) 5796-5803.

[37] C. Ungermann *et al.*, The $\Delta\psi$- and Hsp70/MIM44-Dependent Reaction Cycle Driving Early Steps of Protein Import Into Mitochondria, *EMBO J.* **15** (1996) 735-744.

[38] C. Sirrenberg *et al.*, Import of Carrier Proteins into the Mitochondrial Inner Membrane Mediated By Tim22, *Nature* **384** (1996) 582-585.

[39] C. Sirrenberg *et al.*, Carrier Protein Import into Mitochondria Mediated by the Intermembrane Proteins Tim10/Mrs11 and Tim12/Mrs5, *Nature* **391** (1998) 912-915.

[40] O. Kerscher *et al.*, The Tim54p-Tim22p Complex Mediates Insertion of Proteins into the Mitochondrial Inner Membrane, *J. Cell Biol.* **139** (1997) 1663-1675.

[41] A. Adam *et al.*, Tim9, a New Component of the TIM22-54 Translocase in Mitochondria, *EMBO J.* **18** (1999) 313-319.

[42] C. Koehler *et al.*, Tim9p, an Essential Partner Subunit of Tim10p for the Import of Mitochondrial Carrier Proteins, *EMBO J.* **17** (1998) 6477-6486.

[43] A. van Loon *et al.*, The Presequences of Two Imported Mitochondrial Proteins Contain Information for Intracellular and Intramitochondrial Sorting, *Cell* **44** (1986) 801-812.

[44] P. Böhni *et al.*, Import of Proteins into Mitochondria. Partial Purification of a Matrix-Located Protease Involved in the Cleavage of Mitochondrial Precursor Proteins, *J. Biol. Chem.* **258** (1983) 4937-4943.

[45] G. Hawlitschek *et al.*, Mitochondrial Protein Import: Identification of Processing Peptidase and of PEP, a Processing Enhancing Protein, *Cell* **53** (1988) 795-806.

[46] G. von Heijne, Mitochondrial Targeting Sequences May Form Amphipathic Helices, *EMBO J.* **5** (1986) 1335-1342.

[47] M. Brunner *et al.*, The Mitochondrial Processing Peptidase. In: G. von Heijne (ed.), Signal Peptidases. CRC Press, Boca Raton, 1994, pp. 73-86.

[48] M. Schleyer *et al.*, Requirement of a Membrane Potential for the Posttranslational Transfer of Proteins into Mitochondria, *Eur. J. Biochem.* **125** (1982) 109-16.

[49] H. Fölsch *et al.*, C- to N-terminal Translocation of Preproteins into Mitochondria, *EMBO J.* **22** (1998) 6508-6515.

[50] T. Hase *et al.*, A 70-Kd Protein of the Yeast Mitochondrial Outer Membrane is Targeted and Anchored via its Extreme Amino Terminus, *EMBO J.* **3** (1984) 3157-3164.

[51] H. McBride *et al.*, A Signal-Anchor Sequence Selective for the Mitochondrial Outer Membrane, *J. Cell Biol.* **119** (1992) 1451-1457.

[52] M. Nguyen *et al.*, Targeting of Bcl-2 to the Mitochondrial Outer Membrane by a COOH-Terminal Signal Anchor Sequence, *J. Biol. Chem.* **268** (1993) 25265-25268.

[53] N. Rodriguez-Cousiño *et al.*, An Import Signal in the Cytosolic Domain of the Neurospora Mitochondrial Outer Membrane Protein TOM22, *J. Biol. Chem.* **273** (1998) 11527-11532.

[54] M. Kiebler *et al.*, The Mitochondrial Receptor Complex: a Central Role of MOM22 in Mediating Preprotein Transfer From Receptors to the General Insertion Pore, *Cell* **74** (1993) 483-492.

[55] H. Fölsch *et al.*, Internal Targeting Signal of the BCS1 protein: a Novel Mechanism of Import into Mitochondria, *EMBO J.* **15** (1996) 479-487.

[56] I. Arnold *et al.*, Two Distinct and Independent Mitochondrial Targeting Signals Function in the Sorting of an Inner Membrane Protein, Cytochrome c_1, *J. Biol. Chem.* **273** (1998) 1469-1476.

[57] D. Xia *et al.*, Crystal Structure of the Cytochrome bc1 Complex from Bovine Heart Mitochondria, *Science* **277** (1997) 60-66.

[58] I. Sadler *et al.*, Sequencing of the Nuclear Gene for the Yeast Cytochrome c_1 Precursor Reveals an Unusually Complex Amino-Terminal Presequence, *EMBO J.* **3** (1984) 2137-2143.

[59] J. Römisch *et al.*, The Primary Structure of Cytochrome c_1 from *Neurospora Crassa*, *Eur. J. Biochem.* **164** (1987) 111-115.

[60] E. Guélin *et al.*, Sequence of the AFG3 Gene Encoding a New Member of the Ftsh/Yme1/Tma Subfamily of the AAA-Protein Family, *Yeast* **10** (1994) 1389-1394.

[61] R. Schnall *et al.*, Identification of a Set of Yeast Genes Coding for a Novel Family of Putative ATPases with High Similarity to Constituents of the 26S Protease Complex, *Yeast* **10** (1994) 1141-115.

[62] R. Tauer *et al.*, Yta10p, a Member of a Novel ATPase Family in Yeast, is Essential for Mitochondrial Function, *FEBS Lett.* **353** (1994) 197-200.

[63] A. Tzagoloff *et al.*, A New Member of a Family of ATPases is Essential for Assembly of Mitochondrial Respiratory Chain and ATP Synthetase Complexes in *Saccharomyces Cerevisiae*, *J. Biol. Chem.* **269** (1994) 26144-26151.

[64] G. Mashkevich *et al.*, SHY1, the Yeast Homolog of the Mammalian SURF-1 Gene, Encodes a Mitochondrial Protein Required for Respiration, *J. Biol. Chem.* **272** (1997) 14356-14364.

[65] M. Saraste and J. Walker, Internal Sequence Repeats and the Path of Polypeptide in Mitochondrial ADP/ATP Translocase, *FEBS Lett.* **144** (1982) 250-254.

[66] F. Palmieri, Mitochondrial Carrier Proteins, *FEBS Lett.* **346** (1994) 48-54.

[67] T. Sedman *et al.*, A DNA Helicase Required for Maintenance of the Functional Mitochondrial Genome in Yeast *Saccharomyces Cerevisiae*, Manuscript submitted.

[68] A. Gorbalenya and E. Koonin, Helicases: Amino Acid Sequence Comparisons and Structure-Function Relationships, *Curr. Opin. Struct. Biol.* **3** (1993) 419-429.

[69] G. Taucher-Scholtz *et al.*, Functions of DNA Helicases in Escherichia Coli. In: N. Cozzarelli (ed.), Mechanism of DNA Replication and Recombination. Liss, New York, 1983, pp. 65-76.

[70] A. Lahaye *et al.*, PIF1 DNA Helicase from Saccharomyces Cerevisiae. Biochemical Characterization of the Enzyme, *J. Biol. Chem.* **268** (1993) 26155-26161.

[71] M. Fujiki and K. Verner, Coupling of Protein Synthesis and Mitochondrial Import in a Homologous Yeast In Vitro System, *J. Biol. Chem.* **266** (1991) 6841-6847.

[72] M. Fujiki and K. Verner, Coupling of Cytosolic Protein Synthesis and Mitochondrial Protein Import in Yeast. Evidence for Cotranslational Import In Vivo, *J. Biol. Chem.* **268** (1993) 1914-1920.

Protein translocation into the mitochondrial matrix and inner membrane

Andreas Geissler, Nikolaus Pfanner and Kaye Truscott

Institut für Biochemie und Molekularbiologie, Universität Freiburg, Hermann-Herder-Str. 7, D-79104 Freiburg, Germany

Keywords: protein import, mitochondria, Tim proteins, *Saccharomyces cerevisiae*

1. Introduction

Mitochondria contain up to a thousand different proteins. Among these, only eight polypeptides in *Saccharomyces cerevisiae* and thirteen in humans are encoded by the mitochondrial genome. All other mitochondrial proteins belonging to these organisms are nuclear-encoded. Following synthesis, as precursors on cytosolic ribosomes, mitochondrial proteins are targeted to and imported into the organelle. The translocation process across the outer and inner membranes of mitochondria is mediated by essential proteins located in both mitochondrial membranes and in the aqueous intermembrane space separating the two lipid bilayers (For comprehensive reviews see [1, 2, 3]).

Recently the nomenclature for proteins involved in translocation was standardised [4]. Proteins belonging to the translocation machinery located in the outer membrane are named Tom (translocase of the outer membrane) while those of the inner membrane are named Tim (translocase of the inner membrane). In addition a number is added indicating the molecular mass (kDa) of the protein. Most members of both the TOM and TIM families exist in large complexes (see figure 1). Each membrane complex is able to act independently of the other [5, 6, 7, 8, 9] and up to now there is no indication that a direct association between the membrane associated TOM and TIM machineries exists. However translocation components are present in the intermembrane space that function in the import of a subset of preproteins to the inner membrane and are therefore also referred to as Tims. These proteins provide a link between the outer and inner membrane translocases for import of inner membrane proteins. Since several new Tim components have been identified recently this review focuses on the translocation processes that take place at the inner mitochondrial membrane and the intermembrane space.

2. The TOM complex

The first step in the import of mitochondrial preproteins is their binding to the receptor components of the TOM complex located on the cytosolic side of the outer mitochondrial membrane. At this stage of import preproteins can already be roughly divided into two groups. One class comprises preproteins which carry internal targeting sequences but lack presequences. Members of this class bind primarily to the import receptor Tom70 [10, 11]. The targeting sequences of these preproteins are not very well characterised and usually occur in hydrophobic integral membrane proteins such as the ATP/ADP or phosphate carriers. The other class of preproteins bind to the import receptor Tom20 and carry a typical N-terminal positively charged presequence which is proteolytically removed following translocation into the mitochondrial matrix [12, 13, 14].

Following binding to the receptor components, preproteins are transferred to the GIP (general import pore), a large complex in the outer membrane consisting of Tom40, Tom 22 and the small Tom proteins Tom5, Tom6 and Tom7. Each of these complexes forms up to three channels as revealed by electron micrographs of the complex reconstituted in liposomes [15]. The channel through which preprotein pass is mainly formed by Tom40 and has an effective pore diameter of about 2.2 nm [16].

Although import across the outer membrane has been intensively studied, the force that drives preprotein movement along the components of the Tom machinery and thereby across the membrane has not been elucidated. However experimental evidence has been gathered which supports "the acid chain hypothesis" a model which suggests that positively charged preproteins follow a chain of negatively charged clusters within the Tom machinery.

3. The TIM complexes

3.1 The import pathways diverge in the intermembrane space and at the inner membrane

The import pathways of the two classes of preproteins, which bind to different receptors components of the TOM machinery but utilise the GIP, diverge in the intermembrane space (IMS) (see figure1). Proteins which carry internal targeting signals bind to the Tim9/10 complex after emerging at the *trans* side of the Tom complex. This complex of small soluble proteins located in the IMS directs the bound proteins to the Tim12/22/54 complex in the inner membrane which in turn guides the insertion of integral proteins into the membrane [17,

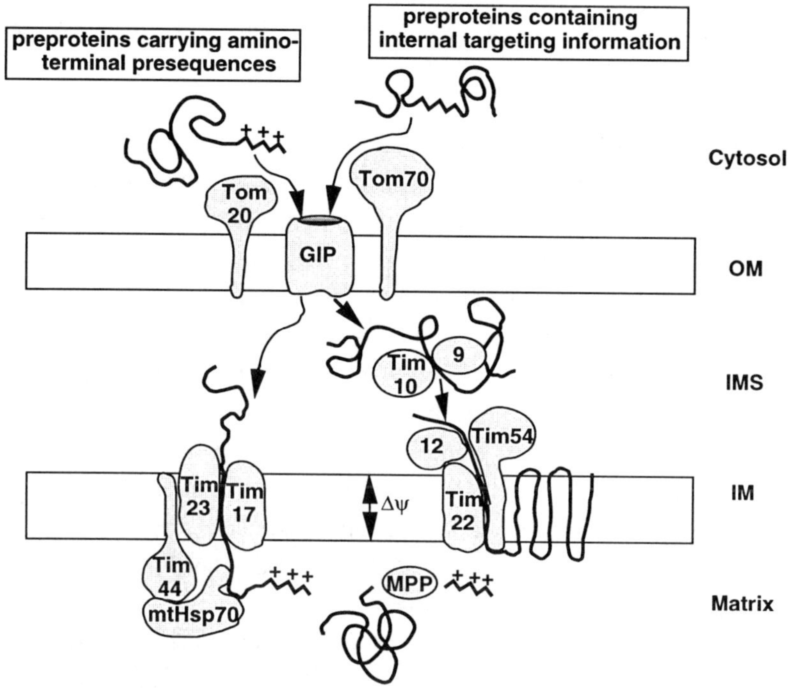

Figure 1: Components of the mitochondrial protein import machinery. Preproteins with N-terminal presequences bind to the Tom20 receptor whereas preproteins carrying internal targeting information bind to the Tom70 receptor. After translocation of both classes of preproteins through the general import pore (GIP) located in the outer membrane (OM) preproteins with N-terminal targeting sequences are imported into the matrix by the Tim17/23/44/mtHsp70 complex, whereas preproteins carrying internal targeting information bind to the Tim9/10 hetero-oligomer at the *trans* side of the outer membrane. Following transfer across the inter membrane space (IMS), the final insertion into the inner membrane is mediated by the Tim12/22/54 complex. $\Delta\psi$ is essential for translocation of proteins across and into the inner membrane (IM).

18]. Proteins carrying an N-terminal positively charged presequence cross the intermembrane space in a Tim9/10 independent manner and are inserted into the Tim17/23 complex. At this location once again the import pathway splits into different routes. Most proteins that utilise the Tim17/23 machinery are imported into the matrix (their final location), where the targeting sequence is cleaved off by the processing peptidase MPP. However, proteins that are sorted to IMS are also imported via this complex. Preproteins which are sorted at the inner membrane may follow one of two mechanisms of translocation. Preproteins following the so called conservative sorting pathway, like the Rieske Fe/S protein, are completely translocated into the matrix where the bipartied presequence is cleaved off by MPP and an octapeptidase [19]. After this processing step the mature protein is reexported to the IMS. In contrast, the import of the cytochrome b_2 precursor follows a non-conservative sorting mechanism [20, 21, 22]. According to this mechanism only the first part of the N-terminal presequence enters the matrix where it is also cleaved off by MPP. The mature portion of the protein remains on the IMS side where the remaining sorting signal is removed by the inner membrane protein peptidase 1. In addition the membrane bound translocation subunit Tim54 which carries a N-terminal, non-cleavable presequence is also imported via Tim17/23 using a mechanism which is yet to be characterised [23].

3.2 The Tim17/23/44/mtHsp70 complex

After crossing the IMS preproteins with N-terminal targeting sequences enter the matrix via the Tim17/23/44/mtHsp70 translocation complex in a strictly membrane potential dependent manner (see Figure1). The membrane potential $\Delta\psi$ (negative inside) exerts an electrophoretic force on the positively charged presequence which drives the initial insertion step. Subsequent translocation of the mature portion of preproteins is facilitated by the ATP-dependent action of mtHsp70, independently of $\Delta\psi$. Following insertion into the matrix targeting sequences are usually cleaved by MPP.

The inner membrane translocation components Tim17 and Tim23 were both identified in parallel by different genetic screens. Tim23, first named Mas6; Mpi3 [24, 25] is an essential, nuclear-encoded protein containing four putative membrane-spanning α-helices and a topology in which both the C- and N-termini face the IMS. Biochemical data confirmed that the newly described 23 kDa polypeptide was an integral membrane protein [24, 25]. A direct interaction of Tim23 with preproteins in transit across the inner mitochondrial membrane was shown using chemical crosslinking experiments [26, 27, 28]. Like Tim23, the essential integral membrane protein Tim17 (originally named Mpi2 or Sms1) is nuclear-encoded [29, 30] and also contains four putative membrane spanning α-helices. It addition it has 46% homology to the C-terminus of Tim23. Homologues of Tim23 and Tim17 have been identified in plants and mammals suggesting that the protein import machinery of the inner membrane is conserved throughout eukaryotes [31, 32]. Since these integral membrane proteins are in direct contact with translocating polypeptides it is likely that both proteins, at least, participate in the formation of the inner membrane import channel [24, 25, 27, 33].

Preproteins of both Tim17 and Tim23 are synthesised without a cleavable presequence and recent data indicates that they are imported via the Tim12/22/54 complex [34, 35]. Subunits of Tim17/23 assemble into large complexes that may contain additional unidentified proteins involved in translocation [36, 28, 37]. As shown by blue native electrophoresis [38] most of Tim17/23 seems to be present in a complex of 90 kDa while smaller amounts of Tim17/23 have been identified in complexes of 140 and 250 kDa respectively [37]. The interaction of the two proteins in these complexes seems to be mediated by hydrophobic portions of the proteins [39]. The 90kDa complex can be shifted to 600kDa when an *invitro* imported preprotein is artificially arrested across the import sites of the inner and outer membranes [37]. This complex contains all of Tim23 and some components of the TOM machinery, which supports the idea that Tim17/23 constitutes the import pore. The molar ratio of Tim23 to the outer membrane channel protein Tom40 is 1:4 [37, 40]. The substiochometric amounts of the inner membrane putative pore forming components relative to Tom40 may limit or regulate the rate of protein import into the matrix although there is no direct evidence for this.

Although Tim17 and Tim23 share common structural features and high sequence homology, they are not able to substitute functionally for each other [29, 30, 41]. The first N-terminal α-helix of Tim23, located in the IMS, is the most obvious feature which is distinct from Tim17. The N-terminal portion can mediate the dimerisation of fusion proteins consisting of the N-terminal domain of Tim23 and a soluble passenger protein [33]. Tim23 dimers in intact mitochondria are disrupted in the presence of mitochondrial presequences. The loss of the membrane potential also weakens the formation of Tim23 dimers [33]. The N-terminal domain of Tim23 may act either as a receptor for matrix targeting presequences or as a lid that seals the inner membrane since, in contrast to the outer membrane, a membrane potential must be maintained across the inner membrane.

After the initial membrane potential dependent insertion of preproteins into the matrix the subsequent translocation of the mature portion of proteins is driven energetically by ATP hydrolysis through the action of mtHsp70. Although some of the total ATP required for the translocation of proteins is consumed in the cytosol, the majority is hydrolysed in the mitochondrial matrix by mtHsp70 [42, 43, 44]. Since proteins arrive in the mitochondrial matrix in an unfolded state they expose structural elements recognised by the chaperone mtHsp70 [45]. The direct binding of mtHsp70 to preproteins is essential for the translocation of the proteins into the matrix [46, 47, 48]. At the inner membrane mtHsp70 is recruited by Tim44 which serves as a membrane anchor for the chaperone thereby positioning it in close proximity to the incoming polypeptides [49, 50, 51]. Only a minor amount of total matrix located mtHsp70 is bound in a 1:1 complex with Tim44 [50], the rest is located in the matrix in an unbound form. The formation of the Tim44/mtHsp70 complex is influenced by the nucleotide bound state of mtHsp70. It has been shown that mtHsp70 binds to Tim44 when it is in an ATP-bound state and hydrolyses the ATP rapidly after interaction with Tim44 [52, 53, 54, 55]. The interaction of mtHsp70 with the nucleotide is controlled by its co-chaperone Mge1, a homologue of the prokaryotic GrpE, which promotes the release of ADP from mtHsp70 [55, 56].

Tim44 was the first translocase component of the inner membrane identified [57]. An essential gene (first named *Mpi1*) was found in a genetic screen that later led to the identification of Tim17 and Tim23. It was also identified independently by biochemical means following precipitation with antibodies raised against crosslink-products of translocation intermediates [58]. The nuclear gene encodes a precursor of 49 kDa which is processed to the 44 kDa mature protein. The sequence of this hydrophilic protein indicates a high content of α-helices with no evidence of membrane-spanning segments. Fractionation experiments and carbonate extractions indicate that Tim44 is a peripheral membrane protein located at the inner surface of the inner mitochondrial membrane [59, 50]. However others suggest that a portion of the protein is exposed to the outer side of the inner membrane or is at least deeply inserted into this membrane [57, 58, 60]. Tim44 can be cross-linked to translocating proteins demonstrating its close proximity to the import sites of the inner membrane

[59, 27, 28, 60]. The possibility that Tim44 directly participates in the formation of the protein import channel cannot be excluded. However when Tim44 is functionally depleted in temperature-sensitive mutants, the insertion of preproteins into the translocation channel is not inhibited [61]. As the number of import sites was the same compared to mitochondria with intact Tim44, it seems unlikely that Tim44 is a component of the import channel.

The most striking effects on protein translocation in temperature-sensitive mutants of Tim44 were observed with preproteins harbouring tightly folded domains [61]. Proteins with tightly folded domains like the heme-binding domain of cytochrome b_2 are known to require a highly active import machinery for import and unfolding of these domains [62, 63, 64]. Tim44 seems to provide a membrane anchor for mtHsp70 allowing the transmission of energy from the hydrolysis of ATP to the movement of the preprotein in the import site. For preproteins with only loosely folded domains Tim44-independent binding of mtHsp70 to the translocating polypeptide seems to be sufficient to drive import. The mtHsp70 binding site of Tim44 has not been clearly defined but a segment of 18 amino acids between residues 185 and 202 may be involved. This region of Tim44 is essential for viability of yeast and although mutant Tim44 lacking this segment assembles into the Tim complex, the interaction between Tim44/mtHsp70 is strongly impaired [65]. This region showa limited similarity to helix II of the J-domain in DnaJ or Sec63 [50] which is involved in their interaction with Hsp70s.

Although the role of mtHsp70 in the translocation process has been intensively studied, the mechanisms by which import into the matrix is promoted by this essential translocation component is still unclear and generates considerable controversy within the field of mitochondrial import. The model of passive trapping is based on the high mobility of translocating proteins in the import channel of the inner membranes [66]. Since Brownian motion can move the polypeptide chain back and forth in the import channel mtHsp70 can bind to the segments newly protruding into the matrix space and thereby prevent a backsliding of the protein [67, 66, 68]. Several rounds of forward movement and trapping result in the full translocation of a polypeptide. Recent data indicate that this "trapping alone" mechanism is not sufficient to drive translocation of preproteins [69]. A second model emphasises that import reactions are strictly ATP-dependent and involves conformational changes of mtHsp70 that are capable of overcoming energy barriers associated with tightly folded domains in preproteins. The model of active pulling suggests that Tim44 bound mtHsp70 performs power strokes that pull proteins into the mitochondrial matrix [70, 71, 69].

3.3 The Tim9/10 and Tim12/22/54 complexes

Unlike proteins with N-terminal presequences, proteins with internal targeting sequences destined for the inner membrane are not able to cross the intermembrane space without the aid of water soluble translocation components located in this compartment. When precursors of inner membrane proteins emerge at the *trans* side of the outer membrane they bind to the Tim9/10 complex and shuttle across the intermembrane space were the preproteins are handed over to the carrier translocase complex consisiting of Tim12, Tim22 and Tim54 (see figure1). The preproteins are then inserted into the inner membrane in a membrane potential dependent step [17,18, 35, 72, 73, 74].

Tim10 (previously named Mrs11) was originally identified as a suppresser of mitochondrial RNA splicing mutants [75] but more recently its role in protein import has been described [17, 18]. Tim9 was identified by database analysis, as a suppresser of a temperature sensitive mutant of Tim10 and by coprecipitation with Tim10 [73, 74]. Tim9 and Tim10 are members of a highly homologous family of small translocation components of the intermembrane space that also includes Tim8, Tim12 and Tim13. These proteins contain a zinc finger motif with four conserved cysteines. Zinc ions were shown to be essential at least for the function of Tim10 and Tim12 [17, 18, 73, 74, 76]. Biogenesis of the small Tim subunits does not seem to involve the receptor components Tom20 and Tom70 or translocase complexes located in the inner membrane or the mitochondrial matrix [23]. Tim9 and Tim10 are both encoded by essential nuclear genes and are targeted to the intermembrane space where they fold to their function state [18, 17, 73, 74, 77]. Analysis by native-electrophoresis and gel filtration revealed that together Tim9/10 form a hetero-hexameric or octameric complex of 70 kDa. Both proteins in this complex interact directly with translocating proteins, as observed by crosslinking of the ADP/ATP carrier preprotein while in transit across the IMS [17, 18, 73]. Very recently two new small proteins, Tim8 and Tim13 were identified. Unlike the other members of this family these intermembrane proteins are not essential. Tim8 and Tim13 together with Tim9 form a complex of 70 kDa. Direct binding of the non-essential Tim oligomer to translocating preproteins has not been observed so far, however genetic data indicate that an interaction with the import machinery does occur. Deafness dystonia protein (DPP), the human homologue of Tim8 is involved in the generation of the neurodegenerative disorder human deafness dystonia [76].

After crossing the intermembrane space the Tim9/10 complex docks to a 300 kDa complex in the inner membrane which consists mainly of Tim12, Tim22 and Tim54. This process facilitates the transfer of preproteins to the carrier translocase of the inner membrane and the subsequent insertion of preproteins into the bilayer. Due to the transfer action of Tim9/10 small amounts of these proteins are associated with the inner membrane translocase complex [17, 18, 73, 74]. Tim12 (first named Mrs5) was initially found in the same screen for suppressors of mitochondrial RNA splicing defects that later led to the identification of Tim10 [75]. As Tim12 (a peripheral membrane protein) coprecipitates with Tim10 and Tim22 , it is thought to establish the binding site for the Tim9/10 complex at the IMS side of the inner membrane [17, 18, 75]. Up until the point at which preproteins bind to Tim12 the translocation of proteins with internal targeting sequences is independent of $\Delta\psi$ [17, 18, 73, 74].

Insertion of proteins into the inner membrane is mediated by Tim22 and Tim54 in a strictly membrane potential dependent manner [17, 18, 72, 34, 35, 72, 73, 74]. The mechanism by which hydrophobic carrier proteins of the inner membrane insert into the membrane is not known. It has been shown that functional inactivation of either Tim22 or Tim54 abolishes or at least strongly inhibits the insertion of proteins with internal targeting sequences into the inner membrane, but not the import of matrix targeted preproteins [72, 34]. Tim22 and Tim54 physically interact with each other forming a 300 kDa complex but an association with the Tim17/23 complex in the inner membrane has not been observed [72, 34]. In contrast to the translocation of matrix targeted preproteins import studies with *ssc1* mutants show that mtHsp70 does not participate in the import pathway of proteins utilising the Tim22/54 complex for import [72, 23].

Tim22 was the first essential component of the carrier translocation pathway to be identified. A yeast genome sequence analysis revealed an open reading frame encoding a protein with a sequence and predicted topology very similar to Tim17 and Tim23 [72]. Interestingly Tim54 was originally identified in a screen for interacting partners of Mmm1, a protein of the outer membrane required for the maintenance of mitochondrial shape [78]. It is unclear if the interaction of Tim54 with the outer membrane is of functional significance or just experimentally artifactual. Secondary structure predictions reveal that Tim54 contains two putative membrane spanning segments although a large part of the C-terminus protrudes into the IMS [34].

Suprisingly the biogenesis of these two essential, nuclear-encoded proteins follows variations of the typical import pathways of mitochondrial proteins described so far [23, 34, 72]. Tim54 contains a positively charged amino-terminus similar to mitochondrial presequences which is not cleaved following translocation. Unlike other preproteins with positively charged presequences Tim54 binds to the Tom70 receptor of the outer membrane. Typically when preproteins bind Tom70 they subsequently utilise the Tim12/22/54 translocase in the inner membrane for import. In fact experiments with *tim23* and *ssc1* mutants show that Tim54 is imported into the inner membrane by the Tim17/23/44 complex [23]. Although Tim22 does not contain an N-terminal targeting sequence it seems to primarily interact with the Tom20 receptor of the outer membrane which is normally used by preproteins containing presequences. Generally Tom20 mediated import is followed by an involvement of the Tim17/23/44 translocase in the final steps of the import at the inner membrane however insertion of Tim22 is mediated by the Tim22/54 translocase. The involvement of Tim22/54 in the import of Tim22 is consistent with the import pathway of homologous proteins Tim17 and Tim23 [23]. It seems that Tim22 and Tim54 utilise aspects of both import routes described for preproteins with N-terminal and internal targeting sequences. The significance of the newly described variations of the classical import pathways remains to be determined but regulation of mitochondrial biogenesis may be a reason for this special import scheme.

4. Conclusions and Perspectives

In our current view proteins destined for the mitochondrial matrix or the inner membrane are translocated by two major import pathways that diverge after the preproteins have crossed the outer membrane through the GIP (general import pore). Preproteins carrying an N-terminal, cleavable presequence are imported via the Tim17/23/44 complex in the inner membrane without interacting with translocation components located in the IMS. Preproteins with internal targeting sequences (typically carrier proteins of the inner membrane) bind to the Tim9/10 complex which shuttles these hydrophobic preproteins across the aqueous phase of the IMS to the inner membrane where they are subsequently inserted into the membrane by the Tim12/22/54 complex. The membrane potential of the inner membrane is required for both translocases to provide the power for translocation or insertion of preproteins. As an additional source of energy ATP-hydrolysis by mtHsp70 is required for the final translocation steps mediated by the Tim17/23/44 machinery.
Are the subunits of the translocation machinery identified so far not only essential but also sufficient for the import process? It is quite possible that future studies will reveal new components of the respective machineries. However there are many gaps in our current knowledge of the translocation machinery. Structural information is lacking and such data would provide considerable insight into the mechanisms that mediate translocation processes. With this knowledge perhaps the tantalising question how the Tim17/23/44 complex is able to form a pore through which the preproteins can cross the inner membrane on the one hand and yet maintain the membrane potential on the other hand by sealing it against leakage, will be answered. Currently the characteristics of the internal targeting signals of inner membrane protein are unknown as is the mechanism by which proteins of the IMS are imported via the Tim17/23/44 complex and sorted to their final location. An intensive investigation over the next few years will hopefully produce answers to at least some of these questions.

5. Acknowledgements

We whould like to thank Drs Joachim Rassow and Michael T. Ryan for critically reading the manuscript. Work of the authors' laboratory was supported by the Deutsche Forschungsgemeinschaft, the Sonderforschungsbereich 388, the Fonds der chemischen Industrie and a long-term fellowship by the Alexander-von-Humboldt Stiftung to K.T..

6. References

[1] Schatz, G. (1996) The protein import system of mitochondria. *J. Biol. Chem.* **271**, 31763-31766.
[2] Neupert, W. (1997) Protein import into mitochondria. *Annu. Rev. Biochem.* **66**, 863-917.

[3] Pfanner, N., Craig, E.A., & Hönlinger, A. (1997) Mitochondrial preprotein translocase. *Annu. Rev. Cell Dev. Biol.* **13**, 25-51.
[4] Pfanner, N., Douglas, M.G., Endo, T., Hoogenraad, N.J., Jensen, R.E., Meijer, M., Neupert, W., Schatz, G., Schmitz, U.K. & Schore, G.C. (1996) Uniform nomenclature for the protein transport machinery of the mitochondrial membranes. *Trends Biochem. Sci.* **21**, 51-52.
[5] Hwang, S., Jascur, T., Vestweber, D., Pon, L., & Schatz, G. (1989) Disrupted yeast mitochondria can import precursor proteins directly through their inner membrane. *J. Cell Biol.* **109**, 487-493.
[6] Hwang, S.T., Wachter, C., & Schatz, G. (1991) Protein import into yeast mitochondrial matrix. A new translocation intermediate between the two mitochondrial membranes. *J. Biol. Chem.* **266**, 21083-21089.
[7] Jascur, T., Goldenberg, D.P., Vestweber, D., & Schatz, G. (1992) Sequential translocation of an artificial precursor protein across the two mitochondrial membranes. *J. Biol. Chem.* **267**, 13636-13641.
[8] Rassow, J. & Pfanner, N. (1991) Mitochondrial preproteins en route from the outer membrane to the inner membrane are exposed to the intermembrane space. *FEBS Lett.* **293**, 85-88.
[9] Horst, M., Hilfiker-Rothenfluh, S., Oppliger, W., & Schatz, G. (1995) Dynamic interaction of the protein translocation systems in the inner and outer membranes of yeast mitochondria. *EMBO J.* **14**, 2293-2297.
[10] Schlossmann, J., Dietmeier, K., Pfanner, N. & Neupert, W. (1994) Specific recognition of mitochondrial preproteins by the cytosolic domain of the import receptor MOM72. *J. Biol. Chem.* **272**, 11893-11901.
[11] Brix, J., Dietmeier, K. & Pfanner, N. (1997) Differential recognition of preproteins by the purified cytosolic domains of the mitochondrial import receptors Tom20, Tom22, and Tom70. *J. Biol. Chem.* **273**, 20730-20735
[12] Söllner, T., Griffiths, G., Pfaller, R., Pfanner, N. & Neupert, W. (1989) MOM19, am import receptor for mitochondrial precursor proteins. *Cell* **59**, 1061-1070
[13] Moczko, M., Gärtner, F. & Pfanner, N. (1993) The protein import receptor MOM19 of yeast mitochondria. *FEBS Lett.* **326**, 251-254.
[14] Harkness, T.A., Nargang, F.E., van der Klei, I., Neupert, W. & Lill, R. (1994) A crucual role of the mitochondrial protein import receptor MOM19 for the biogenesis of mitochondria. *J. Cell. Biol.* **124**, 637-648.
[15] Künkele, K.P., Heins, S., Dembrowski, M., Nargang, F.E., Benz, R., Thieffry, M., Walz, J., Lill, R., Nussberger, S., & Neupert, W. (1998) The preprotein translocation channel of the outer membrane of mitochondria. *Cell* **93**, 1009-1019.
[16] Hill, K., Model, K., Ryan, M.T., Dietmeier, K., Martin, F., Wagner, R., & Pfanner, N. (1998) Tom40 forms the hydrophilic channel of the mitochondrial import pore for preproteins. *Nature* **395**, 516-521.
[17] Koehler, C.M., Jarosch, E., Tokatlidis, K., Schmid, K., Schweyen, R.J., & Schatz, G. (1998a) Import of mitochondrial carriers mediated by essential proteins of the intermembrane space. *Science* **279**, 369-373.
[18] Sirrenberg, C., Endres, M., Fölsch, H., Stuart, R.A., Neupert, W., & Brunner, M. (1998) Carrier protein import into mitochondria mediated by the intermembrane proteins Tim10/Mrs11 and Tim12/Mrs5. *Nature* **391**, 912-915.
[19] Hartl, F.U., Schmidt, B., Wachter, E., Weiss, H. & Neupert, W. (1986) Transport into mitochondria and intramitochondrial sorting of the Fe/S protein of ubiquinol-cytochrome c reductase. *Cell* **47**, 939-951.
[20] van Loon, A.P.G.M., Brändli, A.W. & Schatz, G. (1986) The presequences of two imported mitochondrial porteins contain information for intracellular and intramitochondrial sorting. *Cell* **44**, 801-812.
[21] Gärtner, F., Bömer, U., Guiard, B. & Pfanner, N. (1995) The sorting signal of cytochrome b_2 promotes early divergence from the general mitochondrial pathway and restricts the unfoldase activity op martrix Hsp70. *EMBO J.* **14**, 6043-6057.
[22] Bömer, U., Meijer, M., Guiard, B., Dietmeier, K., Pfanner, N. & Rassow, J. (1997) The sorting route of cytochrome b_2 branches from the general mitochondrial import pathway at the preprotein translocase of the inner membrane. *J. Biol. Chem.* **272**, 30439-30446.
[23] Kurz, M., Martin, H., Rassow, J., Pfanner, N. & Ryan, M.T. (1999) Biogenesis of Tim proteins of the mitochondrial carrier import pathway: Differential targeting mechanism and crossing over with the main import pathway. *Mol. Biol. Cell* , in press
[24] Emtage, J.L.T. & Jensen, R.E. (1993) *MAS6* encodes an essential inner membrane component of the yeast mitochondrial protein import pathway. *J. Cell Biol.* **122**, 1003-1012.
[25] Dekker, P.J.T., Keil, P., Rassow, J., Maarse, A.C., Pfanner, N., & Meijer, M. (1993) Identification of MIM23, a putative component of the protein import machinery of the mitochondrial inner membrane. *FEBS Lett.* **330**, 66-70.

[26] Ryan, K.R. & Jensen, R.E. (1993) Mas6p can be cross-linked to an arrested precursor and interacts with other proteins during mitochondrial protein import. *J. Biol. Chem.* **268**, 23743-23746.
[27] Kübrich, M., Keil, P., Rassow, J., Dekker, P.J.T., Blom, J., Meijer, M., & Pfanner, N. (1994) The polytopic mitochondrial inner membrane proteins MIM17 and MIM23 operate at the same preprotein import site. *FEBS Lett.* **349**, 222-228.
[28] Berthold, J., Bauer, M. F., Schneider, H.-C., Klaus, C., Dietmeier, K., Neupert, W. & Brunner, M. (1995) The MIM complex mediates preprotein translocation across the mitochondrial inner membrane and couples it to the mt-Hsp70/ATP driving system. *Cell,* **81**, 1085-1093.
[29] Maarse, A.C., Blom, J., Keil, P., Pfanner, N., & Meijer, M. (1994) Identification of the essential yeast protein MIM17, an integral mitochondrial inner membrane protein involved in protein import. *FEBS Lett.* **394**, 215-221.
[30] Ryan, K.R., Menold, M.M., Garrett, S., & Jensen, R.E. (1994) *SMS1*, a high-copy suppressor of the yeast *mas6* mutant, encodes an essential inner membrane protein required for mitochondrial protein import. *Mol. Biol. Cell* **5**, 529-538.
[31] Bömer, U., Rassow, J., Zufall, N., Pfanner, N., Meijer, M., & Maarse, A. (1996) The preprotein translocase of the inner mitochondrial membrane: evolutionary conservation of targeting and assembly of Tim17. *J. Mol. Biol.* **262**, 389-395.
[32] Ishihara, N. and Mihara, K. (1998) Identification of the protein import components of the rat mitochondrial inner membrane, rTIM17, rTIM23, and rTIM44. *J. Biochem.* **123**, 722-732.
[33] Bauer, M.F., Sirrenberg, C., Neupert, W. & Brunner, M. (1996) Role of Tim23 as voltage sensor and presequence receptor in protein import into mitochondria. *Cell* **87**, 33-41.
[34] Kerscher, O., Holder, J., Srinivasan, M., Leung, R.S., & Jensen, R.E. (1997) The Tim54p-Tim22p complex mediates insertion of proteins into the mitochondrial inner membrane. *J. Cell Biol.* **139**, 1663-1675.
[35] Káldi, K., Bauer, M.F., Sirrenberg, C., Neupert, W., & Brunner, M. (1998) Biogenesis of Tim23 and Tim17, integral components of the TIM machinery for matrix-targeted preproteins. *EMBO J.* **17**, 1569-1576.
[36] Kübrich, M., Dietmeier, K., & Pfanner, N. (1995) Genetic and biochemical dissection of the mitochondrial protein-import machinery. *Curr. Genet.* **27**, 393-403.
[37] Dekker, P.J.T., Martin, F., Maarse, A.C., Bömer, U., Müller, H., Guiard, B., Meijer, M., Rassow, J., & Pfanner, N. (1997) The Tim core complex defines the number of mitochondrial translocation contact sites and can hold arrested preproteins in the absence of matrix Hsp70/Tim44. *EMBO J.* **16**, 5408-5419.
[38] Schägger, H. & von Jagow, G. (1991) Blue native electrophoresis for isolation of membrane protein complexes in enzymatically active form. *Analyt. Biochem.* **199**, 223-231.
[39] Ryan, K.R., Leung, R.S., & Jensen, R.E. (1998) Characterization of the mitochondrial inner membrane translocase complex: the Tim23p hydrophobic domain interacts with Tim17p but not with other Tim23p molecules. *Mol. Cell. Biol.* **18**, 178-187.
[40] Sirrenberg, C., Endres, M., Becker, K., Bauer, M.F., Walther, E., Neupert, W., & Brunner, M. (1997) Functional cooperation and stoichiometry of protein translocases of the outer and inner membranes of mitochondria. *J. Biol. Chem.* **272**, 29963-29966.
[41] Blom, J., Kübrich, M., Rassow, J., Voos, W., Dekker, P.J.T., Maarse, A.C., Meijer, M. & Pfanner, N. (1993) The essential yeast protein MIM44 (encoded by MPI1) is required at an early step of preprotein translocation across the mitochondrial inner membrane. *Mol. Cell. Biol.* **13**, 7364-7371.
[42] Pfanner, N., Tropschug, M., & Neupert, W. (1987) Mitochondrial protein import: nucleoside triphosphates are involved in conferring import-competence to precursors. *Cell* **49**, 815-823.
[43] Wachter, C., Schatz, G., & Glick, B.S. (1994) Protein import into mitochondria: the requirement for external ATP is precursor-specific whereas intramitochondrial ATP is universally needed for translocation into the matrix. *Mol. Biol. Cell.* **5**, 465-474.
[44] Stuart, R.A., Gruhler, A., van der Klei, I., Guiard, B., Koll, H., & Neupert, W. (1994b) The requirement of matrix ATP for the import of precursor proteins into the mitochondrial matrix and intermembrane space. *Eur. J. Biochem.* **220**, 9-18.
[45] Rassow, J., Hartl, F.U., Guiard, B., Pfanner, N., & Neupert, W. (1990) Polypeptides traverse the mitochondrial envelope in an extended state. *FEBS Lett.* **275**, 190-194.
[46] Kang, P.J., Ostermann, J., Shilling, J., Neupert, W., Craig, E.A., & Pfanner, N. (1990) Requirement for hsp70 in the mitochondrial matrix for translocation and folding of precursor proteins. *Nature* **348**, 137-143.
[47] Scherer, P.E., Krieg, U.C., Hwang, S.T., Vestweber, D., & Schatz, G. (1990) A precursor protein partly translocated into yeast mitochondria is bound to a 70 kD mitochondrial stress protein. *EMBO J.* **9**, 4315-4322.

[48] Ostermann, J., Voos, W., Kang, P.J., Craig, E.A., Neupert, W., & Pfanner, N. (1990) Precursor proteins in transit through mitochondrial contact sites interact with hsp70 in the matrix. *FEBS Lett.* **277**, 281-284.

[49] Schneider, H.-C., Berthold, J., Bauer, M.F., Dietmeier, K., Guiard, B., Brunner, M., & Neupert, W. (1994) Mitochondrial Hsp70/MIM44 complex facilitates protein import. *Nature* **371**, 768-774.

[50] Rassow, J., Maarse, A.C., Krainer, E., Kübrich, M., Müller, H., Meijer, M., Craig, E.A., & Pfanner, N. (1994) Mitochondrial protein import: biochemical and genetic evidence for interaction of matrix hsp70 and the inner membrane protein MIM44. *J. Cell Biol.* **127**, 1547-1556.

[51] Kronidou, N.G., Oppliger, W., Bolliger, L., Hannavy, K., Glick, B.S., Schatz, G. & Horst, M. (1994) Dynamic interaction between Isp45 and mitochondrial hsp70 in the protein import system of the yeast mitochondrial inner membrane. *Proc. Natl. Acad. Sci. USA* **91**, 12818-12822.

[52] von Ahsen, O., Voos, W., Henninger, H., & Pfanner, N. (1995) The mitochondrial protein import machinery. Role of ATP in dissociation of the Hsp70-Mim44 complex. *J. Biol. Chem.* **270**, 29848-29853.

[53] Horst, M., Opplinger, W., Feifel, B., Schatz, G., & Glick. B.S. (1996) The mitochondrial protein import motor: dissociation of mitochondrial hsp70 from its membrane anchor requires ATP binding rather than ATP hydrolysis. *Protein Science* **5**, 759-767.

[54] Ungermann, C., Guiard, B., Neupert, W., & Cyr, D.M. (1996) The $\Delta\Psi$- and Hsp70/MIM44-dependent reaction cycle driving early steps of protein import into mitochondria. *EMBO J.* **15**, 735-744.

[55] Schneider, H.-C., Westermann, B., Neupert, W., & Brunner, M. (1996) The nucleotide exchange factor MGE exerts a key function in the ATP-dependent cycle of mt-Hsp70-Tim44 interaction driving mitochondrial protein import. *EMBO J.* **15**, 5796-5803.

[56] Dekker, P.J.T. & Pfanner, N. (1997) Role of mitochondrial GrpE and phosphate in the ATPase cycle of matrix Hsp70. *J. Mol. Biol.* **270**, 321-327.

[57] Maarse, A.C., Blom, J., Grivell, L.A., & Meijer, M. (1992) MPI1, an essential gene encoding a mitochondrial membrane protein, is possibly involved in protein import into yeast mitochondria. *EMBO J.* **11**, 3619-3628.

[58] Scherer, P.E., Manning-Krieg, U.C., Jenö, P., Schatz, G., & Horst, M. (1992) Identification of a 45-kDa protein at the protein import site of the yeast mitochondrial inner membrane. *Proc. Natl. Acad. Sci. USA* **89**, 11930-11934.

[60] Kanamori, T., Nishikawa, S.-I., Shin, I., Schultz, P.G., & Endo, T. (1997) Probing the environment along the protein import pathways in yeast mitochondria by site-specific photocrosslinking. *Proc. Natl. Acad. Sci. USA* **94**, 485-490.

[61] Bömer, U., Maarse, A.C., Martin, F., Geissler, A., Merlin, A., Schönfisch, B., Meijer, M., Pfanner, N., & Rassow, J. (1998) Separation of structural and dynamic functions of the mitochondrial translocase: Tim44 is crucial for the inner membrane import sites in translocation of tightly folded domains, but not of loosely folded preproteins. *EMBO J.* **17**, 4226-4237.

[62] Glick, B. S., Wachter, C., Reid, G.A., & Schatz, G. (1993) Import of cytochrome b_2 to the mitochondrial intermembrane space: the tighlty folded heme-binding domain makes import dependent upon matrix ATP. *Protein Sci.* **2**, 1901-1917.

[63] Voos, W., Gambill, D., Guiard, B., Pfanner, N., & Craig, E.A. (1993) Presequence and mature part of preproteins strongly influence the dependence of mitochondrial protein import in heat shock protein 70 in the matrix. *J. Cell Biol.* **123**, 119-126.

[64] Voos, W., von Ahsen, O., Müller, H., Guiard, B., Rassow, J., & Pfanner, N. (1996) Differential requirement for the mitochondrial Hsp70-Tim44 complex in unfolding and translocation of preproteins. *EMBO J.* **15**, 2668-2677.

[65] Merlin, A., Voos, W., Maarse, A.C., Meijer, M., Pfanner, N & Rassow, J. (1999) The J-related segment of Tim44 is essential for cell viability: A mutant Tim44 remains in the mitochondrial import site, but inefficiently recruits mtHsp70 and impairs protein translocation. *J. Cell Biol.* **145**, 961-972.

[66] Ungermann, C., Neupert, W., & Cyr, D.M. (1994) The role of Hsp70 in conferring unidirectionality on protein translocation into mitochondria. *Science* **266**, 1250-1253.

[67] Simon, S.M., Peskin, C.S., & Oster, G.F. (1992) What drives the translocation of proteins? *Proc. Natl. Acad. Sci. USA* **89**, 3770-3774.

[69] Voisine, C., Craig, E.A., Zufall, N., von Ahsen, O., Pfanner, N. & Voos, W. (1999) The protein import motor of mitochondria: Unfolding and trapping of preproteins are distinct and separable functions of matrix Hsp70. *Cell* **97**, 565-574.

[70] Glick, B.S. (1995) Can Hsp70 proteins act as force-generating motors? *Cell* **80**, 11-14.

[71] Rassow, J. & Pfanner, N. (1995) Molecular chaperones and intracellular protein translocation. *Rev. Physiol. Biochem. Pharmacol.* **126**, 199-264.
[72] Sirrenberg, C., Bauer, M.F., Guiard, B., Neupert, W., & Brunner, M. (1996) Import of carrier proteins into the mitochondrial inner membrane mediated by Tim22. *Nature* **384**, 582-588.
[73] Koehler, C.M., Merchant, S., Oppliger, W., Schmid, K., Jarosch, E., Dolfini, L., Junne, T., Schatz, G., & Tokatlidis, K. (1998b) Tim9p, an essential partner subunit of Tim10p for the import of mitochondrial carrier proteins. *EMBO J.*, **17**, 6477-6486.
[74] Adam, A., Endres, M., Sirrenberg, C., Lottspeich, F., Neupert, W. & Brunner, M. (1999) Tim9, a new component of the Tim22·54 translocase in mitochondria. *EMBO J.* **18**, 313-319
[75] Jarosch, E., Tuller, G., Daum, G., Waldherr, M., Voskova, A., & Schweyen, R.J. (1996) Mrs5p, an essential protein of the mitochondrial intermembrane space, affects protein import into yeast mitochondria. *J. Biol. Chem.* **271**, 17219-17225.
[76] Koehler, C.M., Leuenberger, D., Merchant, S., Renold, A., Junne, T. & Schatz, G. (1999) Human deafness dystonia syndrome is a mitochondrial disease. *Proc. Natl. Acad. Sci. USA* **96**, 2141-2146.
[77] Jarosch, E., Rodel, G., & Schweyen, R.J. (1997) A soluble 12-kDa protein of the mitochondrial intermembrane space, Mrs11p, is essential for mitochondrial biogenesis and viability of yeast cells. *Mol. Gen. Genet.* **255**, 157-165.
[78] Burgess, S.M., Delannoy, M. & Jensen, R.E. (1994) MMM1 encodes a mitochondrial outer membrane protein essential for establishing and maintaining the structure of yeast mitochondria. *J. Cell Biol.* **126**, 1375-1391.

Protein, Lipid and Membrane Traffic: Pathways and Targeting
J.A.F. Op den Kamp (Ed.)
IOS Press, 2000

MEMBRANE TRAFFIC EXTRAVAGANZA

TOM KIRCHHAUSEN – Harvard Medical School

1 – COAT FORMATION

1.1 *THE PROBLEM, TRAFFIC WITHOUT DISORDER*

Traffic between intracellular compartments requires a special mechanism to transport proteins, lipids and cargo molecules from a donor to an acceptor organelle. Vesicles that bud off the donor membrane and fuse with an acceptor membrane mediate this process. This solves the problem of selection, because the cargo molecules are concentrated into the vesicles before the vesicle separates from the donor membrane, and solves the problem of disorder, because it prevents the intermixing of membrane compartments.

1.2 *THE SOLUTION, EVOLUTION OF BIOLOGICAL ENGINES TO TRANSPORT MEMBRANE PROTEINS AND CARGO MOLECULES*

The first step in cargo concentration is the binding of cargo molecules to the lumenal domain of a transmembrane receptor. This is followed by the interaction of the cytoplasmic domain of the receptor with a protein complex that is believed to trigger the formation of a protein coat, in turn leading to the formation of a pit in the membrane that will eventually become a vesicle. This coupling of cargo concentration and coat formation ensures efficient cargo loading into the assembling vesicles. The neck of the vesicle then narrows, and the vesicle finally pinches off the membrane. At this point the protein coat is removed, and the vesicle is targeted to the acceptor compartment, eventually fusing with the target membrane. Among the best-understood processes for moving receptors and ligands are clathrin-coated pits and vesicles.

1.3 *TYPES OF VESICLE-MEDIATED TRANSPORT SYSTEM*

• *Clathrin coated system:* first and best studied vesicle-mediated transport system required for traffic from the plasma membrane to the endosome and from the TGN to the endosome. Many, many protein components.

• *COPI coatomers:* required for the vesicle-mediated transport between the ER and Golgi during the biosynthesis of secretory proteins. Discovered because coated vesicles (containing COPI) accumulate when GTPγS was used to inhibit VSV−G traffic between the cis and medial compartments of the Golgi apparatus.
Coatomer (complex) exists in solution.

• *COPII coatomers:* required for the vesicle-mediated transport through out the compartments of the Golgi during the biosynthesis of secretory proteins.

• *Caveolae membrane system*: least understood. Defined operationally by their morphology (little caves, pockets or recess), and biochemically by their resistance to detergent extraction and by their 'contents' (lipids [gangliosides, cholesterol], caveolin, acylated proteins (hetero-trimeric G, src, fyn, lck, folate receptor, prion, GPI-proteins, SOS, Grb2, EGFR, etc. etc). No idea of how they assemble, traffics, fuses, etc.

CLATHRIN

Clathrin heavy chain	(β-propeller)
Clathrin light chain	
AP-1	β1, γ, μ1, σ1
AP-2	β2, α, μ2, σ2
AP-3?	β3, δ, μ3, σ3
Eps15	
Eps15R	
AP180	
Amphyphysin	
Epsin	
Synaptojanin	
Dynamin	
Non-visual arresting	

COPI

α	KKXX binding
β	
β'	KKSS binding; WD-repeats (β-propeller),
γ	
δ	
ε	KKXX binding; mammalian ts available (complements ldl-1)
ζ	
Arf	GTP-binding and hydrolysis (required for association with membranes and coat assembly)

COPII

Sec23p	GAP for Sar1p
Sec24p	
Sec13p	WD-repeats (β-propeller)
Sec31p	WD-repeats (β-propeller)
Sar1p	GTP-binding and hydrolysis (required for association with membranes and coat assembly)

1.4 *ASSEMBLY OF CLATHRIN COATS*

Characteristics
- Assembly is very fast (seconds) *in vivo* and in model systems
- Error free and cooperative
- Involves many assembly units (>100)

How does it work?
- 3-dimensional shape of the clathrin triskelion
- Packing of the triskelion into the clathrin lattice
- Atomic structure of the terminal domain and the linker of clathrin -new
- Atomic structure of the terminal domain of clathrin in complex with the clathrin box of β-arrestin and adaptor complexes -new
- Regulation of assembly/disassembly: Clathrin/adaptor interactions, clathrin / Hsc70 / auxilin interactions

1.5 *UNCOATING OF CLATHRIN COATS*

Characteristics
- Energy dependent
- Should happen in coated vesicles and not in coated pits
- Hsc70, auxilin and ATP: is this enough?

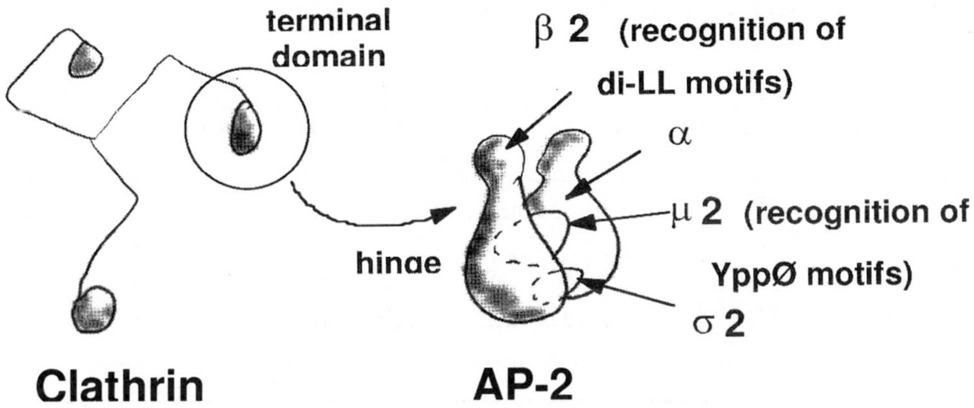

Figure 1. Schematic representation of clathrin and its adaptor complex AP-2.

2 – CARGO SELECTION

2.1 *THE PROBLEM, HOW TO SELECT CARGO*

Endocytic and secretory pathways of eukaryotic cells contain characteristic trans membrane proteins that are targeted to endosomal compartments, to the ER and Golgi compartments, to the trans-Golgi network and to lysosomes.

Targeting can depend on global characteristics of the transported protein (unsaturable) or on discrete signals (saturable) that are recognized by specific receptors (sorting machinery). The selection is exquisite and extremely efficient even though the associations with the sorting machinery are extremely weak.

2.2 *THE SOLUTION, SORTING MACHINES THAT ACT AT VERY HIGH LOCAL CONCENTRATION*

The problem of selection is solved by concentration of cargo in vesicles by (1) binding of cargo to lumenal domain of membrane receptor; by (2) two-dimensional concentration (in the plane of the membrane) of receptor by binding to protein coat component (that may or may not be required for vesiculation). This is referred to as the coupling of cargo loading with vesicle formation.

TABLE 1. Sorting signals recognized in the clathrin pathway

Motif	Sorted proteins containing motif
tyrosine-based (YppØ-type)	CI-M6PR, TfR, EGFR, TGN38/41, gp41,Lamp1
tyrosine based (NPXY-type)	LDLR
dileucine-based	CD4, Glut4, CD3γ, CI-M6PR, Ii, LimpII, Vamp3p (vacuolar t-SNARE)
Acidic clusters	Furin
Dilysine (KKFF-type)	VIP36
Ubiquitin addition	Yeast α-factor, Eps15, GHR

TABLE 2. Sorting signals recognized in the COPI/COPII pathway

Motif	Sorted proteins containing motif
dilysine based (KKXX)	WBP1 (part of yeast N-oligosaccharyl transferase complex), members of Emp24p (cargo receptor ER<>Golgi)
dilysine based (KXKXX)	members of Emp24p (cargo receptor ER<>Golgi)
FF ?	members of Emp24p (cargo receptor Golgi>ER)

2.3 *CHARACTERISTICS OF SORTING SIGNALS*

- Cytoplasmically exposed
- Diverse (YppØ, NPXY, LL, acidic clusters, ubiquitination, KK
- Degenerate
- Weak dependence in preformed structure: atomic structure of $\mu 2$ bound to YppØ - new
- proteins with one sorting signal and proteins with multiple sorting signals

2.4 *CHARACTERISTICS OF SORTING MACHINERY*

- Targeted to specific membrane organelles
- Interaction is saturable
- Interaction is of low affinity (high micromolar/low milimolar) – solved by proximity effects
- Redundant
- High specificity

2.5 *REGULATION OF SORTING*

- By control of exposure of the sorting signal (phosphorylation of signal or adjacent sequences; by association of cytosolic tail with other proteins
- By control of the affinity of sorting machinery (phosphorylation, lipids, interaction with coat proteins)
- By targeting the sorting machinery to the 'right' membrane

2.6 *RECEPTOR SORTING BY CLATHRIN COATED PITS*

A common event in the sorting of many transmembrane proteins is the interaction of a sorting signal in the cytosolic domains and a component of an organellar protein coat, such as the clathrin adaptors AP-1 and AP-2 of clathrin coated pits or the COP-I/COP-II components of coatomers. There are several different classes of sorting signals; some examples are listed in Tables 1and 2.

These sorting signals are diverse and highly degenerate, and they function only in the appropriate context. For example, when the TfR is in the endosomal compartment it is not recognized by AP-2, even though AP-2 recognizes this receptor extremely efficiently when it is at the plasma membrane. Moreover, each sorting signal acts as an addressing code: for example, TGN38 is targeted to the TGN, whereas the TfR, which contains a very similar motif, recycles to the plasma membrane. Switching the sorting motif alters the fate of the protein.

To date, more and more different internalization signals are being discovered. Tyrosine-based signals, however, continue to receive the major attention because they are most commonly found among rapidly internalized proteins. Tyrosine-based signals are characterized by the presence of a critical tyrosine residue within an otherwise degenerate sequence context. The most common tyrosine motifs are NPXY and YppØ, where X represents any amino acid and Ø represents a bulky hydrophobic amino acid. Using combinatorial selection methods and surface plasmon resonance as well as the two-hybrid approach to narrow the requirements for the YppØ interaction with AP-2 and μ2, it has been shown that the optimal sequence for interaction has tyrosine as an anchor, prefers arginine at position +2 and leucine at position +3. In contrast, no preferred sequence surrounding the YppØ is detected indicating that recognition of the YppØ endocytosis signal does not require a pre-folded structure.

Different biochemical and genetic approaches demonstrated interaction of the μ chain of the AP complexes with many different tyrosine signals. The fine specificity of interaction of tyrosine-based signals with μ2 correlates with the interaction of tyrosine-based signals with the complete AP-2 complex. Phosphorylation of the critical tyrosine residue abrogates interaction with μ2/AP-2 (e.g. CTLA4). The avidity for membrane-immobilized AP's might be dramatically increased by the oligomeric state of membrane proteins linking one or more signals. Endocytic receptors might be recruited to pre-formed coated pits since a conformational switch of AP's induced by the interaction with clathrin modulates the affinity.

It has recently been found that phosphoinositide-3'-phosphate (PI3P) enhances the interaction of AP-2 with YppØ -based endocytic signals. Experiments using the PI3-kinase inhibitor Wortmannin, which induces a loss of about 70% of PI3P showed no effect on endocytosis but abolished endosomal traffic between TGN, endosomes and lysosomes. PI3P regulates the affinity of binding of μ2 and YppØ -based sorting signals, whereas inositol has no effect. This observation indicates that there might be a link between signal transduction and membrane traffic, and stresses the importance of phosphoinositide-kinases in vesicular traffic.

2.7 *HOW MANY ADAPTORS?*

AP-1 and AP-2 are hetero-tetramers that clearly co-assemble with clathrin into coated structures primarily at the TGN and at the plasma membrane, respectively. AP-3 and AP-4 are also a hetero-tetramer whose subunits are evolutionary related to those of AP-1 and AP-2. AP-3 and AP-4 are found in perinuclear regions including the TGN. It is less clear whether they involve a clathrin dependent pathway. AP-1, AP-2 and AP-3 recognize YppØ and dileucine based sorting signals. Non-visual β-arrestins are monomers that link G-coupled receptors to the terminal domain of clathrin. Nef, a myristylated protein encoded by HIV, interacts with AP-2 and recruits CD4 to clathrin coated structures.

2.8 *DECODING PROCESS*

Although we cannot yet explain all of the details of the decoding process, some basic principles have recently emerged. At least in the case of the tyrosine YppØ motif and the dileucine motif, distinct subunits of the AP-1 complex recognize the different motifs. Furthermore, different motifs are bound by the AP-1 and AP-2 complexes with different affinities. By varying the site to which the motif binds, and presumably therefore the consequences of the binding event, one can imagine that the sorting machinery would be able to target different motifs to different acceptor membranes.

Multiple coats, different locations and differential but redundant recognition are the constraints of tyrosine-based and dileucine-based signals. The characteristics are differences in avidity of specificity of interaction of YppØ -based signals with AP-1 and AP-2, where AP-2 shows higher/broader specificity. The order of encounter in the biosynthetic pathway is AP-1 at the TGN, AP-3 (AP-4) at the endosome and AP-2 at the plasma membrane.

3 – VESICLE TARGETING AND FUSION

3.1 *THE PROBLEM, HOW TO INSURE THE FIDELITY OF MEMBRANE TRAFFICKING*

Maintenance of the organization of eukaryotic cells requires the non-random and specific shuttle of transport vesicles from donor to acceptor membranes. These form of intracellular traffic requires directionality of transport, addressing of target membranes and efficient fusion.

3.2 *THE SOLUTION, USE OF MOLECULAR MOTORS, TARGETING PROTEINS AND FUSION MOLECULES*

Use of the cytoskeleton (actin) as a highways: delivery to minus-end by dynein; use of myosin-I in traffic involving cortical actin; use of myosin-V for organelle movement; myosin-VI and cycling from endosome to plasma membrane.

SNARE hypothesis, originated with the isolation of a detergent-soluble complex made of three synaptic proteins, VAMP (vesicle associated membrane protein or synaptobrevin; a v-SNARE); SNAP-25 (synaptosome associated protein of 25 kDa; a t-SNARE); syntaxin 1 (a v-SNARE). The complex binds to the soluble NSF and α-SNAP factors. The SNARE complex: specificity of vesicle targeting (original idea) or minimal fusion machine (new idea). Two key discoveries: (1) what turned out to be components of the SNARE complex are proteolytic targets of neurotoxins. (2) NSF, α-SNAP, rabs and homologues of SNAREs are involved in membrane traffic at various cellular locations (ER>Golgi, intra-Golgi, and exocytosis).

<u>Requirements for efficient fusion:</u>
• Docking
• Priming
• Fusion

<u>Fusion molecules:</u>
• Viral proteins – HA and low pH transition; gp41 of HIV: concept of spring-loaded massive conformational change leading to apposition of opposite membranes. Atomic structure - new
• SNARE complex – apposition of opposite membranes. Atomic structure – new
• lipids

REFERENCES

1. Kirchhausen, T.: Coated pits and coated vesicles - sorting it all out. *Curr Opin Struc Biol,* 1993, 3: 182-188.
2. Bednarek, S. Y., Orci, L., and Schekman, R.: Traffic COPs and the formation of vesicle coats. *Trends Cell Biol,* 1996, 6: 468-473.
3. ter Haar, E., Musacchio, A., Harrison, S. C., and Kirchhausen, T.: Atomic structure of clathrin: A β-propeller terminal domain joins an α-zigzag linker. *Cell,* 1998, 95: 563-573.
4. Kirchhausen, T., Bonifacino, J. S., and Riezman, H.: Linking cargo to vesicle formation - receptor tail interactions with coat proteins. *Current Opinion in Cell Biology,* 1997, 9: 488-495.
5. Marks, M. S., Ohno, H., Kirchhausen, T., and Bonifacino, J. S.: Protein sorting by tyrosine-based signals: Adapting to the Ys and wherefores. *Trends Cell Biol,* 1997, 7: 124-128.
6. Greenberg, M., DeTulleo, L., Rapoport, I., Skowronski, J., and Kirchhausen, T.: A dileucine motif in HIV-1 Nef is essential for sorting into clathrin-coated pits and for down-regulation of CD4. *Current Biology,* 1998, 8: 1239-1242.
7. Bednarek, S. Y., Orci, L., and Schekman, R.: Traffic COPs and the formation of vesicle coats. *Trends Cell Biol.,* 1996, 6: 468-473.
8. Rizo, J., and Sudhof, T. C.: Mechanics of membrane fusion. *Nature Structural Biology,* 1998, 5: 839-842.
9. Weissenhorn, W., Dessen, A., Harrison, S. C., Skehel, J. J., and Wiley, D. C.: Atomic Structure of the Ectodomain From Hiv-1 Gp41. *Nature,* 1997, 387: 426-430.

Overexpression of rabGDI induces vesiculation of the Golgi apparatus

Guenther Bode and Thomas Mayer#*

Dept. Exp. Tumorbiology, University of Muenster, Badestr. 9, 48149 Muenster, Germany

#Present address: Program of Cellular Biochemistry and Biophysics, Memorial Sloan - Kettering Cancer Center, 1275 York Avenue RRL 537, New York 10021, USA

**Corresponding Author: phone: +1 212 639 6177*
Fax: +1 212 717 3604

Abstract

Members of the rab/YPT family perform essential functions as molecular switches at multiple steps of the endo- and exocytotic pathway. Rab proteins pass through cycles of membrane-associated, active, GTP bound and cytosolic, inactive, GDP bound forms. The cytosolic form of rab proteins is complexed with the rabGDI (rab guanosine dissociation inhibitor). Overexpression of rabGDI and a certain rab6 mutant results in vesiculation of the Golgi apparatus.

To characterize the vesiculated Golgi apparatus we used cotransfection experiments and immunofluorescence analyses. Vesicles induced by rabGDI contain VSV G as a marker for exocytic membranes and Golgi specific carbohydrate structures as shown by labeling with wheat germ agglutinin. While rabGDI induces reorganization of the Golgi apparatus, changes in the intermediate compartment structure are less prominent.

1. Introduction

Protein transport along the exocytotic and endocytotic pathways is mediated by transport vesicles that bud from a donor compartment and fuse with a target membrane [1]. This process is tightly controlled by rab/YPT proteins, a branch of the superfamily of small

GTPases [2]. Up to now over 30 rab proteins have been described in mammalian cells and, in addition, more than eight YPT proteins have been identified in *Saccharomyces cerevisia.*

Small GTPases act as molecular switches with an "on" and an "off" conformation. Rab proteins undergo a cycle in which the active GTP-bound form on the membrane is converted by GTP hydrolysis into an inactive GDP-bound form. Each individual step of such cycles is regulated by a set of proteins, and requires a conformational change in the corresponding rab protein [for review see 3 and 4]. GTP-bound forms of the rab proteins interact with specific downstream effectors. After GTP hydrolysis the rab protein separates from these effectors, becomes detached from the membrane in its GDP-bound form and is recycled to the cytosol. Membrane extraction and membrane relocation of rab proteins is achieved by the members of the guanosine nucleotide dissociation inhibitor (GDI) family. Thus GDI acts as a recycling factor [5]. RabGDI constitutes a small family of at least three highly related proteins GDIα, GDIß and GDIγ. GDIα is limited in distribution to brain tissue, while the GDIß isoform is expressed ubiquitously.

RabGDI proteins control rab proteins by two mechanisms: 1. They extract GDP-bound rab proteins from membranes and 2. They form cytosolic complexes with rabGDP proteins, which enables them to deliver rab polypeptides back to membranes [6,7]. By controlling membrane translocation and extraction, rabGDI polypeptides are major regulating elements of rab activity in mammalian cells. Excess GDI has been shown to block both ER-to-Golgi and intra-Golgi-transport events *in vitro*. It was shown that rabGDI acts on every rab protein, but individual rab proteins are differentially sensitive to GDI-induced membrane extraction [8]. In an in *vitro* ER-to-Golgi transport assay, it was demonstrated, that the docking of transport vesicles is sensitive to excess levels of Sec19p, a GDI homologue in yeast [9]. Previous studies used cell free systems to describe the effect of rabGDI on isolated steps of vesicular transport. Aim of our study was to analyze the effects of rabGDI overexpression in the intact mammalian cell.

2. Results

To study the effects of the rab/rabGDI system on the internal membrane structure we overexpressed either mutant rab6 T27N or rabGDI in mammalian cells. The rab6 T27N polypeptide is an inactive form which preferentially binds to GDP. In the cell this form binds to the GDI protein to form a cytosolic complex and thereby reduces the intracellular pool of free GDI protein. As an additional experimental approach we also overexpressed the GDI protein in CV1 cells. A myc epitope tag was added to both rab6 T27N and GDI cDNAs. The tagged cDNAs were inserted into the mammalian expression vector pSVSPORT. In this plasmid the expression of inserted cDNA is facilitated by the SV40 promoter (figure 1a and b). 48 hr after transfection cells were fixed and double-stained for the expression of rab6 T27N with a myc epitope specific antibody 9E10 as well as for Golgi membranes with fluorescein coupled lectin wheat germ agglutinin. Rab6 T27N shows a diffuse intracellular distribution indicating that it binds nonspecifically to a whole variety of intracellular membranes (fig. 2b). Golgi staining revealed that these transfected cells display an altered Golgi morphology with an enlarged and vesiculated structure (fig. 2a). In addition, vesicle-like structures are formed in transfected but not in nontransfected cells. Since these vesicle-like structures can be labeled with wheat germ agglutinin, they are most likely derived from Golgi membranes. Expression of rabGDI protein in CV1 cells elicits a much stronger effect on the Golgi apparatus. In nontransfected cells, the Golgi apparatus has its normal perinuclear cap-like structure as revealed by WGA staining. In contrast the rabGDI transfected CV1 cells have lost their perinuclear Golgi staining, instead the lectin identifies vesicle like-structures throughout the cells (fig. 2c and d). These structures spread from the perinuclear region to the

A Structure of the bovine GDI cDNA in pSVSport

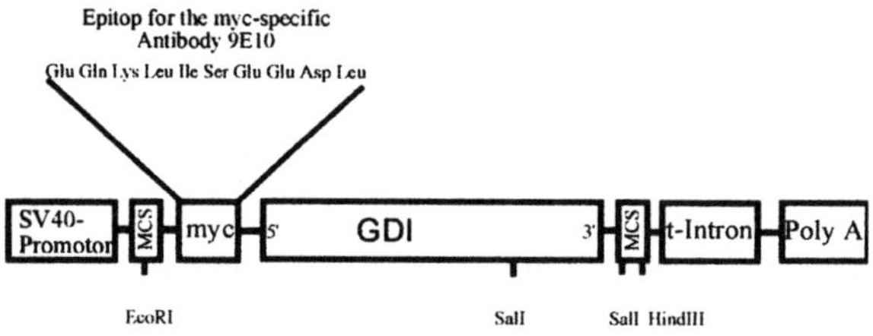

B Structure of the rab6 T27N cDNA in pSVSport

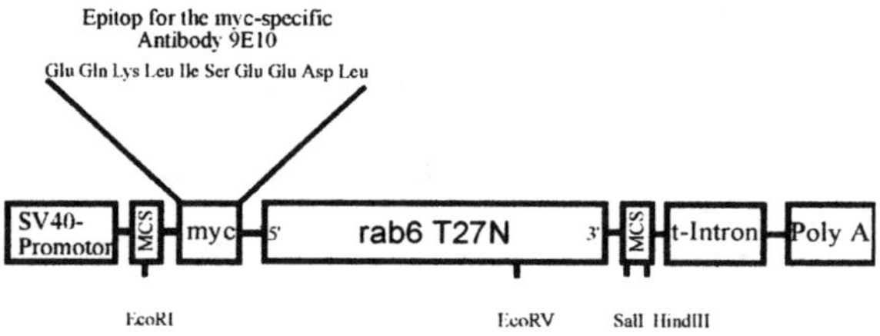

Figure 1 : Expression vectors for rab6 T27N and rabGDI expression

Bovine rabGDI cDNA and human rab6T27N cDNAs were tagged with the myc epitope at the NH_2 - terminus and inserted into the expression plasmid pSV Sport.

periphery. In most cells an accumulation of such Golgi-derived vesicles under the plasma membrane can be observed.

At this point it was not clear whether these structures are breakdown products of the Golgi cisternae or instead are related to intra Golgi transport. To prove that these structures represent transport intermediates we cotransfected CV1 cells with rabGDI and an expression vector containing the cDNA for VSV G protein.

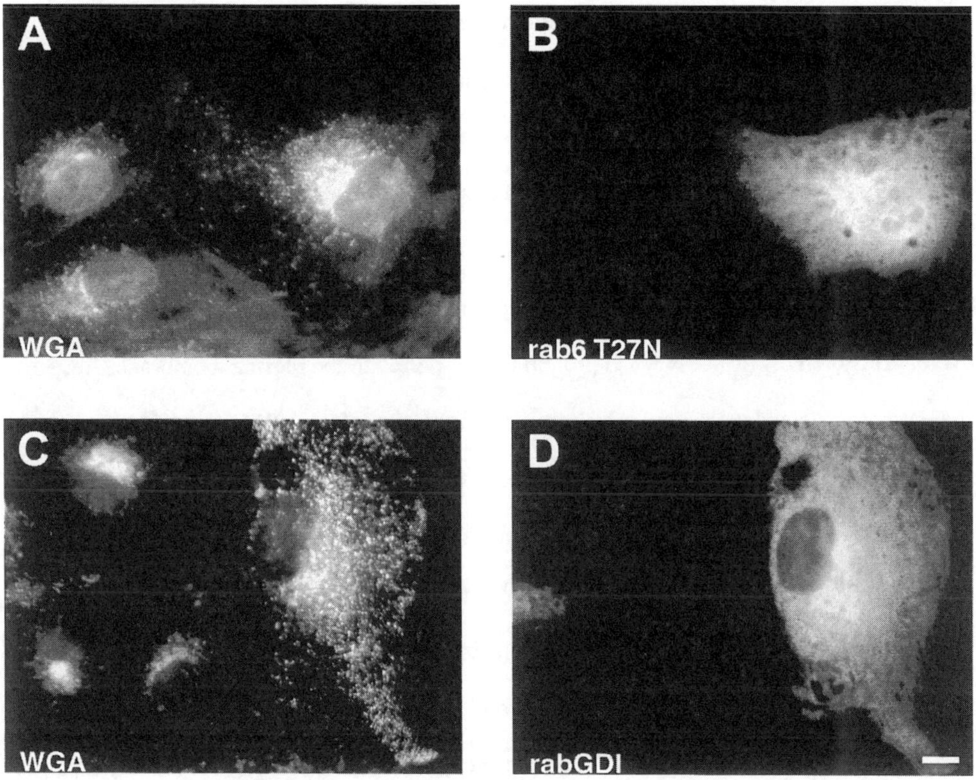

Figure 2: Expression of rab6T27N and rabGDI alters Golgi morphology in CV1 cells

CV1 cells were transfected with rab6 T27N or rabGDI cDNAs. 48hr after transfection cells were fixed and permeabilized. Cells were double labeled with a monoclonal antibody (9E10) against the myc epitope and a secondary anti-mouse antibody coupled to Texas Red (B and D). Golgi membranes were stained with wheat germ agglutinin (A and C). Panels A and B show cells transfected with rab6 T27N, panels C and D show cells transfected with rabGDI. Bar represents 10 μm.

The VSV G protein is a well-characterized marker for anterograde transport in the mammalian cell. Fig. 3b shows VSV G distribution in CV1 cells. In transfected cells most of the VSV G protein localizes to the Golgi cisternae and plasma membrane.

After cotransfection with VSV G protein and rabGDI a different distribution results. The G protein colocalizes with WGA labeled Golgi membranes (fig. 3c and d). Again a significant portion of the vesicles accumulate under the plasma membrane. In the absence of GDI overexpression most of the VSV G protein is transported to the plasma membrane (fig. 3b). In GDI-cotransfected cells, however, only a minor portion of the VSV polypeptides reaches the plasma membrane (fig. 3d). This indicates that rabGDI overexpression inhibits vesicular transport from the ER to the plasma membrane. To analyze whether transport is inhibited at the Golgi level we used a well established 15°C temperature block assay. In this assay budding from the intermediate compartment is inhibited by incubating the cells at 15°C. CV1 cells were transfected with VSV G and rabGDI expression vectors. 48 hr after transfection, cells were shifted to 15°C and further protein synthesis was blocked by addition of cycloheximide to the cells. As demonstrated in figure 4b these conditions cause the VSV G protein to accumulate in perinuclear regions of the CV1 cells. These regions correspond to the intermediate compartment. When the block was released by shifting the cells back to 37°C, distribution of the VSV G protein in the CV1 cells changed. The VSV G could now be observed in vesiculated structures throughout the cells (fig. 4d).

These data show that GDI overexpression induces vesiculation of the Golgi apparatus.

In addition, rabGDI overexpression induced at most minor intermediate compartment changes, while drastically disturbing Golgi morphology.

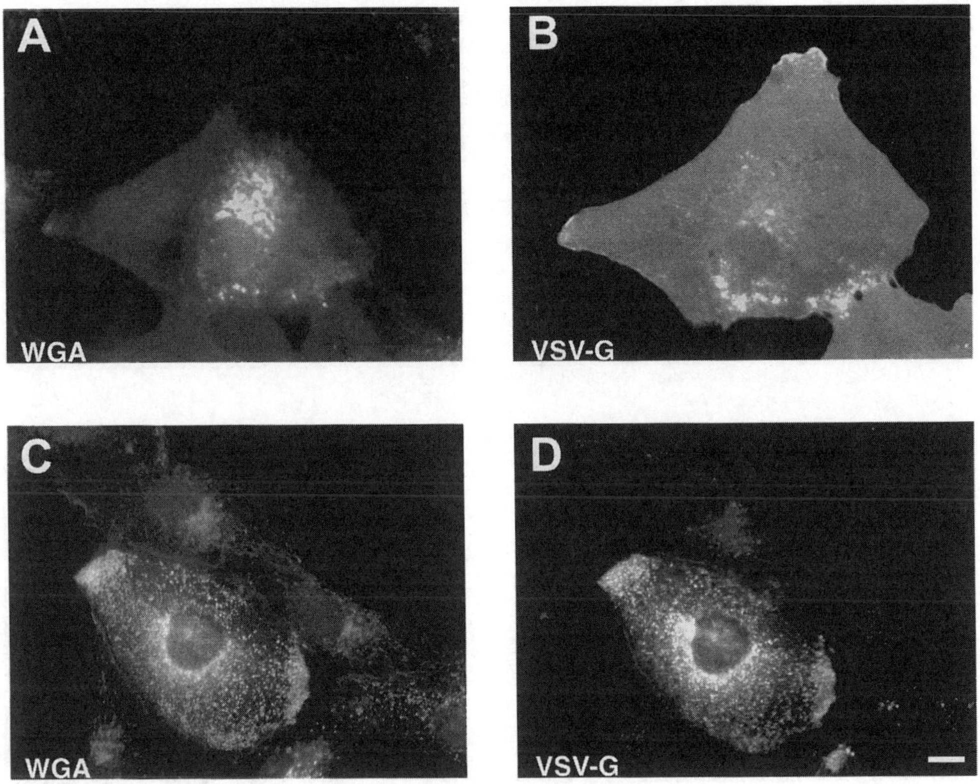

Figure 3: VSV G protein accumulates in Golgi-derived vesicles

CV1 cells were transfected with the VSV G expression vector alone or cotransfected with the rabGDI expression vector. 48hr after transfection cells were fixed and permeabilized. Cells were double labeled with a monoclonal antibody against the VSV G protein and a secondary anti-mouse antibody coupled to Texas Red (B and D). Golgi membranes were stained with wheat germ agglutinin (A and C). A and B show cells transfected with VSV G alone , C and D cells cotransfected with rabGDI. Bar represents 10 μm.

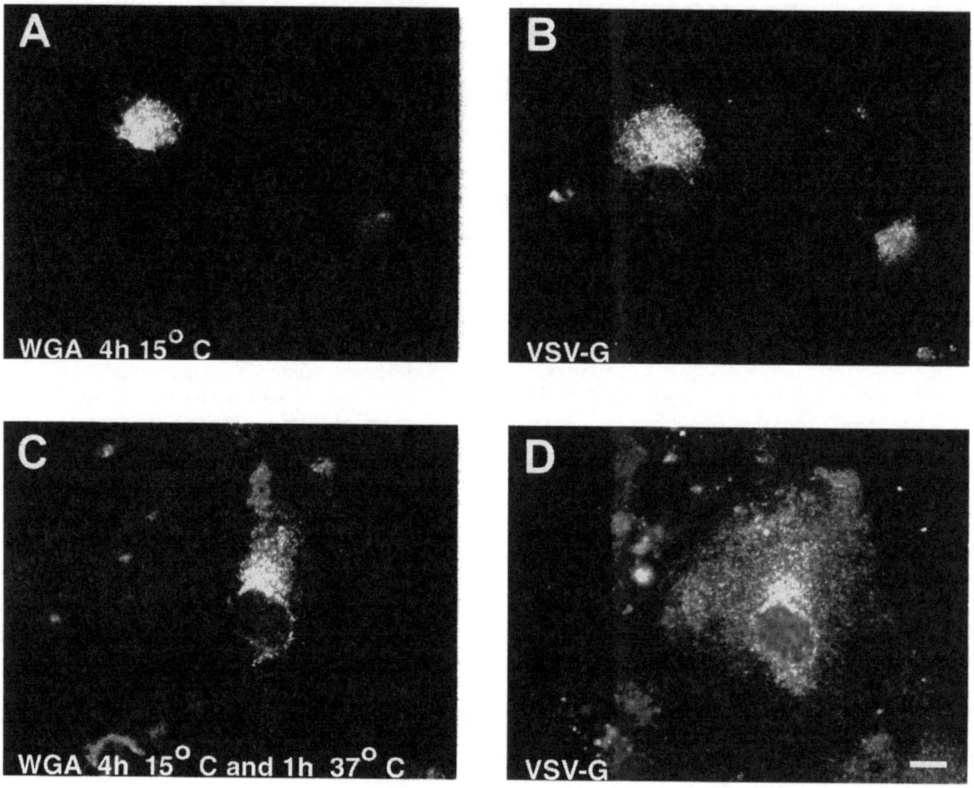

Figure 4: Accumulation of VSV G protein is blocked at 15°C

CV1 cells were cotransfected with VSV G and rabGDI expression vectors. 48hr after transfection cells were fixed and permeabilized. Cells were double-labeled with a monoclonal antibody against VSV G and a secondary anti-mouse antibody coupled to Texas Red (B and D). Golgi apparatuses were stained with wheat germ agglutinin (A and C). A and B show cells incubated in the presence of 50 μg/ml cycloheximide at 15°C for 4hr. In C and D cells were shifted back to 37°C for 1hr. Bar represents 10 μm.

3. Summary and Conclusion

Overexpression of rabGDI alters the structure of the Golgi apparatus. The underlying molecular mechanism for this effect remains to be elucidated. Rab proteins are needed at multiple steps of eukaryotic vesicular transport [for review see 2]. They are active at steps related to vesicular fusion, and are thought to thereby control the specificity of the vesicle-acceptor membrane interaction [for review see 10]. There is an intracellular steady state between active, membrane bound and inactive, cytosolic forms of rab proteins.

The rab GDI polypeptide is one of the major regulators of rab protein activity: it extracts rabs from membranes and keeps them in their inactive, GDP-bound form. Overexpression of rabGDI therefore locks most of the rab proteins into inactive, GDP bound conformations. The resulting low intracellular concentrations of active rabGTP may result in reduced vesicular fusion. If this hypothesis is correct, low intracellular concentrations of active rab would not hinder vesicular budding from donor compartments or vesicular transport to acceptor membranes, but would impair only the vesicular fusion required as initiation of exocytosis. We have shown before that elevated amounts of GDI inhibit protein transport through the Golgi apparatus in cell free systems [6, 11]. Here we show that high intracellular levels of rabGDI also cause accumulation of vesicles in living cells. In our current experiments, vesicular accumulation at the plasma membrane coincided with destruction of typical Golgi cisternal structures. This could have resulted from continued export of membrane material without replenishment of membrane pools.

While transport from the intermediate compartment (between the Golgi cisternae and from Golgi cisternae to the plasma membrane) is inhibited, the intermediate compartment seems not to be affected to the same extent. Two explanations are possible for such a selective inhibition of transport to the plasma membrane: 1. Preferential interaction of GDI with certain rab proteins has already been described [8]. If GDI preferentially inactivated rabs that mediate transport from Golgi cisternae to the plasma membrane, and only to a lesser degree rabs needed for intermediate compartment transport, intermediate compartment structure would remain relatively preserved in the setting of GDI overexpression. 2. If the intermediate compartment is formed by homotypic fusion of vesicles, homotypic fusion might occur independent of rab protein mediation, since no control of vesicle/acceptor membrane specificity is required for this fusion event.

The exact molecular mechanism of GDI activity on Golgi cisternal structure remains to be elucidated.

Acknowledgements

We thank Annette Janning for technical assistance and Drs Klaus van Leyen and Gisela Banauch for their comments on the manuscript. This article contains part of the Ph.D. Thesis of Guenther Bode. This work supported by a grant from the Deutsche Forschungsgemeinschaft (Sonderforschungsbereich 310, Teilprojekt A9) to T.M.

References

1. Rothman, E. R. and Wieland, F. T. (1996) Protein sorting by transport vesicles. Science, 272, 227-234

2. Mayer, T. (1998) Rab proteins: Molecular switches at multiple steps of vesicular transport. CHEMTRACTS-Bioche. Mol. Biol., 11, 459-479

3. Pfeffer, S. R.(1994) Rab GTPases: Master regulators of membrane trafficking. Curr. Opin. Cell Biol., Vol. 6, 522-526

4. Olkonen, M.V. and Stenmark, H. (1997) Role of rab GTPases in membrane traffic. Int. Rev. Cytol. , Vol. 176, 1-85

5. Novick, P., Zerial, M. (1997) The diversity of rab proteins in vesicle transport. Curr. Opin . Cell Biol., 9, 496-504

6. Elazar, Z., Mayer, T., Rothman, JE (1993) Removal of rab GTP-binding proteins from the Golgi membranes by GDP dissociation inhibitor factor inhibits inter-cisternal transport in the Golgi-stacks. J. Biol. Chem., 269, 794-797

7. Ullrich, O., Horiuchi, H., Bucci, C., Zerial, M. (1994) Membrane association of Rab5 mediated by GDP-dissociation inhibitor and accompanied by GDP/GTP exchange. Nature 368, 157-160

8. Chen, W., Feng, Y., Chen, D., Wandinger-Ness, A. (1999) Rab11 is required for trans-Golgi network-to-plasma membrane transport and a preferential target for GDP Dissociation Inhibitor. Mol. Biol.Cell, 9, 3241-3257

9. Cao, X., Ballew, N., Barlowe, C. (1998) Initial docking of ER-derived vesicles requires Uso1p and Ypt1 but is independent of SNARE proteins. EMBO J. 17, 2156-2165

10. Claque, M.J. (1999) Membrane transport: Take your fusion partners. Curr. Biol. 9, 258-260

11. Mayer, T., Touchot, N., Elazar, Z. (1996) Transport between cis and medial Golgi Cisternae requires the function of the ras-related protein rab6. J. Biol. Chem., 271, 16097-16103

The SRP and SecA/SecB targeting pathways in *E. coli* exhibit distinct specificities in substrate recognition

Konstanze Beck, Hans-Georg Koch, Ute Schäfer and Matthias Müller
Institut für Biochemie und Molekularbiologie, Universität Freiburg, Hermann-Herder-Strasse 7, D-79104 Freiburg, Germany, Tel. ++49-761-203-5265; FAX ++49-761-203-5274

Abstract. SecA and SecB mediate posttranslational targeting of secretory proteins to the SecY translocon at the inner membrane. The SRP (signal recognition particle) of *E. coli* mediates the insertion of inner membrane proteins. By use of *in vitro* assay systems for the SRP and Sec components and by chemical and site-specific cross-linking strategies, SecA/SecB and SRP are now demonstrated to be constituents of two separate targeting pathways in prokaryotic protein export. SRP cotranslationally binds to the polytopic membrane protein MtlA. In contrast, binding of SRP and SecA to ribosome-associated nascent chains of the secretory protein OmpA is prevented by the ribosome-borne peptidyl-prolyl isomerase, trigger factor (TF). After release from the ribosome, SecA and SecB bind selectively OmpA, but not MtlA, prior to translocation across the membrane via the SecY translocon. During its membrane passage OmpA is in contact with SecY, SecA and SecD. Following SRP mediated targeting, MtlA when still attached to the ribosome contacts SecY and phospholipids in the inner membrane. Nascent polypeptides therefore are selectively bound by SRP or TF at an early stage at the ribosome to initiate distinct targeting mechanisms for membrane and secretory proteins.

1. Two targeting pathways in protein transport in *E. coli*: SecA/SecB and SRP

A distinct set of Sec proteins is involved in the export of secretory, i.e. periplasmic and outer membrane proteins across the inner membrane of the Gram-negative bacterium *E. coli*. Posttranslational membrane targeting and chaperoning of signal sequence-containing precursors of secretory proteins is mediated by SecA and SecB [1]. The ATPase SecA is also the motor for the translocation across the inner membrane. In addition *E. coli* contains a SRP (signal recognition particle) consisting of 4.5S RNA and Ffh (P48), structural homologues of the 7S RNA and the 54kDa subunit of the mammalian SRP, and a SRP receptor, FtsY, which is homologous to the mammalian SRα (α-subunit of the SRP receptor; [2-5]). In eukaryotic cells, SRP and its receptor function in targeting precursor and membrane proteins to the membrane of the endoplasmic reticulum (reviewed by Rapoport *et al.* [6]).

An involvement of the bacterial SRP/SR in the biogenesis of inner membrane proteins was first demonstrated *in vivo* for lactose permease [7] and later confirmed for several other membrane proteins [8-10]. This is consistent with recent cross-linking studies revealing a preferential recognition by SRP of the more hydrophobic signal sequences, such as the signal-anchor sequences of membrane proteins [11, 12].

The concept of SRP-dependent integration of *E. coli* inner membrane proteins has largely been derived from studies with conditional mutants of the SRP/SRP receptor components. The results therefore had to remain ambiguous as long as secondary effects due to the SRP/SRP receptor mutations were not ruled out as causatives of the integration defects observed. This obstacle has now been overcome by establishing an *in vitro* assay system for the activities of the bacterial SRP [13]. By expressing polytopic membrane proteins by a cell-free transcription/translation system simultaneously deficient in Ffh (P48), 4.5S RNA, FstY, SecA and SecB we have shown that integration of mannitol permease (MtlA) and SecY into inside-out inner membrane vesicles was strictly dependent on the bacterial SRP/SRP receptor components. Moreover, SecA, SecB, and $\Delta\mu H^+$, all of which are found to be required for the efficient translocation of pOmpA under identical experimental conditions, are completely inactive in supporting membrane integration. Nor were they found to be stimulatory to the SRP-catalyzed integration process. On the other hand, no requirement for SRP components could be demonstrated for the translocation of the secretory protein OmpA, neither when the full-length protein nor when ribosome-associated nascent chains (RANCs) were analyzed. These results indicate the occurrence of distinct targeting routes for secretory and polytopic membrane proteins. Contradictory to the principle of a dual targeting, other membrane proteins were found to be dependent also on SecA for integration *in vivo* [14] or cross-linked to SecA *in vitro* [15].

To prove different specificities in the recognition of secretory and membrane proteins resulting in non-overlapping SecA/SecB and SRP targeting pathways respectively, we have analyzed early events in protein targeting of the polytopic inner membrane protein mannitol permease (MtlA) and a protein translocated across the plasma membrane (OmpA) by different cross-linking approaches.

2. *In vitro* cross-linking to study the substrate specificity of targeting factors

^{35}S-radiolabelled ribosome-associated nascent chains (RANCs) of MtlA and of OmpA were synthesized *in vitro*. According to a recently described method, complementary oligodeoxynucleotides were added to a cell-free transcription/translation system from *E. coli* to create truncated mRNAs allowing the synthesis of RANCs with defined lengths [16]. RANCs were isolated by cosedimentation with ribosomes through a sucrose cushion.

Unspecific cross-linking was performed with the water soluble heterobifunctional chemical cross-linker Sulfo-HSAB (N-hydroxysulfosuccinimidyl 4-azidobenzoate; PIERCE) and the hydrophobic homobifunctional cross-linking reagent DSS (disuccinimidyl suberate; PIERCE).

Another, site-specific cross-linking method was adopted to an *in vitro* system from *E. coli.* The photoactivable amino acid derivative Tmd-Phe (L-4'-[3-(trifluoromethyl)-3H-diazirin-3-yl] phenylalanine) was site-specifically incorporated into RANCs of MtlA by *in vitro* suppression of TAG stop codons engineered at selective positions by use of a suppressor tRNA which had chemically been loaded with Tmd-Phe [17]. UV-irradiation of such Tmd-Phe mutants generates a highly reactive carbene and leads to covalent cross-linking with any adjacent molecule.

Covalently linked molecules gave rise to additional radiolabelled products which were visualized by autoradiography following SDS-PAGE. Binding partners were identified by co-immunoprecipitation with specific antibodies in the case of proteins. Adducts resulting from interactions of membrane bound radiolabelled substrates with phospholipids were identified due to their sensitivity against digestion with phospholipase A2.

3. Discrimination between SRP- and SecA/SecB dependent substrates involves selective recognition of nascent chains by SRP and TF

In a first step RANCs of MtlA and pOmpA were synthesized *in vitro* and analyzed for early cotranslational interactions with targeting factors in a ribosome bound state. Release from the ribosome by puromycin allowed the analysis of posttranslational binding of targeting factors to N-terminal peptides in a cytoplasmic environment prior to interactions of transported proteins with membranes.

3.1 The polytopic membrane protein MtlA is specifically and exclusively recognized by SRP

MtlA is a polytopic membrane protein lacking major translocated periplasmic loops. It consists of a N-terminal membrane domain with 6 transmembrane helices and a C-terminal cytoplasmic domain [18]. Site-specific cross-linking with Tmd-Phe at different positions in the first signal-anchor sequence and chemical cross-linking of RANCs of MtlA containing the N-terminal 2 or 4 transmembrane domains respectively showed that its signal-anchor sequence is recognized by SRP via its protein part Ffh (P48). TF, a peptidyl-prolyl isomerase, associated with prokaryotic ribosomes [19] which had been found cross-linked to various other cytosolic, secretory and membrane proteins in heterologous and homologous *in vitro* systems [11, 15, 20], was not cross-linked to RANCs of MtlA. Even in the absence of functional SRP only an extremely faint cross-linking of TF with RANCs of MtlA was detectable by co-immunoprecipitation.

After release of the nascent chains from the ribosome by puromycin, SRP was no longer found associated with MtlA. In particular, under no conditions could SecA and SecB be identified as cross-linking partners of this polytopic membrane protein. This is consistent with previous results obtained *in vitro* and *in vivo* unanimously classifying polytopic membrane proteins as SecA and SecB independent proteins [13, 21]. Because neither TF nor SecA and SecB were found to bind to MtlA in its ribosome-associated as well as its free form, these polytopic membrane proteins must lack the still unknown recognition sites for TF, SecA, and SecB.

3.2 TF prevents recognition of secretory proteins by SRP; they are posttranslationally recognized by SecA and SecB

As a representative of a classical secretory protein, pOmpA was tested by *in vitro* cross-linking for recognition by targeting factors. The translocation of pOmpA *in vitro* in inside-out inner membrane vesicles requires SecA and SecB but is independent of SRP/SR [13]. Nascent chains of pOmpA are also transported in a SecA/SecB-dependent manner but only after their release from the ribosome [16].

In our cross-linking assay, RANCs of pOmpA were found associated with TF when ribosome-associated. They did not cross-link to Ffh (P48), nor to SecA and SecB. After release from the ribosome they were, however, recognized by SecA and SecB. Only in a TF-deficient background, Ffh (P48) and SecA could be cross-linked to RANCs of pOmpA, a phenomenon that was suppressed by adding TF back.

Thus, SRP can in principle also bind to a secretory protein emerging from the ribosome. Consistently, SRP was shown to bind only to the secretory protein PhoE when added in high concentrations in the presence of trigger factor [11]. These findings point to an important function of TF in keeping SRP away from secretory proteins thereby interfering with a cotranslational targeting of secretory proteins to the translocon. Such a situation could be envisioned to drastically reduce the translocation activity of a cell now being limited by the number of translocons. A similar scenario could be created if SecA and SecB were not prevented by TF from binding cotranslationally to RANCs.

4. The SecY translocon: a protein pore for insertion and translocation

Once targeted to the membrane and kept in a translocation competent form by SecA and SecB, pOmpA like other secretory proteins is translocated across the plasma membrane through the translocon whose essential components appear to be SecY and SecE [1]. The precursor of the M13 major coat protein, termed procoat, inserts spontaneously into the plasma membrane of *E. coli* [12, 22] by means of a so-called insertion domain consisting of two transmembrane segments and a short intervening sequence. In the past, these findings had invoked the question of a spontaneous insertion even of larger membrane proteins if they were considered to be composed of a string of independent insertion domains [23]. Recent data suggest that the unassisted insertion of procoat is rather an exception and that membrane proteins insert in a SRP-dependent manner via the SecYE transport pore [12, 15, 24]. This is convincingly shown by *in vitro* experiments in which SecY is jammed by interrupting the transport of a secretory protein in a translocon-spanning stage. To this end, a chimera consisting of a pOmpA peptide and a C-terminally coupled tRNA was created according to Matlack *et al.* [25]. When membrane vesicles were incubated with this OmpA-tRNA construct, re-isolated, and then added to cell-free synthesis reactions, both translocation of pOmpA as well as integration of MtlA decreased to the same extent clearly indicating that both processes depend on the same translocon [13].

4.1 A secretory protein when spanning the translocon is in immediate proximity to SecA, SecY and SecD

When cross-linking was performed using this transport intermediate of pOmpA, SecY, the core component of the translocon, and SecA which inserts together with portions of the transported secretory proteins into the translocon to drive their membrane passage, were identified as adducts. Immunoprecipitation of cross-linked products also identified SecD as binding partner for translocating OmpA. The precise function of SecD in protein translocation has so far remained elusive while several findings suggest a function of SecD in supporting that of SecA. In addition a yet unidentified protein of about 60kD is in the vicinity of this transport intermediate. Following complete translocation into the periplasmic environment of inside-out inner membrane vesicles, OmpA is cross-linked to Skp, a periplasmic chaperone involved in the folding of outer membrane proteins. Skp interacts with newly translocated non-native outer membrane proteins thus facilitating their release from the plasma membrane into the periplasmic space. Skp is further required for the maintenance of the solubility of periplasmic assembly intermediates of outer membrane proteins [26].

4.2 Interaction of nascent MtlA with SecY and phospholipids reflects cotranslational targeting and insertion of polytopic membrane proteins in E. coli

SRP binding to RANCs of MtlA occurs only in the context of the ribosome and is abolished in the presence of *E. coli* membranes [12, 15, our own results] suggesting a cotranslational mode of membrane insertion. Therefore, to search for interaction partners during membrane insertion of MtlA, no construction of arrested transport intermediates like in the case of OmpA was needed; membrane vesicles could simply be added to *in vitro* synthesized RANCs of MtlA. SRP-mediated cotranslational targeting should directly lead to an association of RANCs of MtlA with membrane structures.

In these conditions, DSS cross-linking resulted in a covalent adduct of RANCs of MtlA with SecY which could only be observed if MtlA was still attached to ribosomes. Several Tmd-Phe mutants of MtlA gave rise to cross-linking products of the first transmembrane domain of MtlA with phospholipids and with a so far unidentified 52kD protein in the plasma membrane.

5. Secretory and membrane proteins are directed by distinct targeting pathways in *E. coli*

In summary, analysis of functional activities and cross-linking approaches demonstrate a clear distinction in substrate specificity for the SRP and the SecA/SecB mediated targeting pathways in *E. coli*. These results are supportive of the previous *in vivo* studies demonstrating a SRP-involvement in the integration of membrane proteins. A cotranslational mode of integration mediated by SRP would represent an efficient means to prevent premature denaturation of hydrophobic membrane proteins whereas the chaperoning activity of SecB and possibly of SecA guarantees maintenance of translocation competence in the posttranslational targeting pathway of secretory proteins.

In contrast to the clear-cut discrimination of SRP and SecA/SecB between polytopic membrane and secretory proteins, other membrane proteins containing large periplasmic domains were found to be dependent on SRP and SecA [11, 14, 15]. A plausible explanation of these findings is that these proteins need SecA for the translocation of their extended periplasmic domains across the lipid bilayer in addition to SRP which they might use for cotranslational targeting. To prove this idea, studies with chimeric constructs composed of exclusively SecA/SecB- and SRP-dependent proteins are required. The initiated approach of site-specific cross-linking is likely to elucidate specific binding sites for SecA, SecB and trigger factor on secretory proteins.

References

[1] F. Duong, J. Eichler, A. Price, M.R. Leonard, and W. Wickner, Biogenesis of the gram-negative bacterial envelope, *Cell* **91***(1997)* pp. 567-573

[2] J. C. R. Struck, H.Y. Toschka, T. Specht, and V.A. Erdmann, Common structural features between eukaryotic 7SL RNAs, eubacterial 4.5S RNA and scRNA and archaebacterial 7S RNA, *Nucleic Acids Res.* **16** (1988) p. 7740

[3] M. A. Poritz, K. Strub, and P. Walter, Human SRP RNA and *E. coli* 4.5S RNA contain a highly homologous structural domain, *Cell* **55** (1988) pp. 4-6

[4] K. Römisch, J. Webb, J. Herz, S. Prehn., R. Frank, M. Vingron, and B. Dobberstein, Homology of the 54K protein of signal recognition particle, docking protein , and two *E. coli* proteins with putative GTP-binding domains, *Nature* **340** (1989) pp. 478-482

[5] H.D. Bernstein, M.A. Poritz, K. Strub, P.J Hoben, S. Brenner, and P. Walter, Model for signal sequence recognition from amino-acid sequence of 54K subunit of signal recognition particle, *Nature* **340** (1989) pp. 482-486

[6] T. A. Rapoport, B. Jungnicker, and U. Kutay, Protein transport across the eukaryotic endoplasmic reticulum and bacterial inner membranes, *Annu. Rev. Biochem.* **65** (1996) pp.271-303

[7] J. MacFarlane and M. Müller, The functional integrationof a polytopic membrane protein in *Eschichia coli* is dependent on the bacterial signal-recognition particle, *Eur. J. Biochem.* **233** (1995) pp. 766-771

[8] J.-W.L. de Gier, P. Mansournia, Q.A. Valent, G.J. Phillips, J. Luirink, and G. von Heijne, Assembly of a cytoplasmic membrane protein in *Escherichia coli* is dependent on the signal recognition particle, *FEBS Lett.* **399** (1996) pp. 307-309

[9] A. Seluanov and E. Bibi, FtsY, the prokaryotic signal recognition partcle receptor homologue, is essential for biogenesis of membrane proteins, *J. Biol. Chem.* **272** (1997) pp. 2053-2055

[10] N.D. Ulbrandt, J.A. Newitt, and H.D. Bernstein, The *E. coli* signal recognition particle is required for the insertion of a subset of inner membrane proteins, *Cell* **88** (1997) pp. 187-196

[11] Q.A. Valent, J.-W.L. de Gier, G. von Heijne, D.A. Kendall, C.M. ten Hagen-Jongman, B. Oudega, and J. Luirink, Nascent membrane and presecretory proteins synthesized in *Escherichia coli* associate with signal recognition particle and trigger factor, *Mol. Microbiol.* **25** (1997) pp. 53-64

[12] J.-W.L. de Gier, P.A. Scotte, A. Sääf, Q.A. Valent, A. Kuhn, and J. Luirink, Differential use of the signal recognition particle translocase targeting pathway for inner membrane protein assembly in *Escherichia coli*, *Proc. Natl. Acad. Sci. USA* **95** (1998) pp. 14646-14651

[13] H.-G. Koch, T. Hengelage, C. Neumann-Haefelin, H.K. Hoffschulte, K.-L. Schimz, and M. Müller, *In vitro* studies with purified components reveal signal recognition particle (SRP) and SecA/SecB as constituents of two independent targeting pathways of *Escherichia coli*, *Mol. Biol. Cell*, in press

[14] H.-Y. Qi and H.D. Bernstein, SecA is required for the insertion of inner membrane proteins targeted by the *Escherichia coli* signal recognition particle, *J. Biol. Chem.* **274** (1999) pp. 8993-8997

[15] Q.A. Valent, P.A. Scotti, S. High, J.-W.L. de Gier, G. von Heijne, G. Lentzen, W. Wintermeyer, B. Oudega, and J. Luirink, The *Escherichia coli* SRP and SecB targeting pathways converge at the translocon, *EMBO J.* **17** (1998) pp. 2504-2512

[16] M. Behrmann, H.-G. Koch, T. Hengelage, Wieseler, B., H.K. Hoffschulte, and M. Müller, Requirements for the translocation of elongation-arrested, ribosome-associated OmpA across the plasma membrane of *Escherichia coli*, *J. Biol. Chem.* **273** (1998) pp. 13898-13904

[17] R. Graf, J. Brunner, B. Dobberstein, and B. Martoglio, Probing the molecular environment of proteins by site-specific photocrosslinking. In: J. E. Celis (ed.), Cell Biology - A Laboratory Handbook, second edition, Academic Press, San Diego, 1997, pp. 495-501

[18] J.E. Sugiyama, S. Mahmoodian, and G.R. Jacobson, Membrane topology analysis of *Escherichia coli* mannitol permease by using a nested-deletion method to create mtlA-phoA fusions, *Proc. Natl. Acad. Sci. USA* **88** (1991) pp. 9603-9607

[19] G. Stoller, K.P. Rücknagel, H.H. Nierhaus, F.X. Schmid, G. Fischer, and J.-U. Rahfeld, A ribosome-associated peptidyl-prolyl *cis/trans* isomerase identified as the trigger factor, *EMBO J.* **14** (1995) pp. 4939-4948

[20] Valent, Q. A., D.A. Kendall, S. High, R. Kusters, B. Oudega, and J. Luirink, Early events in preprotein recognition in *E. coli:* interaction of SRP and trigger factor with nascent polypeptides, *EMBO J.* **14** (1995) pp. 5494-5505

[21] P. Werner, M.H. Saier, and M. Müller, Membrane insertion of the mannitol permease of *Escherichia coli* occurs under conditions of impaired SecA function, *J. Biol. Chem.* **267** (1992) pp. 24523-24532

[22] P.B. Wolfe, M. Rice, and W. Wickner, Effects of two sec genes on protein assembly into the plasma membrane of *Escherichia coli, J. Biol. Chem.* **260** (1985) pp. 1836-1841

[23] M. Müller and J. MacFarlane, Membrane Assembly in Bacteria. In: A. H. Maddy and J. R. Harris (eds.), Subcellular Biochemistry, Vol. 22: Membrane Biogenesis, Plenum Press, New York, 1994, pp. 327-359

[24] J.A. Newitt and H.D. Bernstein, A mutation in *Escherichia coli secY* gene that produces distinct effects on inner membrane protein insertion and protein export, *J. Biol. Chem.* **273** (1998) pp. 12451-12456

[25] K.E.S. Matlack, K. Plath, B. Misselwitz, and T.A. Rapoport, Protein transport by purified yeast Sec complex and Kar2p without membranes, *Science* **277** (1997) pp. 938-941

[26] U. Schäfer, K. Beck, and M. Müller, Skp, a molecular chaperone of gram-negative bacteria, is required for the formation of soluble periplasmic intermediates of outer membrane proteins, submitted for publication

Protein, Lipid and Membrane Traffic: Pathways and Targeting
J.A.F. Op den Kamp (Ed.)
IOS Press, 2000

Sphingosine base requirement for endocytosis in yeast

Sylvie Friant, Bettina Zanolari, Alan Munn[§] and Howard Riezman

Biozentrum of the University of Basel, CH-4056 Basel, Switzerland

[§] *Institute of Molecular Agrobiology, 1 Research Link, The National University of Singapore, Singapore 117604, Republic of Singapore*

Abstract

A synthetic lethal screen used for isolation of endocytosis-deficient (*end*) mutants in yeast resulted in isolation of the *end8-1* mutant. The *end8-1* mutant has a temperature-sensitive growth defect. This mutant is defective in the internalization step of both receptor-mediated and fluid-phase endocytosis at non-permissive temperature. The *end8-1* mutant has been shown to be allelic to *LCB1*. The Lcb1 protein is part of serine palmitoyltransferase, the enzyme required for the first step in *de novo* synthesis of sphingosines, ceramides and sphingolipids in yeast. C2-ceramide and phytosphingosine suppress the endocytic defect in *end8-1=lcb1-100* cells showing that the defect is due to the lack of synthesis of these compounds. In summary, sphingosine base is required for the internalization step of both receptor-mediated and fluid-phase endocytosis in the yeast *Saccharomyces cerevisiae*.

Address all correspondence to Dr. Sylvie Friant, Biozentrum, University of Basel, Klingelbergstrasse 70, CH 4056 Basel, Switzerland. Tel.: 41 61 267 2160. Fax.: 41 61 267 2149
Email.: friant@ubaclu.unibas.ch

Introduction

Endocytosis is a process performed by all eukaryotic cells in which various molecules are internalized from the extracellular space to the cell interior. The internalization step of endocytosis requires clathrin, actin, proteins that are required for proper actin cytoskeleton organization, such as fimbrin, calmodulin, type I myosin, amphiphysin, and a large set of proteins associated to these major components [1, 2]. Recently, not only proteins but also lipids have been implicated in signal transduction and in several stages of membrane trafficking, but the role of lipids in vesicle budding and fusion in living cells is poorly understood. It is known that sphingolipid synthesis is necessary for trafficking of glycosylphosphatidylinositol (GPI)-anchored proteins from the endoplasmic reticulum to the Golgi apparatus [3, 4]. Phospholipids, in particular phosphorylated derivatives of phosphatidylinositol (PtdIns) also appear to play a critical role in regulating transport events [5]. A requirement for phosphoinositides in Golgi to vacuole trafficking in yeast was established by the characterization of the Vps34p kinase responsible for PtdIns(3)P synthesis [6]. More recently phospholipids were also implicated in cargo-selective sorting into the vacuole lumen [7]. We report here that *de novo* sphingosine base synthesis is required for the internalization step of both receptor-mediated and fluid-phase endocytosis in the yeast, *Saccharomyces cerevisiae*.

Results

Isolation of endocytosis-deficient mutant *end8-1*

Yeast mutants defective in both acidification of the lysosome-like vacuole and endocytosis are inviable, based on this result new endocytosis-deficient mutants (*end* mutants) were isolated. The *VMA2* gene encodes the 60-kD vacuolar H^+-ATPase (V-ATPase) regulatory subunit (subunit B). In cells bearing a disruption of *VMA2* (*vma2Δ*), the V-ATPase does not assemble on the vacuolar membrane and consequently the lumen of the vacuole fails to become acidified [8]. Yeast *vma2Δ* mutants are able to grow if the external medium is around pH 5.5, but are unable to grow if the medium is at pH 7 or above. The *vma2Δ* strain is viable when the cells are grown on acidic plates, because the cells are able to take up the protons that are in the medium by fluid-phase endocytosis. Based on this result, a screen for isolation of endocytosis-deficient (*end*) mutants

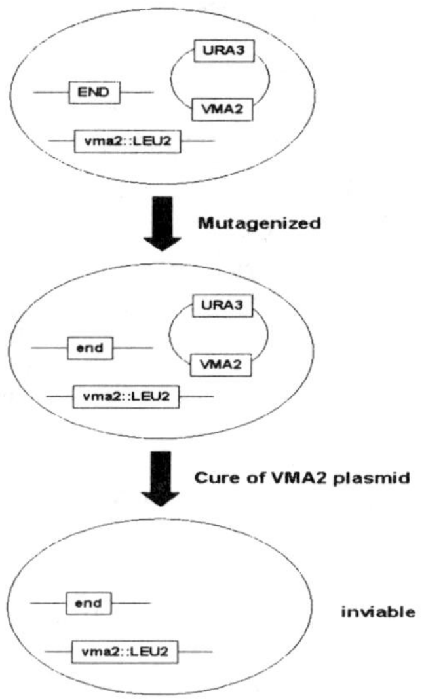

Figure 1: The *vma2*-synthetic lethality mutant screen. In the starting yeast strain, carrying the wild-type *VMA2* and *URA3* genes on a plasmid, the *VMA2* chromosomal gene was disrupted by *LEU2*. After mutagenesis some of the cells will gain a new mutation (*end*) which prevents growth without a wild-type copy of *VMA2*. 5FOA is used to select between the *end* cells and the others, since only the Ura^- cells can grow on the selective medium.

was developed. This screen was based on the isolation of mutants that are synthetically lethal with *vma2Δ*. The principle of this screen is presented in figure 1. This screen resulted in isolation of the *end8-1* mutant among others [9].

The yeast *end8-1* mutant is defective in the internalization step of endocytosis

In the yeast, *Saccharomyces cerevisiae*, the internalization step of receptor-mediated endocytosis can be followed by using the α-factor pheromone that is bound and internalized by its G-protein coupled receptor, Ste2p [10]. Internalized α-factor is transported through two intermediate compartments to the vacuole where it is degraded by resident vacuolar proteases. [^{35}S] α-factor uptake assays are performed on mid-log phase cells using the continuous presence protocol [11]. The assays were carried out at 24°C and 37°C with a 15 min preshift before adding [^{35}S]α-factor. Internalization was calculated by dividing the internalized counts (pH 1-resistant) by the total cell-associated counts (pH 6-resistant) for each time point (Fig. 2). The yeast *end8-1* mutant has a temperature-sensitive growth defect. This mutant was assayed for α-factor internalization at 24°C and 37°C. At 24°C, it exhibits α-factor uptake kinetics that are slightly slower than wild-type

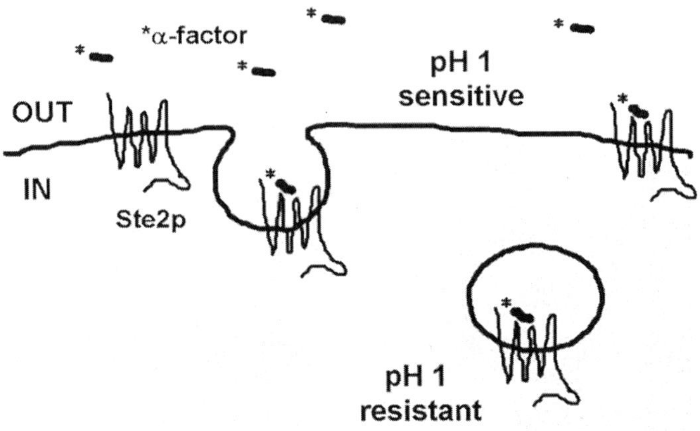

Figure 2: α-factor internalization assay. ^{35}S-α-factor binds to his receptor Ste2p and is internalized. The internalization rate is determined by assessing the fraction of radiolabeled α-factor that cannot be removed from the cells by an acid-wash protocol.

cells, but at 37°C this mutant is strongly defective for α-factor internalization (Fig. 3). The ability of *end8-1* cells to carry out fluid-phase endocytosis at 24°C and 37°C was also tested. The fluid-phase endocytosis can be visualized by accumulation of a fluorescent dye, the lucifer yellow-carbohydrazide (LY), in the vacuole upon endocytosis. This mutant shows a few small vacuoles when viewed by Nomarski optics and is defective for accumulation of LY in the vacuole at 24°C and 37°C (data not shown). In summary, the *end8-1* mutant is defective in the internalization step of both receptor-mediated and fluid-phase endocytosis at non-permissive temperature.

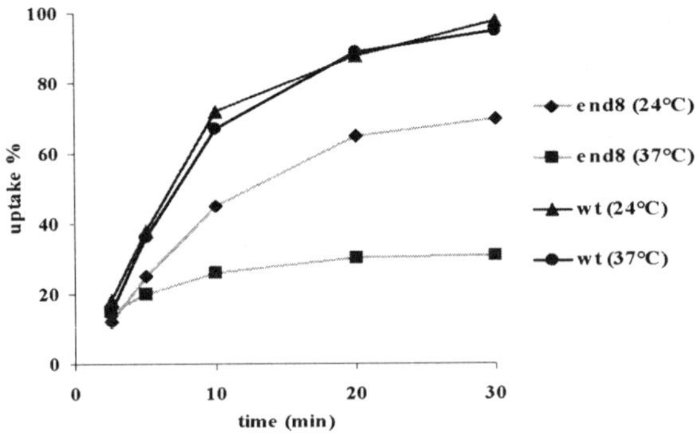

Figure 3: *end8* mutant is defective in α-factor internalization. Continuous presence α-factor uptake assays were performed at 24°C and 37°C on the wild-type strain (*END*) and the mutant *end8*. *end8* cells are defective in internalization even at early time points at 37°C, but uptake is less defective at 24°C.

De novo sphingosine base synthesis is required for endocytosis

The cloned *END8* gene has been shown to be allelic to *LCB1*. The Lcb1 protein is part of serine palmitoyltransferase, the enzyme required for the first step of *de novo* sphingosine, ceramide, and sphingolipid synthesis (Fig. 4) [12]. These findings suggest that sphingosine bases could play an important role in the internalization step of endocytosis. To test this hypothesis, water-soluble analogs of sphingolipids were added to the *end8-1=lcb1-100* strain and the rate of receptor-mediated endocytosis was determined. Cells were grown at 24°C to mid-log phase and radiolabeled α-factor was added to the cells after a 15 min incubation either at 24°C or 37°C. N-acetyl-ceramide (C2-ceramide) and phytosphingosine (PHS) suppress the endocytic defect observed in *lcb1-100* cells at 37°C, but N-hexyl-ceramide (C6-ceramide) does not, even though both were incorporated into cells (data not shown). An explanation for the differences obtained with C2-ceramide and C6-ceramide could be that C2-ceramide suppresses the *lcb1-100* defect in endocytosis at 37°C, because it mimics the active molecule better than C6-ceramide. C2-ceramide carries only a short fatty acid chain and therefore is more similar to sphingosine base than to ceramide that carries a C-26 long chain fatty acid (Fig. 4). In yeast, the sphingosine bases are 3-ketosphinganine (KDS), dihydrosphingosine (DHS) and phytosphingosine (PHS) [13]. The effect of sphingosine base suppression is very fast, within 5 min incubation of the *lcb1-100* cells in presence of PHS the endocytosis defect at 37°C is completely restored. Given all these results, we conclude that *de novo* sphingosine base synthesis is required for the internalization step of endocytosis in yeast.

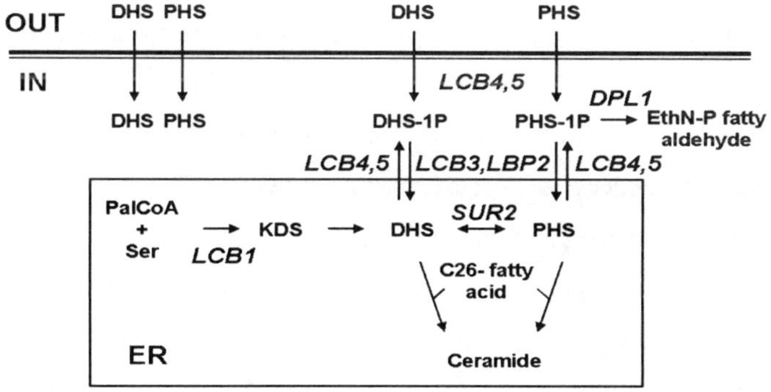

Figure 4: The sphingosine bases biosynthetic pathway in *Saccharomyces cerevisiae*. Known pathway intermediates, substrates and genes are indicated.

Discussion

In this paper we present evidence that sphingosine base is required for the internalization step of both receptor-mediated and fluid-phase endocytosis in yeast. This evidence comes from isolation of the *lcb1-100* temperature-sensitive mutant that is defective in endocytosis. The *LCB1* gene is an essential gene that encodes subunit of the serine palmitoyl-transferase enzyme (SPT). SPT catalyses the first step of *de novo* sphingolipid synthesis, the condensation of serine and palmitoyl-CoA to yield the sphingosine base 3-ketodihydrosphingosine (KDS) (Fig. 4). In yeast, sphingosine bases are generated by *de novo* synthesis in the endoplasmic reticulum.

Here we show that addition of sphingosine bases (PHS) or C2-ceramide suppress the internalization defect of *lcb1-100* mutant, whereas addition of C6-ceramide does not. C2-ceramide is unlikely to be a good analog of natural ceramide in yeast because it is structurally distant and less hydrophobic than ceramide. C2-ceramide is more similar to sphingosine bases whereas C6-ceramide can be compared to natural yeast ceramide. Natural ceramides cannot be used because they are highly insoluble in water. In summary, the *lcb1-100* defect in endocytosis can be suppressed by addition of sphingosine or sphingosine-related bases. The use of genetic approaches should help to clarify the sphingosine base compound (KDS, DHS or PHS) that is required for endocytosis in yeast. Sphingosine bases and their phosphorylated derivatives (DHS-1P and PHS-1P) are thought to be signaling molecules for regulating a variety of mammalian cellular processes including cell growth, motility and death [14-16]. The recent identification of two *Saccharomyces cerevisiae* genes encoding sphingosine kinases, *LCB4* and *LCB5*, should allow us to determine if phosphorylated sphingosine bases are required to trigger endocytosis in yeast [17].

Sphingosine bases when added to the culture medium suppress the *lcb1-100* mutant endocytic defect and this effect on endocytosis occurs extremely rapidly. This could be consistent with sphingosine bases activating a signal transduction pathway that is needed for the internalization step of endocytosis. In yeast, sphingolipids have been implicated in stress response [18, 19]. Strains mutated in either *LCB1* or *LCB2* genes are unable to grow in stress conditions such as heat stress or low pH [19, 20]. In the heat stress response, the addition of sphinganine or other sphingosine bases to the culture medium mimics the effect of heat and induces trehalose accumulation by transcriptional activation of the *TPS2* gene [21]. The *TPS2* gene is one of the heat inducible trehalose biosynthetic genes. The ability of these sphingosine bases to mimic the effect of a 40°C heat shock is consistent with the hypothesis of activation of a signal transduction pathway leading to induction of the *TPS2* transcription.

Finally this paper shows the first evidence for sphingosine base requirement for endocytosis in *Saccharomyces cerevisiae*, and the use of genetic approaches should help to clarify this pathway in yeast.

References

1. Riezman, H., Munn, A., Geli, M. I. & Hicke, L. (1996) Actin-, myosin- and ubiquitin-dependent endocytosis, *Experientia. 52*, 1033-41.
2. Geli, M. I. & Riezman, H. (1998) Endocytic internalization in yeast and animal cells: similar and different, *J Cell Sci. 111*, 1031-7.
3. Horvath, A., Sutterlin, C., Manning-Krieg, U., Movva, N. R. & Riezman, H. (1994) Ceramide synthesis enhances transport of GPI-anchored proteins to the Golgi apparatus in yeast, *Embo J. 13*, 3687-95.
4. Sutterlin, C., Doering, T. L., Schimmoller, F., Schroder, S. & Riezman, H. (1997) Specific requirements for the ER to Golgi transport of GPI-anchored proteins in yeast, *J Cell Sci. 110*, 2703-14.
5. De Camilli, P., Emr, S. D., McPherson, P. S. & Novick, P. (1996) Phosphoinositides as regulators in membrane traffic, *Science. 271*, 1533-9.
6. Schu, P. V., Takegawa, K., Fry, M. J., Stack, J. H., Waterfield, M. D. & Emr, S. D. (1993) Phosphatidylinositol 3-kinase encoded by yeast VPS34 gene essential for protein sorting, *Science. 260*, 88-91.
7. Wurmser, A. E. & Emr, S. D. (1998) Phosphoinositide signaling and turnover: PtdIns(3)P, a regulator of membrane traffic, is transported to the vacuole and degraded by a process that requires lumenal vacuolar hydrolase activities, *Embo J. 17*, 4930-42.
8. Yamashiro, C. T., Kane, P. M., Wolczyk, D. F., Preston, R. A. & Stevens, T. H. (1990) Role of vacuolar acidification in protein sorting and zymogen activation: a genetic analysis of the yeast vacuolar proton- translocating ATPase, *Mol Cell Biol. 10*, 3737-49.
9. Munn, A. L. & Riezman, H. (1994) Endocytosis is required for the growth of vacuolar H(+)-ATPase- defective yeast: identification of six new END genes, *J Cell Biol. 127*, 373-86.

10. Zanolari, B. & Riezman, H. (1991) Quantitation of alpha-factor internalization and response during the Saccharomyces cerevisiae cell cycle, *Mol Cell Biol. 11*, 5251-8.

11. Dulic, V., Egerton, M., Elguindi, I., Raths, S., Singer, B. & Riezman, H. (1991) Yeast endocytosis assays, *Methods Enzymol. 194*, 697-710.

12. Nagiec, M. M., Baltisberger, J. A., Wells, G. B., Lester, R. L. & Dickson, R. C. (1994) The LCB2 gene of Saccharomyces and the related LCB1 gene encode subunits of serine palmitoyltransferase, the initial enzyme in sphingolipid synthesis, *Proc Natl Acad Sci U S A. 91*, 7899-902.

13. Dickson, R. C. & Lester, R. L. (1999) Yeast sphingolipids, *Biochim Biophys Acta. 1426*, 347-57.

14. Spiegel, S. (1999) Sphingosine 1-phosphate: a prototype of a new class of second messengers, *J Leukoc Biol. 65*, 341-4.

15. Hannun, Y. A. (1997) Sphingolipid second messengers: tumor suppressor lipids, *Adv Exp Med Biol*, 305-12.

16. Obeid, L. M. & Hannun, Y. A. (1995) Ceramide: a stress signal and mediator of growth suppression and apoptosis, *J Cell Biochem. 58*, 191-8.

17. Nagiec, M. M., Skrzypek, M., Nagiec, E. E., Lester, R. L. & Dickson, R. C. (1998) The LCB4 (YOR171c) and LCB5 (YLR260w) genes of Saccharomyces encode sphingoid long chain base kinases, *J Biol Chem. 273*, 19437-42.

18. Skrzypek, M. S., Nagiec, M. M., Lester, R. L. & Dickson, R. C. (1999) Analysis of phosphorylated sphingolipid long-chain bases reveals potential roles in heat stress and growth control in Saccharomyces, *J Bacteriol. 181*, 1134-40.

19. Jenkins, G. M., Richards, A., Wahl, T., Mao, C., Obeid, L. & Hannun, Y. (1997) Involvement of yeast sphingolipids in the heat stress response of Saccharomyces cerevisiae, *J Biol Chem. 272*, 32566-72.

20. Pinto, W. J., Srinivasan, B., Shepherd, S., Schmidt, A., Dickson, R. C. & Lester, R. L. (1992) Sphingolipid long-chain-base auxotrophs of Saccharomyces cerevisiae: genetics, physiology, and a method for their selection, *J Bacteriol. 174*, 2565-74.

21. Dickson, R. C., Nagiec, E. E., Skrzypek, M., Tillman, P., Wells, G. B. & Lester, R. L. (1997) Sphingolipids are potential heat stress signals in Saccharomyces, *J Biol Chem. 272*, 30196-200.

Endocytosis meets Exocytosis

Vivienne Laird
Dept. Biochemistry, Biozentrum, Klingelbergstrasse 70, 4056 Basel, Switzerland

Abstract

Trafficking of proteins through the cell has been the topic of immense study since the initial characterisation of the secretory pathway [1]. Endosomal compartments have long been known to be complicated sorting organelles, differentiating between recycling proteins from the cell surface or newly synthesised proteins from the TGN *en route* to the lysosome. Recently, a new pathway from the TGN to the cell surface via endosomal compartments has been suggested [2, 3]. To further investigate this pathway, I have designed an assay which showed that newly synthesised protein from the TGN formed immunocomplexes with antibody which had been internalised by receptor-mediated endocytosis. Coimmunofluorescence confirmed that internalised antibody filled organelles of the endocytic pathway. These results confirm the presence of this pathway, and this assay gives us a biochemical tool to further study the trafficking of proteins to the cell surface via endosomes.

Endosomal Targeting Signals

The most common targeting motif within the endocytic pathway consists of a short, linear stretch of amino acids which centres on a tyrosine residue. This motif is found within the cytoplasmic domain of the protein and commonly has the sequence NPXY or YXXØ, where X is any amino acid and Ø is a bulky hydrophobic residue (extensively reviewed by [4] and [5]). Tyrosine sorting signals have been shown, by several different methods, to interact with adaptor proteins, of which there are at least four types [5-8]. Within the cell tyrosine motifs play several different sorting roles. Tyrosine-containing motifs mediate endocytosis from the plasma membrane, but can also be responsible for targeting from the TGN to the lysosome without first targeting to the cell surface as well as targeting from the TGN to the basolateral membrane in polarised cells (for details see the reviews by [4] and [9]). Hence, tyrosine signals in the tails of TGN38, lamp-1 and CD68 have been shown to interact specifically with the medium (μ) chain of both adaptor complexes AP-1 (predominantly found at the TGN) and AP-2 (found at the cell surface) [6-8, 10]. In contrast, the tyrosine motif in the transferrin receptor tail only interacted with AP-2 *in vitro* [10]. The reason for the ability of signals to be recognised by several adaptor complexes is unclear. One explanation may be that certain proteins use several different targeting pathways. For example there is conflicting evidence for the routing of lamp-1 from the TGN. Evidence exists for two routes to the lysosome, with [11] or without [12] trafficking via the cell surface. The majority of molecules may be targeted directly to the lysosome via AP-1, but the ability of the tyrosine signals to be recognised also by AP-2 may allow 'escaped' lamp-1 proteins at the cell surface to be retrieved to early endosomes where they can be resorted for targeting to the lysosome.

Added complications arise with the involvement of tyrosine-based signals in basolateral targeting. One well studied protein is the LDL receptor which contains two tyrosine based motifs in its cytoplasmic tail. In this case the distal signal mediates basolateral targeting and recycling of the receptor, both signals requiring a critical tyrosine residue at position 35, whilst the proximal signal is required for apical to basolateral transcytosis [13, 14]. Other examples of critical tyrosine residues mediating both endocytic and basolateral targeting events are for the asialoglycoprotein receptor [15], lysosomal acid phosphatase [16] and influenza virus haemagglutinin [17].

Another type of signal which can mediate different sorting events is the di-leucine motif. Di-leucine signals, like tyrosine-based motifs, have also been shown to interact with adaptor proteins [5] and can mediate both endocytosis and basolateral sorting events e.g. the IgG Fc receptor [14, 18]. It has also been shown to be important for the targeting and internalisation of the invariant chain [19] and to mediate direct targeting to lysosomes of CD3 chains [20] and Limp II [21].

It has been shown that other residues around the 'core' signal may also contribute, or be essential, for specificity e.g. a glycine residue preceding the tyrosine enhances lamp-1 targeting to lysosomes [12, 22] or more generally the presence of acidic residues are important for internalisation, endosomal targeting or basolateral targeting [14, 19]. Recently, a novel diacidic motif for lysosomal targeting was identified in the HIV-1 Nef protein and, to complicate matters further, this motif is suggested to interact with COP-I coat proteins [23]. Another vital determinant, shown for TGN38 and lamp-1, is the spacing of the tyrosine-containing sorting motif from the transmembrane domain [24].

Sorting Machinery at the TGN

Proteins synthesised on membrane-bound ribosomes of the endoplasmic reticulum are transported through organelles of the secretory pathway, and upon reaching the TGN, are then sorted for delivery to various destinations within the cell. These destinations include targeting to the plasma membrane (apical or basolateral membranes in polarised cells), endosomes, lysosomes or incorporation into secretory granules for release at the cell surface. I will not discuss secretory granule formation further, but concentrate on the sorting of proteins to the endosomes and lysosomes, and then touch on sorting in polarised cells.

The best characterised proteins for studying TGN to endosome transport are the mannose 6-phosphate receptors (MPRs) which are responsible for transporting soluble lysosomal proteins which carry the mannose 6-phosphate tag. The receptors themselves do not go to lysosomes but release their cargo in late endosomes and recycle back to the TGN [25, 26]. There are two MPRs, a cation-dependent, 46 kDa receptor and a cation-independent 300 kDa protein, present in the same exit vesicles from the TGN [27]. These receptors are the most abundant proteins to be transported out of the TGN and hence are the best studied. It is clear that transport out of the TGN is mediated by clathrin but there are, however, conflicting data as to whether the receptor tail itself is responsible for coat recruitment or not. One argument is that the

MPRs are concentrated in clathrin coated vesicles with the aid of ARF-1 [28] and that AP-1 recruitment to the membrane depends upon the MPR concentration in the cell [29]. It was proposed, therefore, that ARF activates the MPRs at the TGN thereby generating a high affinity membrane binding site for AP-1 [30]. However, further results when using the same MPR-deficient fibroblasts were conflicting with those found previously and suggested that AP-1 could in fact be recruited to the membrane with high affinity, irrespective of the presence of the MPRs [31]. Trafficking of the MPRs from endosomes back to the TGN has been shown to require a specific cytosolic factor, TIP47 [32] and the normal fusion machinery [33].

The MPRs transport soluble lysosomal hydrolases, whereas sorting of lysosomal membrane proteins is independent of the MPRs. The most studied protein family are the lysosomal membrane proteins or lamps, but the nature of the vesicles which carry these proteins is not yet established. A detailed interaction of the tail of lamp-1 with AP-1 and AP-2 was demonstrated, and colocalisation of lamp-1 with AP-1 shown by electron microscopy [7]. It was concluded that the MPRs and lamp-1 are transported in the same clathrin coated vesicles. However, using an *in vitro* budding assay, vesicles containing lamp-1 or the MPRs were shown to run at different positions on Nycodenz gradients, suggesting that the lamps are sorted away from the MPRs at the level of the TGN and are not in fact in the same vesicles [34].

Basolateral sorting signals have been clearly identified (as described earlier), however, the precise nature of the vesicles involved in polarised traffic and the nature of the sorting machinery remains unclear. It is known, however, that there are differential requirements for basolateral and apical sorting. Basolateral sorting requires the Rab-NSF-SNAP-SNARE machinery, whereas apical sorting does not [35]. Apical targeting of proteins has been proposed to be a lipid mediated sorting event in the TGN. Lipid 'rafts' insoluble to extraction with Triton X-100 have been proposed to provide a mechanism of sorting apical proteins within the TGN [36]. This hypothesis is still the subject of much discussion, and although evidence is mounting [37] the debate continues.

Clathrin is one of the major coat proteins found on the TGN, but other coat components have been observed. Coatomer (or COP-I) is involved in recycling cargo back to the earlier organelles of the secretory pathway [38], but it has also been implicated in the exocytosis of basolaterally targeted vesicles containing VSV-G protein [39]. A lace-like coat, distinct from clathrin or coatomer, was observed by EM [40] and associated protein components may include p200 [41], p62 or rab 6 [42].

Endosomes and Endocytosis

The terminology for endocytic organelles has been based upon their relative kinetics of labeling by endocytic tracers, or morphology. Hence, some organelles have several different names. Here I will use the terms early endosome, multivesicular body, sorting endosome, pericentriolar recycling endosome, late endosome and lysosome. Alternatively, endosomal compartments could be defined by using resident, non cargo proteins as markers. Marker proteins for the endocytic compartments, and many other organelles within the cell, are the Rab proteins, small GTPases which play complicated roles in membrane dynamics [43]. Rab 4 [44, 45], and rab 5 [46] are associated with the early endosome; rab 11 plays a regulatory role through the pericentriolar recycling endosome [47]; rab 7 localises to the late endosome [46]; and both rab 17 [48] and rab 25 [49] associate with the subapical compartment in polarised cells.

Early endosomes are the first structures to be labeled by tracers in incoming vesicles (normally clathrin coated vesicles) and represent a dynamic network of tubules and vesicles dispersed throughout the cytoplasm [50]. In some instances, the endocytic pathway may appear as a large, connected tubular reticulum [51]. The early endosome is filled 2-3 minutes after internalisation [52] and is the site of ligand dissociation from recycling receptors due to the internal pH shift to 6.3-6.8. Tubular extensions extend from the more vesicular body of the early endosome which are the site of concentration of the recycling receptors. These projections are often referred to as the sorting endosome and are positive for both recycling receptors and proteins destined for the lysosome. However, the recycling endosome, which is formed from the tubules of the sorting endosome and migrates towards the microtubule organising centre, is only positive for recycling receptors [53, 54]. Released ligands such as LDL [55] and some receptors destined for the lysosome e.g. the EGF receptor [54] remain in the larger, spherical part of the endosome, which will ultimately form the multivesicular body that is proposed to traffic to late endosomes.

There are currently two models for transfer of endocytosed material from the early to the late endosome. There is the vesicular model, in which it is proposed that the multivesicular bodies budded from early endosomes are the transport intermediates between early and late endosomes. These bodies contain proteins of both early and late endosomes and are non fusogenic with early endosomes but are capable of homotypic and late endosome fusion [56, 57]. Multivesicular bodies are then transported to the late endosome in a dynein and microtubule-dependent manner [56]. The precise nature of the transport coat is unclear, however, clathrin has been visualised on endosomes and multivesicular bodies [58, 59] but there is also extensive evidence for COP-I association with endosomes [23, 60-62].

The second model proposes gradual maturation from early to late endosomes [63, 64]. The data supporting a maturation model are not as numerous as for the vesicular model, and it has been argued that the maturation stages studied may be the maturation of the early endosomes to multivesicular bodies which then fuse with late endosomes [65].

Sorting Machinery at the Endosome

Endosomes play an important role in sorting. Previously it was unclear how the sorting machineries at the TGN and endosome related to each other. At first, the sorting machineries were believed to be different, based upon a study of the polymeric IgG receptor [66]. Serine phosphorylation was required for transcytosis of the receptor from the basolateral to the apical surface. Mutation of the serine to an alanine resulted in a recycling receptor. This suggested that polarised sorting required alternative machinery at the endosome. However, studies of the tyrosine signals in the LDL receptor showed that a similar signal could result in either basolateral targeting or internalisation [13]. This suggested that the sorting machineries were similar. In fact, now it has also been shown that the same amino acids of the polymeric IgG receptor mediate polarised sorting in the exo- and endocytic pathways [67] and that serine phosphorylation is also required for AP-1 recruitment at the TGN [68]. Furthermore, clathrin lattices have been found on the tubular extensions of perinuclear recycling endosomes (positive for both the transferrin receptor and the polymeric immunoglobulin receptor) and have been suggested to contribute to the polarised sorting mechanism [69].

Endosomes therefore have sorting machinery which is similar to the machinery in the TGN. Furthermore, many coats such as clathrin and coatomer are present on both Golgi and endosomal membranes [38]. The endosomal sorting machinery can, in addition, distinguish between cargo which enters from the TGN to be targeted to lysosomes, from endocytosed cargo which can either also be targeted to lysosomes, or be recycled. This suggests that endosomal compartments may be a potential secondary sorting station for exocytic proteins leaving the TGN.

Endosomes and Exocytosis

Proteins leaving the TGN can potentially take several routes, trafficking directly to the cell surface or to internal organelles such as endosomes and lysosomes. Several assays have been developed to study the trafficking of newly synthesised proteins through the endocytic pathway. One type of assay is a 'capture' assay. Proteins, such as antibodies, are internalised via endocytosis thereby loading the endocytic pathway. This internal pool of antibody is therefore present to 'catch' newly synthesised proteins *en route* to their final destination. One of the earliest reports was from Ludwig *et al.* [70] where they were studying the route of newly synthesised lysosomal proteins from the TGN to the lysosome. In this study they performed pulse-chase experiments, at the end of which they internalised anti-cathepsin D antibodies by fluid phase endocytosis for 5 minutes (only filling the early endosomes with antibody). They found that antibody-cathepsin D complexes could form within early endosomes, and this lead them to the conclusion that some lysosomal proteins are transported to the early endosome, before ultimately being targeted to the lysosome [70].

Diesner *et al.* [71] adopted a slightly different approach to investigate the trafficking of arylsulfatase A, again a soluble lysosomal enzyme. In this instance antibodies against arylsulfatase A were conjugated to transferrin and internalised via receptor mediated endocytosis into the early and recycling endosomes. They were also able to detect immune complexes between internalised antibody and labeled arylsulfatase A [71].

Recently, work from several laboratories has demonstrated a new indirect route from the TGN to the cell surface. Certain proteins destined for the cell surface are also believed to pass through an endosomal compartment before arriving at the plasma membrane. In one case the transferrin receptor and horseradish peroxidase were used as model proteins [2]. In this example a different approach was adopted. Gold-conjugated antibodies against the transferrin receptor were internalised, and then loaded endosomes were isolated by gradient centrifugation. Electron microscopy was also performed on treated cells. They showed that less than 5% of the horseradish peroxidase activity was found in the gold-labeled isolated endosomes, and that there were no double-labeled elements upon thin section EM. On the other hand, labeled transferrin receptor released from the 20°C block in the TGN could be coisolated with the gold-labeled, isolated endosomes. A time course study confirmed that the transferrin receptor was present in the endosomes before reaching the cell surface [2].

A crossroads between the exocytic and endocytic pathways of a protein destined for the cell surface has also been demonstrated using the Semliki Forest virus precursor protein p62 [72]. In this case the cleavage of the p62 precursor protein by furin was studied. It was shown that cleavage only occurred upon release of the protein from the TGN, but also that this cleavage could be inhibited by endocytosed antibodies against p62 or furin. Furin is known to be localised mainly in the TGN, but does cycle between the TGN, endosomes and the cell surface. It was therefore proposed that cleavage of p62 occurs in a post TGN compartment, but before the protein reaches the cell surface and that it may be an early endocytic compartment.

Further evidence for an exocytic route to the cell surface via endosomes was demonstrated by immunoisolation techniques [3]. In this case the model proteins used were a basolateral recycling receptor, namely the asialoglycoprotein receptor H1 subunit and the secretory protein α_1-proteinase inhibitor. In this case the proteins were tagged with a tyrosine sulphation site, a stretch of 9 amino acids from rat cholecystokinin precursor protein which allows the protein to be specifically radiolabeled in the TGN of MDCK cells [73]. Using MDCK cell lines stably expressing the tagged proteins, pulse-chase experiments were performed and MPR-positive compartments immunoisolated. It could be shown that the newly synthesised H1 protein coimmunoisolated with the MPR-positive compartments after only 5 minutes chase from the TGN. This time was insufficient for the H1 protein to reach the cell surface and therefore the result was not due to an endocytic event. As expected, the constitutively secreted α_1-proteinase inhibitor did not coisolate with these compartments.

Trafficking of newly synthesised proteins via endosomes seems to be a more general phenomenon, and therefore, to continue to investigate this pathway further, we have designed an assay which combines several of the features from each of the assays previously described. The model protein studied is the H1 subunit of the asialoglycoprotein receptor with the tyrosine sulphation tag (H1TS) which is stably expressed in MDCK cells. Use of this protein allows direct labeling in the TGN with [^{35}S]sulphate. The assay used is along the principle of an antibody capture assay, as used by Ludwig *et al.* [70], but the difference is that the polyclonal antibody against H1TS is internalised via the endocytic recycling of the receptor itself (Laird and Spiess, manuscript in preparation)

The Antibody Capture Assay

Antibodies against H1TS are bound on the surface of the MDCK cells at 4°C. They are then internalised at 37°C for varying times. This pre-internalisation step allows the antibody to preload the endosomal compartments. H1TS is a recycling complex, and therefore the bound antibody would be expected to fill the early, sorting and recycling endosomes. After the various internalisation incubations, the cells are pulse labeled for 7 minutes. This time period is long enough to give a good labeling efficiency, but is short enough to ensure that the majority of labeled protein molecules are still within the TGN [3]. After the pulse, the proteins are then chased out of the TGN for 5 or 20 minutes. This allows the proteins to be chased either into the MPR-positive compartments, or to the cell surface. If required, the surface of the cells is shaved with proteinase K to ensure any antibody/H1TS complexes formed are the result of internal contact only. The cells are then lysed in the presence of cold H1, to ensure no post-lysis mixing, and any immune complexes formed can easily be isolated directly with protein A sepharose.

Using this assay, we can clearly show capture of newly synthesised H1TS by antibodies within preloaded endosomes. Background capture levels can be in the order of 10%, which includes capture due to inefficient proteinase K stripping of the cell surface and complexes formed during lysis. If the antibody is pre-internalised for up to 23 minutes, to allow even the furthest perinuclear recycling endosome to be filled, capture may increase to 15%. This increase most probably represents capture of the small number of molecules which escape out of the TGN during the 7 minute pulse label. Trafficking of recycling receptors back to the TGN has been shown to be very low [74], however, the slight increase in capture may also reflect a small amount of capture in the TGN. When proteins are chased out of the TGN for 5 minutes, capture increases significantly up to 26%. This represents the capture of proteins which are in the MPR-positive compartments. If the proteins are allowed to travel to the cell surface, and not be shaved by proteinase K, a maximal 40-50% capture is achieved. Hence, approximately 68% of the trafficking H1TS molecules can be captured when they are chased out of the TGN for 5 minutes (Laird and Spiess, manuscript in preparation).

In order to confirm the cellular location of the antibody, immunofluorescence studies were performed. Antibody was bound at the surface of H1TS expressing cells and then internalised at 37°C for 30 minutes (equivalent to the total time taken for all internalisation, labeling and chase steps). Clear punctate peripheral and perinuclear structures could be seen, characteristic of the early endocytic and recycling compartments. Extensive cross-linking of H1 with primary and secondary antibodies has been shown to lead to mislocalisation to the lysosome [75], however, colabeling with lamp-1 confirmed that only a very minor fraction of the internalised antibody was present within lysosomes.

The Future

I have reviewed existing data, and described a new assay which has yielded supportive data for this new pathway to the cell surface via endosomes. It is now a fast moving and exciting field to work in, since many questions remain to be answered: How many other proteins also traffic through endosomes *en route* to the cell surface? What are the signals which govern inclusion or exclusion into this pathway? What is the nature of the vesicles e.g. is it clathrin mediated? Where exactly do the endo- and exocytic pathways meet? The list can just continue. One thing is clear, the complexity of the endosomal system has been underestimated, both in the physical nature of the organelles, the dynamics between compartments and the sorting roles played by endosomes. I am sure we will be hearing more about these organelles for many years to come.

Acknowledgments

Thanks go to M. Spiess for critical review of the manuscript. This work was supported by The Wellcome Trust (Fellowship awarded to V. Laird) and Grant 31-43483.95 of the Swiss National Science Foundation (awarded to M. Spiess).

References

1. Palade, G. (1975) *Science* **189,** 347-358
2. Futter, C. E., Connolly, C. N., Cutler, D. F., and Hopkins, C. R. (1995) *J. Biol. Chem.* **270,** 10999-11003
3. Leitinger, B., Hille-Rehfeld, A., and Spiess, M. (1995) *Proc. Natl. Acad. Sci. U. S. A.* **92,** 10109-10113
4. Marks, M. S., Ohno, H., Kirchhausen, T., and Bonifacino, J. S. (1997) *Trends Cell Biol.* **7,** 124-128
5. Heilker, R., Spiess, M., and Crottet, P. (1999) *BioEssays* **21,** in press
6. Boll, W., Ohno, H., Songyang, Z., Rapoport, I., Cantley, L. C., Bonifacino, J. S., and Kirchhausen, T. (1996) *EMBO J.* **15,** 5789-5795
7. Höning, S., Griffith, J., Geuze, H. J., and Hunziker, W. (1996) *EMBO J.* **15,** 5230-5239
8. Ohno, H., Fournier, M. C., Poy, G., and Bonifacino, J. S. (1996) *J. Biol. Chem.* **271,** 29009-29015
9. Sandoval, I. V., and Bakke, O. (1994) *Trends Cell Biol.* **4,** 292-297
10. Ohno, H., Stewart, J., Fournier, M.-C., Bosshart, H., Rhee, I., Miyatake, S., Saito, T., Gallusser, A. s., Kirchhausen, T., and Bonifacino, J. S. (1995) *Science* **269,** 1872-1875
11. Carlsson, S. R., and Fukuda, M. (1992) *Arch. Biochem. Biophys.* **296,** 630-639
12. Harter, C., and Mellman, I. (1992) *J. Cell Biol.* **117,** 311-325
13. Matter, K., Whitney, J. A., Yamamoto, E. M., and Mellman, I. (1993) *Cell* **74,** 1053-1064
14. Matter, K., Yamamoto, E. M., and Mellman, I. (1994) *J. Cell Biol.* **126,** 991-1004
15. Geffen, I., Fuhrer, C., Leitinger, B., Weiss, M., Huggel, K., Griffiths, G., and Spiess, M. (1993) *J. Biol. Chem.* **268,** 20772-20777
16. Prill, V., Lehmann, L., von Figura, K., and Peters, C. (1993) *EMBO J.* **12,** 2181-2193
17. Brewer, C. B., and Roth, M. G. (1991) *J. Cell Biol.* **114,** 413-421
18. Hunziker, W., and Fumey, C. (1994) *EMBO J.* **13,** 2963-2969
19. Pond, L., Kuhn, L. A., Teyton, L., Schutze, M., Tainer, J. A., Jackson, M. R., and Peterson, P. A. (1995) *J. Biol. Chem.* **270,** 19989-19997
20. Letourneur, F., and Klausner, R. D. (1992) *Cell* **69,** 1143-1157
21. Sandoval, I. V., Arredondo, J. J., Alcalde, J., Gonzalez Noriega, A., Vandekerckhove, J., Jimenez, M. A., and Rico, M. (1994) *J. Biol. Chem.* **269,** 6622-6631
22. Höning, S., and Hunziker, W. (1995) *J. Cell Biol.* **128,** 321-332
23. Piguet, V., Gu, F., Foti, M., Demaurex, N., Gruenberg, J., Carpentier, J. L., and Trono, D. (1999) *Cell* **97,** 63-73
24. Rohrer, J., Schweizer, A., Russell, D., and Kornfeld, S. (1996) *J. Cell Biol.* **132,** 565-576
25. Griffiths, G., Hoflack, B., Simons, K., Mellman, I., and Kornfeld, S. (1988) *Cell* **52,** 329-341
26. Hunziker, W., and Geuze, H. J. (1996) *Bioessays* **18,** 379-389
27. Klumperman, J., Hille, A., Veenendaal, T., Oorschot, V., Stoorvogel, W., von Figura, K., and Geuze, H. J. (1993) *J. Cell Biol.* **121,** 997-1010
28. Le Borgne, R., Griffiths, G., and Hoflack, B. (1996) *J. Biol. Chem.* **271,** 2162-2170
29. Le Borgne, R., and Hoflack, B. (1997) *J. Cell Biol.* **137,** 335-345
30. Ludwig, T., Le Borgne, R., and Hoflack, B. (1995) *Trends Cell Biol.* **5,** 202-206
31. Zhu, Y., Traub, L. M., and Kornfeld, S. (1999) *Mol. Biol. Cell* **10,** 537-549
32. Diaz, E., and Pfeffer, S. R. (1998) *Cell* **93,** 433-443
33. Itin, C., Rancano, C., Nakajima, Y., and Pfeffer, S. R. (1997) *J. Biol. Chem.* **272,** 27737-27744
34. Karlsson, K., and Carlsson, S. R. (1998) *J. Biol. Chem.* **273,** 18966-18973
35. Ikonen, E., Tagaya, M., Ullrich, O., Montecucco, C., and Simons, K. (1995) *Cell* **81,** 571-580
36. Simons, K., and Ikonen, E. (1997) *Nature* **387,** 569-572
37. Varma, R., and Mayor, S. (1998) *Nature* **394,** 798-801
38. Robinson, M. S. (1997) *Trends Cell Biol.* **7,** 99-102
39. Simon, J. P., Shen, T. H., Ivanov, I. E., Gravotta, D., Morimoto, T., Adesnik, M., and Sabatini, D. D. (1998) *Proc. Natl. Acad. Sci. U. S. A.* **95,** 1073-1078
40. Ladinsky, M. S., Kremer, J. R., Furcinitti, P. S., McIntosh, J. R., and Howell, K. E. (1994) *J. Cell Biol.* **127,** 29-38
41. Müsch, A., Cohen, D., and Rodriguez-Boulan, E. (1997) *J. Cell Biol.* **138,** 291-306
42. Jones, S. M., Crosby, J. R., Salamero, J., and Howell, K. E. (1993) *J. Cell Biol.* **122,** 775-788

43. Novick, P., and Zerial, M. (1997) *Curr. Opin. Cell Biol.* **9,** 496-504
44. Daro, E., van der Sluijs, P., Galli, T., and Mellman, I. (1996) *Proc. Natl. Acad. Sci. U. S. A.* **93,** 9559-9564
45. van der Sluijs, P., Hull, M., Webster, P., Male, P., Goud, B., and Mellman, I. (1992) *Cell* **70,** 729-740
46. Chavrier, P., Parton, R. G., Hauri, H. P., Simons, K., and Zerial, M. (1990) *Cell* **62,** 317-329
47. Ullrich, O., Reinsch, S., Urbe, S., Zerial, M., and Parton, R. G. (1996) *J. Cell Biol.* **135,** 913-924
48. Zacchi, P., Stenmark, H., Parton, R. G., Orioli, D., Lim, F., Giner, A., Mellman, I., Zerial, M., and Murphy, C. (1998) *J. Cell Biol.* **140,** 1039-1053
49. Casanova, J. E., Wang, X., Kumar, R., Bhartur, S. G., Navarre, J., Woodrum, J. E., Altschuler, Y., Ray, G. S., and Goldenring, J. R. (1999) *Mol. Biol. Cell* **10,** 47-61
50. Helenius, A., Mellman, I., Wall, D., and Hubbard, A. (1983) *Trends Cell Biol.* **8,** 245-250
51. Hopkins, C. R., Gibson, A., Shipman, M., and Miller, K. (1990) *Nature* **346,** 335-339
52. Schmid, S., Fuchs, R., Male, P., and Mellman, I. (1988) *Cell* **52,** 73-83
53. Yamashiro, D. J., Tycko, B., Fluss, S. R., and Maxfield, F. R. (1984) *Cell* **37,** 789-800
54. Felder, S., Miller, K., Moehren, G., Ullrich, A., Schlessinger, J., and Hopkins, C. R. (1990) *Cell* **61,** 623-634
55. Ghosh, R. N., Gelman, D. L., and Maxfield, F. R. (1994) *J. Cell Sci.* **107,** 2177-2189
56. Aniento, F., Emans, N., Griffiths, G., and Gruenberg, J. (1993) *J. Cell Biol.* **123,** 1373-1387
57. Emans, N., Gorvel, J. P., Walter, C., Gerke, V., Kellner, R., Griffiths, G., and Gruenberg, J. (1993) *J. Cell Biol.* **120,** 1357-1369
58. Louvard, D., Morris, C., Warren, G., Stanley, K., Winkler, F., and Reggio, H. (1983) *EMBO J.* **2,** 1655-1664
59. Stoorvogel, W., Oorschot, V., and Geuze, H. J. (1996) *J. Cell Biol.* **132,** 21-33
60. Whitney, J. A., Gomez, M., Sheff, D., Kreis, T. E., and Mellman, I. (1995) *Cell* **83,** 703-713
61. Aniento, F., Gu, F., Parton, R. G., and Gruenberg, J. (1996) *J. Cell Biol.* **133,** 29-41
62. Gu, F., Aniento, F., Parton, R. G., and Gruenberg, J. (1997) *J. Cell Biol.* **139,** 1183-1195
63. Stoorvogel, W., Strous, G. J., Geuze, H. J., Oorschot, V., and Schwartz, A. L. (1991) *Cell* **65,** 417-427
64. Dunn, K. W., and Maxfield, F. R. (1992) *J. Cell Biol.* **117,** 301-310
65. Gruenberg, J., and Maxfield, F. R. (1995) *Curr. Opin. Cell Biol.* **7,** 552-563
66. Casanova, J. E., Breitfeld, P. P., Ross, S. A., and Mostov, K. E. (1990) *Science* **248,** 742-745
67. Aroeti, B., and Mostov, K. E. (1994) *EMBO J.* **13,** 2297-2304
68. Orzech, E., Schlessinger, K., Weiss, A., Okamoto, C. T., and Aroeti, B. (1999) *J. Biol. Chem.* **274,** 2201-2215
69. Futter, C. E., Gibson, A., Allchin, E. H., Maxwell, S., Ruddock, L. J., Odorizzi, G., Domingo, D., Trowbridge, I. S., and Hopkins, C. R. (1998) *J. Cell Biol.* **141,** 611-623
70. Ludwig, T., Griffiths, G., and Hoflack, B. (1991) *J. Cell Biol.* **115,** 1561-1572
71. Diesner, F., Sommerlade, H. J., and Braulke, T. (1993) *Biochem. Biophys. Res. Comm.* **197,** 1-7
72. Sariola, M., Saraste, J., and Kuisman, E. (1995) *J. Cell Sci.* **108,** 2465-2475
73. Leitinger, B., Brown, J. L., and Spiess, M. (1994) *J. Biol. Chem.* **269,** 8115-8121
74. Snider, M. D. (1989) *Methods Cell Biol.* **32,** 339-350
75. Bischoff, J., Libresco, S., Shia, M. A., and Lodish, H. F. (1988) *J. Cell Biol.* **106,** 1067-1074

Protein, Lipid and Membrane Traffic: Pathways and Targeting
J.A.F. Op den Kamp (Ed.)
IOS Press, 2000

Budding of retrograde transport vesicles from the cis-Golgi in yeast *Saccharomyces cerevisiae*

Anne Spang and Randy Schekman
Department of Molecular and Cell Biology
Howard Hughes Medical Institute
University of California, Berkeley
401 Barker Hall
Berkeley, California 94720, USA

Retrograde transport from the Golgi to the ER is an essential process. The most important functions of retrograde transport may be the delivery of endocytosed proteins to the ER, the retrieval of proteins that have escaped the ER, and the recycling of transport factors, such as SNAREs, for another round of anterograde transport. This transport step is mediated by the heptameric protein complex, coatomer, and the small ras-like GTPase Arf1p, which together form the COPI-coat. The recruitment of the COPI-coat proteins to the cis-Golgi is thought to be initiated by the binding of Arf1p to the membrane. Arf1p is myristoylated at its C-terminus. In the GDP-bound (inactive) state the fatty acid modification is buried within the molecule and thus Arf1p is soluble in the cytoplasm. Upon exchange of GDP for GTP on Arf1p, the C-terminus is exposed and anchors Arf1p into the membrane. The activation of Arf1p is mediated by a guanine nucleotide exchange factor (GEF). After the association of Arf1p with the membrane, coatomer and cargo can be recruited into budding complexes resulting in the formation of COPI-coated vesicles that pinch from the donor membrane (Figure 1).

Although this scenario gives a nice overview, it does not allow to understand the mechanisms and regulation underlying the COPI-vesicle budding process from the Golgi.

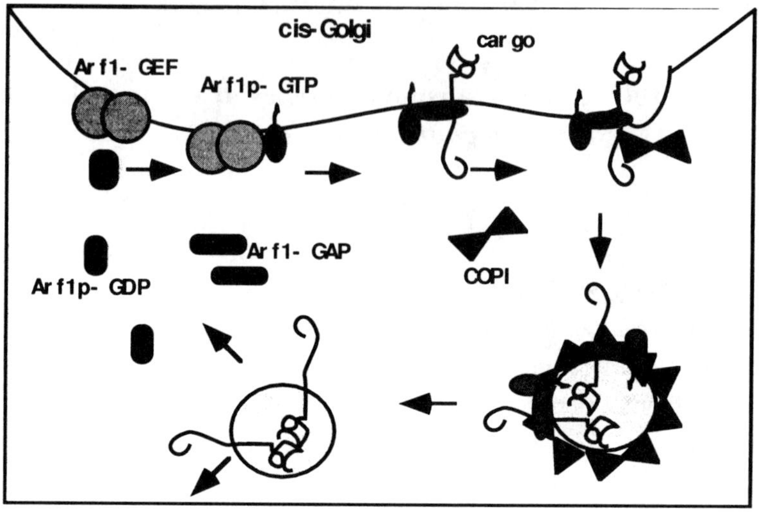

Figure 1: Schematic drawing of COPI-vesicle budding from the cis-Golgi. For explanation see text.

What are the prerequisites for vesicle formation at the donor membrane? Do cargo molecules have to be in close proximity to the emerging budding site or can a vesicle form in the absence of cargo? In a very reductionist system where biological membranes were replaced by chemically defined synthetic liposomes, COPI-coated vesicles could be formed in the presence of activated Arf1p and coatomer (Spang et al., 1998a). Similar conclusions can be drawn from experiments employing the COPII-coat proteins (Sar1p, Sec23p/Sec24p, Sec13p/Sec31p) that mediate the anterograde transport from the ER to the Golgi. Yeung and co-workers (1995) showed that ER membranes isolated from cells incubated with cycloheximide form COPII vesicles in vitro, at a normal rate and yield, that are depleted of biosynthetic cargo molecules. Furthermore COPII-coated vesicles bud from chemically defined synthetic liposomes, formulated to reflect the composition of microsomal membranes, upon addition of a non-hydrolyzable analogue of GTP and the COPII components (Matsuoka et al., 1998). Thus the inclusion of cargo does not seem to be required for the vesicle formation per se. If so, what defines a budding site? The activation of Arf1p is thought to be the starting point in COPI bud formation and subsequent vesicle budding. In order for this to occur, the cytosolic guanine nucleotide exchange factor (Gea1p or Gea2p, Peyroche et al., 1996) must bind to the cis-Golgi membrane. How this is achieved remains unclear. Protein receptor molecules as well as a certain lipid environment are equally likely to provide a recruiting mechanism.

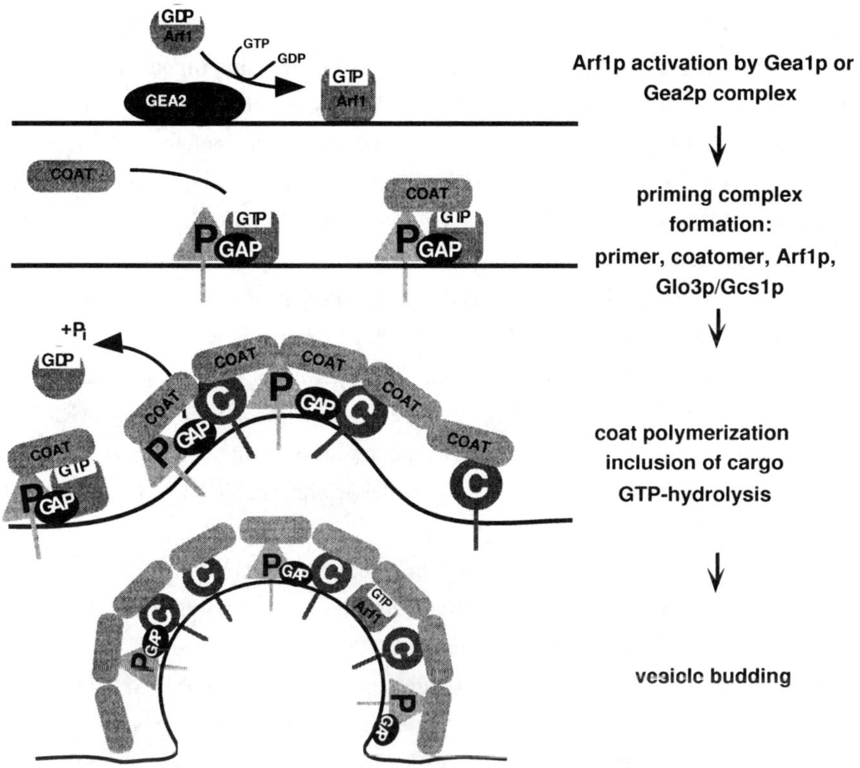

Figure 2: Model for vesicle budding mechanism from the cis-Golgi (adapted from Springer, 1999). For explanation see text.

Once Arf1p is activated and embedded into the Golgi membrane, it is likely to move around until it encounters a membrane protein or a membrane protein complex whose packaging is essential to vesicle biogenesis (Springer et al., 1999, Figure 2). We have speculated that certain essential cargo proteins called primers may initiate coat assembly. Primers for coatomer may contain the KKXX (or KXKXX) sequence motif known to be required for retrieval of certain proteins from the Golgi to the ER. The KKXX motif can bind directly to coatomer (Cosson and Letourneur, 1997) and might recruit coatomer to the membrane. Coatomer can also bind directly to Arf1p. The binding sites in coatomer for

Arf1p and KKXX-motifs are different, thus leading to local concentration of proteins (refs). Other sequences that interact with coatomer have been identified and new ones remain to be found, but proteins that interact with both with coatomer and Arf1p remain to be discovered.

A candidate for a primer protein is Erd2p, the HDEL-receptor. This Golgi protein recognizes soluble ER-resident proteins that have escaped their compartment and delivers them into COPI-vesicles destined for retrieval to the ER (Dean and Pelham, 1990). In mammalian cells, it has been shown that the loaded Erd2p interacts with the ARF-GTPase activating protein (ARF-GAP) (Aoe et al., 1997). This interaction ensures the inclusion of loaded Erd2p and ARF-GAP into COPI-coated vesicles whereas unloaded receptor is left behind in the Golgi. In *S. cerevisiae*, retrograde transport from the Golgi is mediated by two ARF-GAPs, Gcs1p and Glo3p, with overlapping function (Poon et al., 1999). Because ARFs possess a low intrinsic GTPase activity a GAP is likely to be included into the polymerizing coat to promote ARF-mediated GTP-hydrolysis. It is not known, if the loaded Erd2p is included in vesicles through the interaction with ARF-GAP or if ARF-GAP interacts with multiple primers to ensure its incorporation in the polymerizing coat. The predictions that result from these different hypotheses are testable and will clarify the issue.

It is unclear if v-SNAREs, which are cycling between the ER and the Golgi, act as primers in retrograde transport. Springer and Schekman have demonstrated that Bet1p, a v-SNARE in the ER-Golgi shuttle, interacts with the GTPase Sar1p and the Sec23p/Sec24p complex, which is part of the COPII-coat (Springer and Schekman, 1998). Given the similarity between the two systems and the absolute requirement for recycling of the v-SNAREs it seems plausible to suggest that a retrograde v-SNARE acts as primer in the formation of retrograde COPI vesicle.

Primer, Arf1p and coatomer form a priming complex, or micro domains, which are prone to form vesicles. Additional membrane cargo that may interact weakly with coatomer may diffuse into these polymerizing priming complexes and become incorporated into a coated bud. In these emerging coated buds, ARF-GAP would be in close proximity to Arf1p and stimulate the GTPase activity of Arf1p. After GTP-hydrolysis, Arf1p-GDP would diffuse out of the coat into the cytoplasm for another round of activation. Alternatively a timing mechanism, which tells ARF-GAP when to act on Arf1p, may control cargo recognition by a proofreading process. Although there is no evidence for such a timing mechanism Glo3p, one of yeast ARF-GAPs, is a component of COPI-coated vesicles derived from enriched Golgi membranes (Poon et al., 1999).

How can the requirement of primers as discussed above be reconciled with results from synthetic liposomes and budding in the absence of biosynthetic cargo? Although cargo destined for secretion or on its transient journey through the early secretory pathway is not required for vesicle formation to occur, primers may be essential for this process in vivo. The conditions in the reductionist synthetic liposome systems may bypass this requirement. Nevertheless, the requirements for COPII primer recruitment, vesicle budding, and cargo selection are recapitulated in a synthetic liposome budding reaction (Matsuoka et al., 1998b). In native ER membranes, primers may re-cycle between ER and Golgi. Therefore in the absence of protein synthesis, non-biosynthetic cargo can be still included into vesicles and sustain transport.

Additional work will be necessary to establish the role of ARF-GAPs in cargo recruitment, the interaction of ARF-GAPs with primers and the priming complex as well as to determine the point of action on Arf1p. A clear role for SNAREs in the nucleation of COPI and COPII coats remains to be established in vivo and reconstituted reactions.

References

[1] A. Spang, K. Matsuoka, S. Hamamoto, R. Schekman and L. Orci, Coatomer, Arf1p, and nucleotide are required to bud coat protein complex I-coated vesicles from large synthetic liposomes, *Proc Natl Acad Sci U S A*, **95** (1998) 11199-204

[2] T. Yeung, C. Barlowe and R. Schekman, Uncoupled packaging of targeting and cargo molecules during transport vesicle budding from the endoplasmic reticulum, *J. Biol.Chem.* **270** (1995) 30567-30570

[3] K. Matsuoka, L. Orci, M. Amherdt, S. Y. Bednarek, S. Hamamoto, R. Schekman and T. Yeung, COPII-coated vesicle formation reconstituted with purified coat proteins and chemically defined liposomes, *Cell*, **93** (1998) 263-75

[4] A. Peyroche, S. Paris and C. L. Jackson, Nucleotide exchange on ARF mediated by yeast Gea1 protein, *Nature*, **384** (1996) 479-481

[5] S. Springer, A. Spang and R. Schekman, A primer on vesicle budding, *Cell*, **97** (1999) 145-8

[6] P. Cosson and F. Letourneur, Coatomer (COPI)-coated vesicles: role in intracellular transport and protein sorting, *Curr Opin Cell Biol*, **9** (1997) 484-7

[7] N. Dean and H. R. B. Pelham, Recycling of proteins from the Golgi compartment to the ER in yeast, *J. Cell Biol.*, **111** (1990) 369-377

[8] T. Aoe, E. Cukierman, A. Lee, D. Cassel, P. J. Peters and V. W. Hsu, The KDEL receptor, ERD2, regulates intracellular traffic by recruiting a GTPase-activating protein for ARF1., *EMBO J.*, **16** (1997) 7305-7316

[9] P. P. Poon, D. Cassel, A. Spang, M. Rotman, E. Pick, R. A. Singer and G. C. Johnston, Retrograde transport from the yeast Golgi is mediated by two ARF GAP proteins with overlapping function, *EMBO J*, **18** (1999) 555-564

[10] S. Springer and R. Schekman, Nucleation of COPII vesicular coat complex by endoplasmic reticulum to Golgi vesicle SNAREs, *Science*, **281** (1998) 698-700

[11] K. Matsuoka, Y. Morimitsu, K. Uchida and R. Schekman, Coat assembly directs v-SNARE concentration into synthetic COPII vesicles, *Mol Cell*, **2** (1998) 703-8

Protein, Lipid and Membrane Traffic: Pathways and Targeting
J.A.F. Op den Kamp (Ed.)
IOS Press, 2000

Investigating the roles of SecB chaperone on HasA secretion via the ABC (ATP binding cassette) dependent pathway

Guillaume Sapriel and Philippe Delepelaire
Unité des Membranes Bactériennes
Bat. des biotechnologies
Institut Pasteur
25-28, rue du Dr Roux
75724 Paris Cedex 15
France

1. Introduction

In Gram-negative bacteria proteins are secreted to the extracellular medium across the cell envelope via at least four different specific pathways. One of them is the ABC dependent secretion system. It is characterized by an apparatus made of three proteins: two inner membrane proteins, the ABC protein and the Membrane Fusion Protein (MFP), believed to make specific contacts with the outer membrane protein (OMP).

Proteins secreted via this pathway belong to many different families as proteases, cytolysins (the model being the α-hemolysin from some uropathogenic *Escherichia coli* isolates), lipases or hemophores and are found in many different species. Usually, these proteins are coded on operons that contain the corresponding genes for specific secretion functions. Secretion apparati functionally cross-complement between members of one family sharing high sequence homology, but rarely do so between members of different families.

In contrast to the case of the General Secretory Pathway, secretion by ABC pathway occurs in one step without any stable periplasmic intermediate and, secreted proteins do not have an N-terminal signal peptide (Wandersman, 1996). In most cases, secreted proteins possess an uncleaved C-terminal secretion signal. The ABC protein specifically recognizes the secretion signal and is responsible for specificity (Binet and Wandersman, 1995). This interaction triggers the assembly of the secretion apparatus proteins in a complex (Létoffé et al., 1996).

2. Involvment of chaperone proteins in ABC pathway

Chaperone proteins and protein translocation across membranes

The C-terminal position of the secretion signal implies that the protein has to be entirely synthesized before interacting with the ABC transporter. This constraint raises the question of the conformational state of the protein before secretion. Proteins could be secreted via this pathway in a folded or in an unfolded state. In this second hypothesis, folding kinetics has to be slower than secretion kinetics. Folding kinetics could be slown down to maintain proteins partially unfolded, in a state competent for secretion. Secretion kinetics could also be enhanced by targeting the protein or by presentation of the secretion signal to the secretion apparatus. It is also possible that some secondary or tertiary structures of the secreted protein, corresponding to a folding intermediate, would be necessary for secretion. All these hypothetical mechanisms could involve chaperone proteins.

Chaperone proteins transiently bind non native structures in polypeptide chains and can thus prevent "illegitimate" interactions, thereby affecting the folding of proteins. Chaperone proteins play an essential role in many different translocation pathways across membranes like in translocation through the endoplasmic reticulum (Rapoport et al., 1996), or the mitochondrion (Lill et al., 1996) and also in export through bacterial cytoplasmic membrane via the general secretory pathway. In *E. coli* , the main chaperone of the general secretory pathway is the SecB protein (Collier, 1993). This chaperone affects the translocation kinetics of many proteins but is not essential for growth. SecB function has been identified by its effect on the export across the cytoplasmic membrane of the periplasmic Maltose-Binding Protein (MBP).

In the case of the ABC pathway, previous studies have established that for α-hemolysin, which is a 107kDa polypeptide, secretion via its specific ABC transporter in *E. coli* does not depend upon the SecB chaperone neither upon GroEL, the main Hsp60 of *E. coli*. It is also independent of SecA, the ATPase of the Sec translocon (Blight and Holland, 1994).(Mackman et al., 1986).

SecB chaperone is necessary to HasA secretion via the ABC pathway

One of our model proteins secreted by the ABC pathway is HasA. It is a 188 amino acids heme binding protein secreted by *Serratia marcescens* that allows the bacteria to use heme or hemoglobin as iron source in iron starvation condition (Létoffé et al., 1994 a; Létoffé et al., 1994 b). We are studying HasA secretion via its specific apparatus reconstituted in *E. coli* and consisting of the three proteins HasD (ABC protein), HasE (Membrane Fusion Protein) and TolC (outer membrane protein).

In previous studies, the chaperone protein SecB has been shown to be necessary for HasA secretion (Delepelaire and Wandersman, 1998). In the reconstituted system in *E. coli*, HasA secretion was strongly reduced in the absence of SecB. Secretion was restored in presence of a plasmid coding for SecB.

The amount of secretion apparatus proteins was not affected in a SecB minus strain and a small C-terminal fragment was not dependent upon SecB for secretion. These data suggest that SecB effect on HasA secretion was not due to an indirect effect on secretion apparatus. Moreover, SecB could be coimunoprecipitated with HasA indicating that SecB must probably interact directly with HasA.

In *S. marcescens* strain, a protein is specifically recognised by anti *E.coli* SecB antibodies indicating the presence of a possible homologue of this protein (de Cock and Tommassen, 1991). Moreover, introduction of a MBP (Maltose Binding Protein) variant that traps SecB affected strongly HasA secretion in *S. marcescens*. These results indicate that a SecB homologue of *S. marcescens* is necessary for HasA secretion.

3. Questions risen by SecB involvment in HasA secretion

Chaperone requirement and specificity

Whereas HasA requires SecB for secretion, proteins such as *Erwinia chrysanthemi* proteases or *E. coli* hemolysin do not. For these proteins, we are investigating the possible involvment of other chaperones such as DnaJ and DnaK. If different chaperones are specific to different secreted proteins, it would be interesting to find what determines this specificity.

It is also possible that some internal regions of secreted proteins have a chaperone function. In this respect, metalloproteases, toxins, lipases secreted by this pathway have a glycin-rich repeat region that is not present in HasA protein. In the case of the metalloproteases, it has been shown that this region can enhance the secretion of passenger proteins (Létoffé and Wandersman, 1992); furthermore in the active folded protein this region folds into a ß-roll to which calcium ions are tightly bound (Baumann et al., 1993). It is at present not known where calcium binding occurs; this could provide a way of regulating the secretion of those proteins.

Possible roles of SecB chaperone in HasA secretion

In the classical Sec export pathway of Gram-negative bacteria, SecB has two different functions: an antifolding activity, that maintains proteins in an export competent state, and a targeting activity, that enhances interaction between precursors and SecA, the

ATPase of the Sec translocon (Fekkes and Driessen, 1999). Specific complexes can be isolated between SecB and SecA and this interaction is strenghtened by the presence of a precursor molecule. A SecB binding site on SecA has been identified in the last C-terminal residues of SecA (Fekkes et al., 1997).

Since SecB has affinity for SecA, we first investigated the possible involvment of SecA in HasA secretion. For this purpose, we used azide, which at low concentrations specifically inhibits SecA ATPase activity (Oliver, 1993). Pulse chase experiment results showed that SecA is not necessary to HasA secretion (Fig.1).

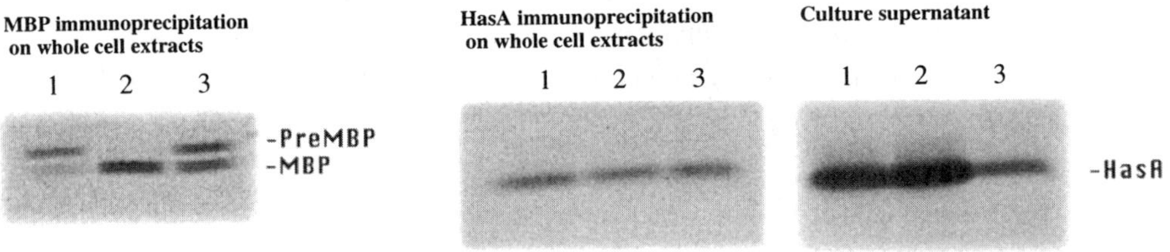

Fig.1 Pulse chase experiment autoradigramme. Cells where transformed with a plasmid pSYAC1 (Ghigo *et al.* 1997) containing *hasA* gene and its specific secretion functions, cultured in M9 minimal medium at 30°C. Maltose Binding Protein synthesis was induced by addition of 0,4% maltose.The cells where pulse-labeled with (S35)methionine for 1 min and processed for immunoprecipitation after 2 min as described by Delepelaire and Wandersman (1998). Supernatant was collected by centrifugation and TCA precipitated after 15 min. 3% of the labelled cells and 30% of the supernatant were loaded on acrylamid gel.
1: strain MC4100 (pSYAC1) + 3mM azide. **2**: strain MC4100 (pSYAC1). **3**: strain MC4100*secB*::Tn*5* (pSYAC1).

Several SecB point mutants have been identified (Kimsey et al., 1995). For some of them, interaction with SecA is weakened whereas complex formation between SecB and preMBP is normal. For the other SecB point mutants, complex formation between SecB and preMBP is low. These data allowed a tentative assignment of SecB residues specifically involved in SecA targeting or in precursor recognition and anti-folding activity. We are now using these mutants to test the involvment of those residues in SecB function in the HasA secretion pathway.

By analogy with the Sec pathway one might assume that SecB has a targeting function for the ABC transporter which is enhanced by the presence of HasA. The use of His-tagged SecB will allow us to test whether SecB recognizes the ABC transporter.

Investigating interaction of HasA with SecB and the ABC transporter

Whereas full length HasA is SecB dependent for secretion, a 56 amino acids C-terminal fragment is not SecB dependent for secretion. This raises the question of the SecB binding sites on HasA. We have previously shown that a truncated form of HasA containing the 132 N-terminal aminoacids leads to an accumulation of MBP precursor, indicative of SecB trapping.

In the case of the general secretory pathway, slow folding MBP variants have been isolated whose translocation across the cytoplasmic membrane is SecB independent (Collier et al., 1988). This suggests that in this pathway, SecB is required to slow down precursor folding kinetics to maintain it in a competent state for translocation. If SecB function is the same for HasA secretion, one might expect similar HasA variants to be isolated. The 56 amino acids C-terminal fragment that is not SecB dependent for secretion could be such a candidate since it does not display secondary stucture in aqueous solution. In order to possibly correlate SecB requirement for secretion and HasA folding kinetics, we are now investigating SecB dependance for secretion of other HasA variants.

In the general secretory pathway, a cooperative effect has been shown between SecB and the N-terminal peptide signal for MBP interaction with SecA. HasA could also contain regions that would enhance interaction between C-teminal secretion signal and the ABC protein. This effect could be direct or indirect by triggering HasA release from SecB. Afinity tests using His-tagged SecB and HasA variants are now being investigated in order to know if some HasA specific regions can enhance interaction with the ABC protein.

HasA tridimensional structure complexed with heme is now known (Arnoux et al., 1999) : it is a unimodular protein. The C-terminal secretion signal in the folded protein has the same structure as isolated in a membrane mimetic environment: it is composed of two α–helices with the very last residues without preferred structure (Izadi-Pruneyre et al., 1999). Several residues of the C-terminal secretion signal make hydrogen bonds with residues from the N-terminal end (Arnoux et al., 1999). Even thought HasA conformational state in cytoplasm is not known, C-terminal secretion signal interaction with other parts of the protein could support the hypothesis of the presentation of this signal by another part of the protein.

Altogether our work on HasA secretion has provided evidence for the involvement of chaperone proteins in the ABC pathway. Specific SecB role in HasA secretion is now being investigated. It also remains to be established which functions play this role in *Serratia.*

References

Arnoux, P., Haser, R., Izadi, N., Lecroisey, A., Delepierre, M., Wandersman, C. and Czjzek, M. (1999). "The crystal structure of HasA, a hemophore secreted by Serratia marcescens." *Nat Struct Biol* **6**(6): 516-520.

Baumann, U., Wu, S., Flaherty, K. M. and McKay, D. B. (1993). "Three-dimensional structure of the alkaline protease of *Pseudomonas aeruginosa* : a two-domain protein with a calcium binding parallel beta roll motif." *EMBO J.* **12**: 3357-3364.

Binet, R. and Wandersman, C. (1995). "Protein secretion by hybrid bacterial ABC-transporters: specific functions of the membrane ATPase and the membrane fusion protein." *EMBO J.* **14**: 2298-2306.

Blight, M. A. and Holland, B. (1994). "Heterologous protein secretion and the versatile *Escherichia coli* haemolysin translocator." *Tibtech* **12**: 450-455.

Collier, D. N. (1993). "SecB: a molecular chaperone of *Escherichia coli* protein secretion pathway." *Advances in Protein Chemistry* **44**: 151-193.

Collier, D. N., Bankaitis, V. A., Weiss, J. B. and Bassford, P. J., Jr. (1988). "The antifolding activity of SecB promotes the export of the E. coli maltose-binding protein." *Cell* **53**(2): 273-283.

de Cock, H. and Tommassen, J. (1991). "Conservation of components of the Escherichia coli export machinery in prokaryotes." *FEMS Microbiology Letters* **64**(2-3): 195-199.

Delepelaire, P. and Wandersman, C. (1998). "The SecB chaperone is involved in the secretion of the *Serratia marcescens* HasA protein trough an ABC transporter." *EMBO J.* **17**: 936-944.

Fekkes, P. and Driessen, A. J. (1999). "Protein targeting to the bacterial cytoplasmic membrane." *Microbiology & Molecular Biology Review (Washington, DC)* **63**(1): 161-173.

Fekkes, P., van der Does, C. and Driessen, A. J. (1997). "The molecular chaperone SecB is released from the carboxy-terminus of SecA during initiation of precursor protein translocation." *EMBO Journal* **16**(20): 6105-6113.

Ghigo, J. M.Létoffé, S. and Wandersman, C. (1997) Anew type of hemophore-dependant heme acquisition system of *Serratia marcescens* reconstituted in *Escherichia coli. J. Bacteriol.*, **179**, 3572-3579.

Izadi-Pruneyre, N., Wolff, N., Redeker, V., Wandersman, C., Delepierre, M. and Lecroisey, A. (1999). "NMR studies of the C-terminal secretion signal of the haem-binding protein, HasA." *European Journal of Biochemistry* **261**(2): 562-568.

Kimsey, H. H., Dagarag, M. D. and Kumamoto, C. A. (1995). "Diverse effects of mutation on the activity of the Escherichia coli export chaperone SecB." *Journal of Biological Chemistry* **270**(39): 22831-22835.

Létoffé, S., Delepelaire, P. and Wandersman, C. (1996). "Protein secretion in gram-negative bacteria: assembly of the three components of ABC protein-mediated exporters is ordered and promoted by substrate binding." *EMBO Journal* **15**(21): 5804-5811.

Létoffé, S., Ghigo, J. M. and Wandersman, C. (1994 a). "Iron acquisition from heme and hemoglobin by *Serratia marcescens* extracellular protein." *Proc. Natl. Acad. Sci. USA* **91**: 9876-9880.

Létoffé, S., Ghigo, J. M. and Wandersman, C. (1994 b). "Secretion of the *Serratia marcescens* HasA protein by an ABC Transporter." *J. Bacteriol.* **176**: 5372-5377.

Létoffé, S. and Wandersman, C. (1992). "Secretion of CyaA-PrtB and HlyA-PrtB Fusions Proteins in *Escherichia coli*: Involvement of Glycine-Rich Repeat Domain of *Erwinia chrysanthemi* Protease B." *J. Bacteriol.* **174**: 4920-4927.

Lill, R., Nargang, F. E. and Neupert, W. (1996). "Biogenesis of mitochondrial proteins." *Current Opinion in Cell Biology* **8**(4): 505-512.

Mackman, N., Nicaud, J. M., Gray, V. and Holland, I. B. (1986). Secretion of hemolysin by *Escherichia coli*. *Current topics in Microbiology and Immunology*. Berlin-Heidelberg, Springer-Verlag. **125:** 159-181.

Oliver, D. B. (1993). "SecA protein: autoregulated ATPase catalysing preprotein insertion and translocation across the Escherichia coli inner membrane." *Molecular Microbiology* **7**(2): 159-165.

Rapoport, T. A., Jungnickel, B. and Kutay, U. (1996). "Protein transport across the eukaryotic endoplasmic reticulum and bacterial inner membranes." *Annual Review of Biochemistry* **65**: 271-303.

Wandersman, C. (1996). Secretion across the bacterial outer membrane. *Escherichia coli and Salmonella tiphymurium cellular and molecular biology*. F. C. Neidhardt. Washington D.C., ASM Press: 955-966.

Protein, Lipid and Membrane Traffic: Pathways and Targeting
J.A.F. Op den Kamp (Ed.)
IOS Press, 2000

Glucose transport in the extremely thermoacidophilic *Sulfolobus solfataricus*

Sonja-V. Albers, Marieke G.L. Elferink, Arnold J.M. Driessen and Wil N. Konings

Department of Microbiology, Groningen Biomolecular Sciences and Biotechnology Institute, University of Groningen, Kerklaan 30, 9751 NN Haren, The Netherlands

Abstract. The archaeon *Sulfolobus solfataricus* grows optimally at a temperature of 80°C and a pH of 2.5-3.5 on carbon sources such as yeast extracts, tryptone and various sugars. Cells rapidly accumulate glucose. This transport activity involves a membrane-bound glucose-binding protein that interacts with its substrate with very high affinity (K_d of 0.43 μM) and retains high glucose affinity at very low pH values (up to pH 0.6). The binding protein was extracted with detergent, and purified to homogeneity as a 65-kDa glycoprotein. The gene coding for the binding protein was identified in the *S. solfataricus* P2 genome. Sequence analysis suggests the protein to be anchored to the membrane via an amino terminal transmembrane segment. Neighbouring genes encode two membrane proteins and an ATP-binding subunit that are transcribed in the reverse direction, whereas a homologous gene cluster in *Pyrococcus horikoshii* OT3 was found to be organised in an operon. These data indicate that *S. solfataricus* utilizes a binding-protein-dependent ATP-binding cassette (ABC) transporter for the uptake of glucose.

1. Introduction

Sulfolobus solfataricus is able to live in very acidic (pH 2-5) and very hot (up to 87°C) environments. It is a member of the third kingdom of life, the Archaea. Most members of this group exhibit the ability to survive extreme environmental conditions such as very high or low temperature, very acidic or alkaliphilic pH values, high salinity or high pressure. Woese et al. [26] postulated the Archaea as a distinct kingdom in 1990 based on studies on 16sRNAs. The analysis of the genome sequences of many different Archaea (*Methanococcus janaschii*, *Methanobacterium thermoautotrophicum*, *Archaeoglobus fulgidus*, *Pyrococcus furiosus*, *Sulfolobus solfataricus*) emphasizes their genetically different position. Two third of the genes found in these Archaea do not have homologues in Bacteria and Eucarya. The membranes of Archaea contain ether lipids in which two phytanyl chains are ether-linked to glycerol or another polyalcohol, nonitol. The lipid chains are composed of isoprene subunits, which contain methyl sidegroups at every fourth carbon atom of the chain backbone. Membrane spanning (bolaform amphiphilic) tetraether lipids are found in extreme thermophiles and acidophiles [6] (Fig.1). These lipids have a C_{40} isoprenoid phytanoyl chain that spans the entire membrane. Freeze-fracturing of liposomes made from tetraether lipids reveals that cleavage between two leaflets as seen with diether or diester lipids does not occur [7]. Liposomes made of *S. acidocaldarius* derived tetraether lipids are characterized by a high stability and a very low proton permeability [9]. These liposomes were already used for the functional reconstitution of several membrane proteins such as cytochrome-c-oxidase of *B. stearothermophilus* and beaf heart, bacteriorhodopsin of *H. halobium* [8].

The archaeon *Sulfolobus solfataricus* grows optimally at a temperature of 80°C and a pH of 2 - 3.5 on carbon sources such as yeast extracts, tryptone and various sugars. As an acidophile *Sulfolobus* faces the problem of a high ΔpH, acid outside. It has been shown that the cytoplasm has a pH value near to neutral [19] so that the pH gradient across the membrane can be maintained only when the proton permeability of the membrane is very low. Extensive studies with liposomes from *Sulfolobus acidocaldarius* lipids revealed that the proton permeability of these membranes at 80°C is the same as for *E. coli* membranes at 20°C [8, 23]. The high ΔpH results in a large proton motive force (PMF), which is slightly reduced by a reversed $\Delta\psi$ (inside positive), but with a value of about -200 mV it is still high [18]. The reversed $\Delta\psi$ seems not to have any consequences for the orientation of the transmembrane segments of membrane proteins in *Sulfolobus* [24]. In neutrophiles membrane proteins carry a net positive charge in their intracellular loops (positive inside rule [25]). Also in acidophiles, the positive inside rule applies, and therefore the $\Delta\psi$ can not be a determinant for membrane protein topology as suggested earlier [3].

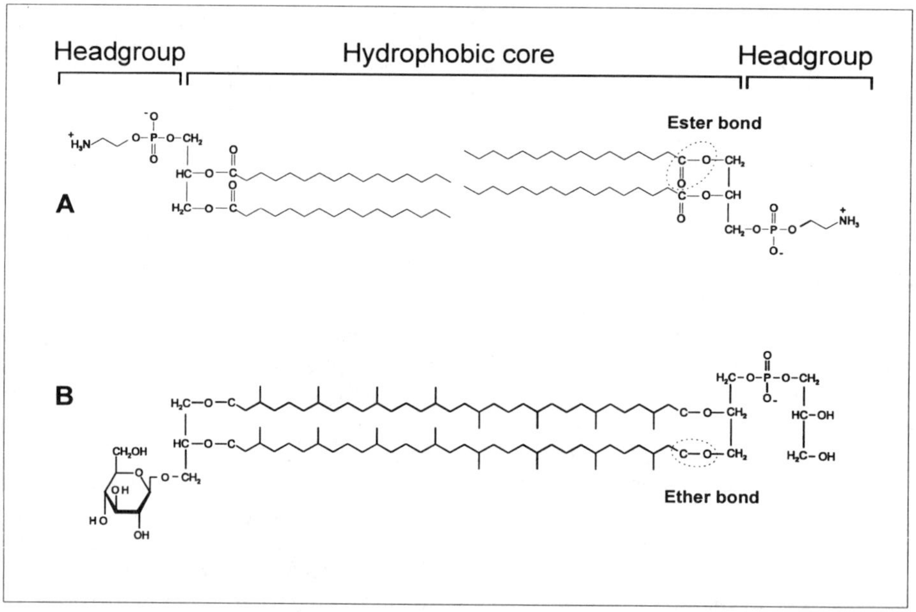

Fig.1. Lipids from Archaea and Bacteria. **A**: bilayer forming lipids in Bacteria: Phosphatidylethanolamine (PE) from *Escherichia coli*. The acyl chain is straight (not in all cases: some Bacterial lipids have a methyl branch, or a cyclohexyl group, at the end of the acyl chain, other lipids have one or more unsaturated bonds). The connection of the acyl chain with the headgroup is an ester. **B**: Monolayer forming lipids in thermo-acidophilic Archaea: Main glycophospholipid (MPL) of *Thermoplasma acidophilum*. The phytanyl chain contains isoprane-like branches. The connection of the phytanyl chain with the headgroup is an ether. Some acidophilic tetraethers contain cyclopentane rings.

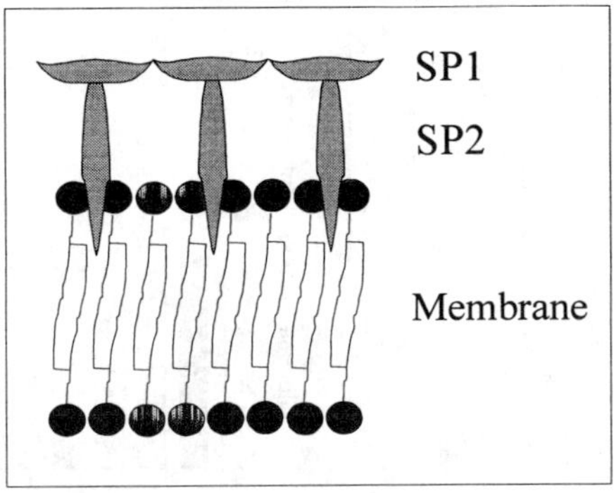

Fig. 2. Model of the membrane of *S. solfataricus*. Lipids shown here are membrane-spanning tetraether lipids.

We are especially interested in the transport proteins of *S. solfataricus*. These proteins have to face a large pH difference and have to be stable at high temperature. Such transport proteins may be good candidates for crystallization and structure elucidation.

2. Approaches

Two different approaches were used. In the first genetic approach attempts were made to express two ORF's, which are coding for putative secondary transporters. Up to now heterologous expression of these proteins in *E. coli* or *L. lactis* was not successful. At this moment this approach suffers from the lack of transformation or endogenous expression systems in *Sulfolobus* itself. Some vector systems, which have been published, were found to be useful only for some cytoplasmic proteins [5].

The second approach was the isolation of membrane vesicles from *S. solfataricus*. The isolation of such closed membrane vesicles from *S. solfataricus* membranes was hampered by the presence of the S-layer, which surrounds these cells. The S-layer consists of two proteins, SP1 and SP2 (Fig.2). SP1 can easily be stripped off, but SP2 sticks into the membranes [12]. SP2 likely prohibits the formation of closed membrane vesicles. Although the isolated membranes could not be used for uptake studies, since no electrochemical proton gradients could be generated across these membranes, the membranes were found to bind glucose.

3. Results

Glucose binding was also detected in Triton X- 100 total membrane protein extracts of *S. solfataricus*. It was concluded that these membranes contain a glucose-binding protein (GBP). GBP is tightly associated with the membrane, because it resists treatment with 22 mM Na_2CO_3, pH 10, and 1% Triton X- 100 is needed to solubilize it. The glucose-binding activity appeared to be associated with a 65 kDa protein band on SDS-PAGE, which gave a positive signal with glycoprotein stain. An affinity chromatography column, Con A, is able to bind glycoproteins and can be used for the first purification step. This step reduces the amount of proteins from a total membrane protein extract to 7-10 proteins, from which GBP can subsequently be separated by anion exchange chromatography. The glycoprotein-fraction also contained sugar-binding proteins for sucrose, maltose, cellobiose and arabinose. The isolated glucose-binding protein exhibited the same pH optimum for glucose binding as was found with crude membranes or solubilized membrane proteins. The optimal pH value was around 1.5, but significant binding activity could even be detected at a pH of 0.6.

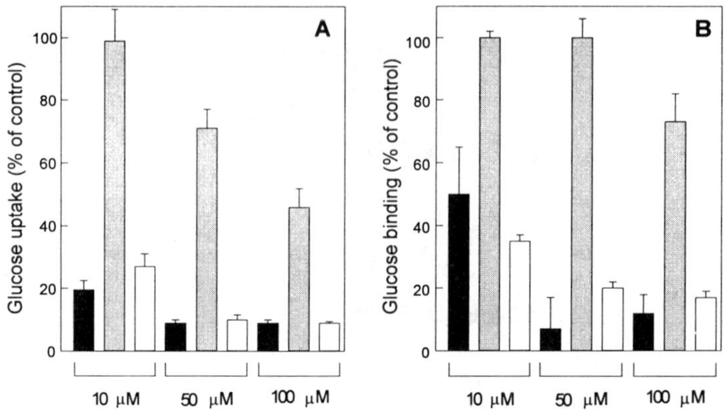

Fig. 3. Glucose uptake by *S. solfataricus* cells (A) and binding of glucose to purified GBP (B) in the presence of competing substrates. Uptake and binding assays were performed at 60°C and 1 μM [^{14}C]glucose in the presence of the indicated concentration of mannose (black), 2-deoxyglucose (grey), and galactose (white).

The K_D for glucose binding was determined to be 430 nM at pH 2 and 60°C. The substrate specificity was tested by adding competing concentrations of non-labelled sugar prior to the binding assay. Binding of glucose to the purified binding protein was strongly inhibited by galactose and mannose, but not by 2-deoxyglucose. The same substrate specificity was observed for glucose transport in intact cells (Fig. 3), indicating that GBP is directly involved in glucose uptake [1].

To identify the gene coding for GBP, the amino-terminal amino acid sequence of the purified protein was determined. A stretch of 31 amino acid residues could be determined and allowed the identification of the complete open reading frame in the genomic database of *S. solfataricus* P2. This gene codes for a protein with a molecular mass of 61 kDa. The size difference with the purified protein (~65 kDa) is most likely due to the glycosylation of the mature protein. The protein contains eleven possible glycosylation sites, while hydropathy analysis revealed strong hydrophobic regions at the amino-terminus and at the carboxyl-terminus of the protein. Both may form a transmembrane segment that anchors the protein to the cytoplasmic membrane.

The amino-terminal amino acid sequence of the purified glucose-binding protein completely matches the predicted sequence from the DNA database, except that the first 12 amino acids are lacking in the purified protein (Fig. 4). One possibility is that the protein is truncated as a result of proteolytic degradation. The other possibility is that the protein bears a signal sequence, which is processed after synthesis and before transport over the cytoplasmic membrane.

Comparisons with known signal sequences showed that archaeal flagellins have very similar characteristics in their N-terminal sequences. Kalmakoff and Jarrel [16] have shown that signal sequences of archaeal flagellins are not homologous to bacterial flagellins, but rather resemble the bacterial proteins of the IV pilin superfamily which includes also DNA uptake and protein excretion systems [13]. The genome of *S. solfataricus* bears a homologue of VirB, a component of a DNA uptake system in *Agrobacterium*. The open reading frames located.downstream of *Methanococcus voltae* flagellins show significant homology to pilB [4], a nucleotide binding protein involved in pilus function in *E. coli* [22]. This demonstrates that type IV secretion systems are also present in Archaea [2].

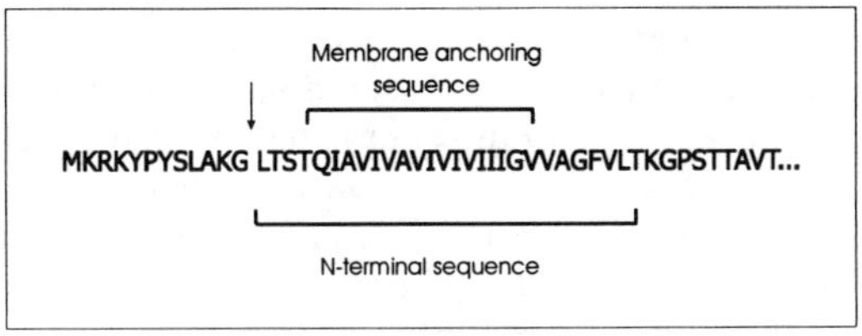

Fig. 4. Amino acid sequence of the N-terminus of GBP. The sequenced 31 amino acids and the putative membrane spanning part are indicated. The signal sequence is cleaved off at the G12. The processing site is indicated by an arrow.

Databank searches revealed that GBP shows the highest homology with the product of the *P. horikoshii* PH1214 gene (24 % identity and 40 % similarity). Downstream of PH1214 of *P. horikoshii*, two other ORFs are located in the same direction of transcription, PH1215 and PH1216 (Fig. 5). These two ORFs code for integral membrane proteins that are homologous to sugar permeases bearing the inner membrane component signature typical for binding-protein-dependent-transport systems. Further analysis of the DNA sequence surrounding the glucose-binding protein gene of *S. solfataricus* revealed three upstream genes, i.e., ORF32, ORF34 and ORF35 that are transcribed in the reverse direction (Fig. 5). ORF34 and ORF35 show homology to binding-protein-dependent sugar permeases, while ORF32 is similar to several ATP-binding proteins. Furthermore, ORF34 and PH1215 (28 % identity, 51 % similarity), and ORF35 and PH1216 (24 % identity, 54 % similarity) are homologous (Fig. 5). This genetic organisation suggests that GBP is a subunit of an ABC-transporter.

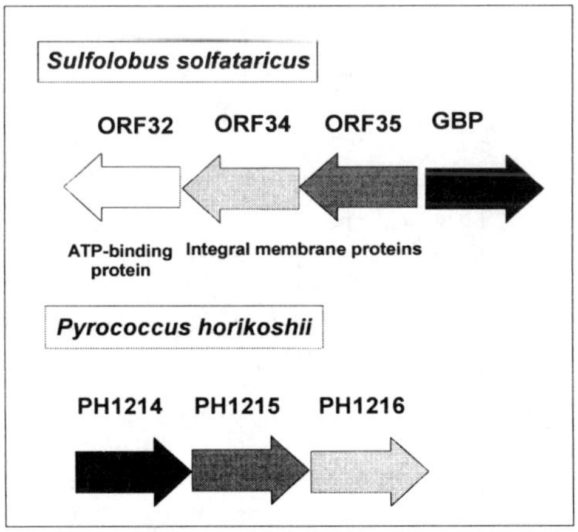

Fig. 5. Organisation of the genomic region around the glucose-binding protein (GBP) of *S. solfataricus* P2 and its homologue, PH1214, of *Pyrococcus horikoshii* OT3. Homologous genes are indicated by the same shading pattern. ORF34, ORF35, PH1215, and PH1216 code for putative membrane proteins, and ORF32 encodes an ATP-binding protein belonging to the ABC superfamily.

Trials for heterologous expression of GBP in *E. coli* have been unsuccessful. This could be due to the hydrophobic stretches, which could integrate into and distort the host membranes. Also the external pH of *E. coli* medium is higher than of *S. solfataricus*. This higher pH could prevent proper folding of GBP. On the other hand ORF32, the ATP-binding protein of the ABC-transporter, can be expressed in the cytoplasm of *E. coli*. This expression can be enhanced strongly by the co-expression of plasmid encoded tRNAs, which are rare for *E. coli*, but frequently used by *S. solfataricus*. *S. solfataricus* has a low G/C content (38%). It prefers codons which are A/T rich and in the cases of arginine, lysine and isoleucine are barely used by *E. coli*. The gene encoding the ATP-binding protein contains 42 of these codons and upon expression of this protein these tRNAs are depleted in *E. coli* resulting in low expression levels.

The ATP-binding protein was expressed with an N-terminal 6 histidine-tag and a Ni-NTA affinity chromatography column was used as a first purification step. Impurities were further removed by cation exchange chromatography. The protein, which has a molecular weight of 41 kDa, hydrolyzes ATP as shown by the release of phosphate. The optimal temperature for ATP hydrolysis is 70°C (Fig. 6).

4. Discussion

Transport of glucose in *S. solfataricus* is mediated by a high affinity binding-protein-dependent system that is specific for glucose, galactose, and mannose [1]. No 2-deoxyglucose binding was found to the membrane vesicles and to the purified binding protein. This suggests that a hydroxyl group at the C2 position of the sugar is critical for binding of the substrate, but GBP does not discriminate between C2 and C4-epimers of glucose.

A short amino-terminal sequence of 12 amino acids predicted on basis of the nucleotide-sequence is not present in the purified GBP. Similar observations have been made for the flagellin proteins in methanogens. This suggests that these proteins are processed and secreted in a similar way [2].

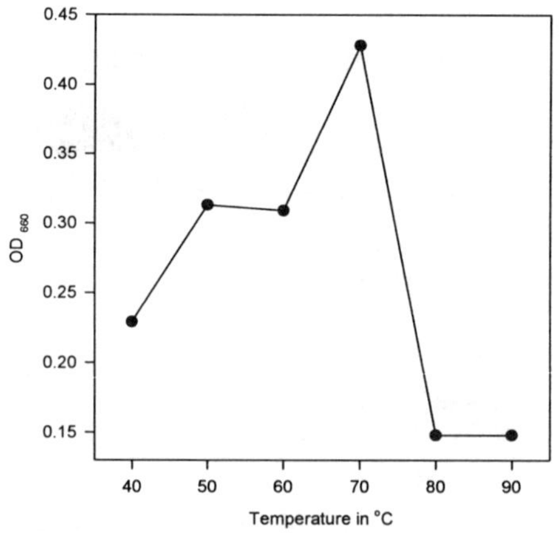

Fig. 6. Temperature optimum of ATP hydrolysis by recombinant ORF32. ATP hydrolysis is shown as release of free phosphate, which is measured by a colorimetric assay using malachite green at a wavelenght of 660 nm.

Another unusual aspect of the glucose-binding protein is its extreme acid resistance. The protein exhibits a very low pH optimum, i.e., around pH 1.5, which is comparable to that of pepsin in the stomach [11]. In this respect, GBP differs from two other extracellular enzymes of *S. solfataricus* that have been analysed, i.e., α-glucosidase and an esterase, which exhibit pH optima of 4.5 and 6, respectively [14, 20]. The purified GBP in detergent solution appeared somewhat more acid susceptible than the membrane-bound enzyme.

In Gram-negative bacteria, binding proteins exist in a soluble form in the periplasm. *S. solfataricus* lacks an outer membrane, and is instead surrounded by an S-layer. This hexagonal-paracrystalline-proteinaceous structure contains large pores of 4-5 nm to allow contact with the external medium [17]. It is thought that this structure serves as a molecular sieve, but it is not known if it can act as a barrier for proteins that are present in the space between the cytoplasmic membrane and the S-layer. The transmembrane segment(s) of GBP most likely serves as an anchor to the membrane like the lipid moiety that retains binding proteins at the cytoplasmic membrane of Gram-positive bacteria. Although the exact membrane topology of the protein is not yet known, the major part of the protein is most likely located at the outer surface of the membrane where it is glycosylated and where it can perform its function as a binding protein.

A transporter with affinity for glucose and galactose has previously been identified in *Brucella abortus* [10]. This system presumably catalyzes a sugar/H^+ symport reaction. Energetically such a mechanism would also be attractive for *S. solfataricus*, since it maintains a very large pmf across its membrane. Uptake of glucose by *S. solfataricus* mediated by a binding protein that is associated with a secondary transport system, is unlikely (15; 21), because the gene encoding the glucose-binding protein is located adjacent to genes encoding two integral membrane proteins and one ATPase subunit that are typical for ABC transporters. These three genes are transcribed in the reverse direction relative to the binding protein and a direct link is not immediately obvious. However, homologues of these genes in the genome of *Pyrococcus horikoshii* OT3 are contained in a single operon-like structure. Moreover, the two integral membrane proteins, ORF34 and ORF35, are also homologous to many other sugar permeases belonging to an ABC transporter. We therefore conclude that the glucose-binding protein is a subunit of an ABC transport system. The binding-protein-dependent maltose/trehalose transporters of *Thermococcus litoralis* [27] and *S. shibatae* [28], and the glucose transporter of *S. solfataricus*, exhibit very high affinities for their sugar substrates, i.e., in the submicromolar range. Moreover, *S. solfataricus* also contains high affinity binding proteins for at least maltose, arabinose, cellobiose and sucrose. For most of these substrates some bacteria can use phospho*enol*pyruvate dependent phosphotransferase systems (PTS). PTS has not yet been found in Archaea. Evidently, Archaea make use of ABC-transporters which activity depends on ATP levels inside of the cell. In contrast to PTS, ABC-transporter involve high-affinity-binding proteins. The high affinity of the binding proteins allows archaeal cells to efficiently utilise carbon sources in substrate poor environments such as the hydrothermal vents in the deep sea or the hot sulfuric pools.

Acknowledgements

This work was supported by a TMR grant of the European Commission (ERBFMBIC971980).

References

[1] Albers, S.-V., Elferink, M.G.L., Charlebois, R.L., Sensen, C.W., Driessen, A.J.M., and Konings, W.N. (1999) *J.Bacteriol.* in press.

[2] Albers, S.-V., Konings, W.N., and Driessen, A.J.M. (1999) *Mol.Microbiol.* **31**, 1595-1596.

[3] Andersson, H. and Von Heijne, G. (1994) *EMBO J.* **13**, 2267-2272.

[4] Bayley, D.P. and Jarrell, K.F. (1998) *J.Mol.Evol.* **46**, 370-373.

[5] Cannio, R., Contursi, P., Rossi, M., and Bartolucci, S. (1998) *J.Bacteriol.* **180**, 3237-3240.

[6] De Rosa, M., Trincone, A., Nicolaus, B., and Gambacorta, A.(1991) in Life under extreme conditions (di Prisco, G., Ed.), pp. 61-87, Springer-Verlag, Berlin Heidelberg.

[7] Elferink, M.G.L., De Wit, J.G., Demel, R., Driessen, A.J.M., and Konings, W.N. (1992) *J.Biol.Chem.* **267**, 1375-1381.

[8] Elferink, M.G.L., De Wit, J.G., Driessen, A.J.M., and Konings, W.N. (1993) *Eur.J.Biochem.* **214**, 917-925.

[9] Elferink, M.G.L., De Wit, J.G., Driessen, A.J.M., and Konings, W.N. (1994) *Biochim.Biophys.Acta* **1193**, 247-254.

[10] Essenberg, R.C., Candler, C., and Nida, S.K. (1997) *Microbiology* **143**, 1549-1555.

[11] Fox, P.F. and Whitaker, J.R. (1977) *Biochem.J.* **161**, 389-398.

[12] Grogan, D.W. (1996) *Can.J.Microbiol.* **42**, 1163-1171.

[13] Hobbs, M. and Mattick, J.S. (1993) *Mol Microbiol* **10**, 233-243.

[14] Huddleston, S., Yallop, C.A., and Charalambous, B.M. (1995) *Biochem.Biophys.Res.Commun.* **216**, 495-500.

[15] Jacobs, M.H., Van der Heide, T., Driessen, A.J.M., and Konings, W.N. (1996) *Proc.Natl.Acad.Sci.U.S.A.* **93**, 12786-12790.

[16] Kalmokoff, M.L. and Jarrell, K.F. (1991) *J.Bacteriol.* **173**, 7113-7125.

[17] Koenig, H. (1988) *Can.J.Microbiol.* **34**, 395-406.

[18] Michels, M. and Bakker, E.P. (1985) *J.Bacteriol.* **161**, 231-237.

[19] Moll, R. and Schäfer, G. (1988) *FEBS Lett.* **232**, 359-363.

[20] Rolfsmeier, M., Haseltine, C., Bini, E., Clark, A., and Blum, P. (1998) *J.Bacteriol.* **180**, 1287-1295.

[21] Shaw, J.G., Hamblin, M.J., and Kelly, D.J. (1991) *Mol Microbiol* **5**, 3055-3062.

[22] Sohel, I., Puente, J.L., Ramer, S.W., Bieber, D., Wu, C.Y., and Schoolnik, G.K. (1996) *J.Bacteriol.* **178**, 2613-2628.

[23] Van de Vossenberg, J.L.C.M., Ubbink-Kok, T., Elferink, M.G.L., Driessen, A.J.M., and Konings, W.N. (1995) *Mol.Microbiol.* **18**, 925-932.

[24] Van de Vossenberg, J.L.C.M., Van der Does, C., Albers, S.-V., Driessen, A.J.M., and Van Klompenburg, W. (1998) *Mol Microbiol* **29**, 1125-1126.

[25] Von Heijne, G. (1986) *EMBO J.* **5**, 3021-3027.

[26] Woese, C.R., Kandler, O., and Wheelis, M.L. (1990) *Proc.Natl.Acad.Sci.USA* **87**, 4576-4579.

[27] Xavier, K.B., Martins, L.O., Peist, R., Kossmann, M., Boos, W., and Santos, H. (1996) *J.Bacteriol.* **178**, 4773-4777.

[28] Yallop, C.A. and Charalambous, B.M. (1996) *Microbiology* **142**, 3373-3380.

Protein, Lipid and Membrane Traffic: Pathways and Targeting
J.A.F. Op den Kamp (Ed.)
IOS Press, 2000

Towards the localization and characterization of the ligand binding sites of the lactose transport protein from *Streptococcus thermophilus*

Liesbeth M. Veenhoff[1], Paul J.R. Spooner[2], Anthony Watts[2] and Bert Poolman[1]
[1]Department of Microbiology, Groningen Biomolecular Sciences and Biotechnology Institute, University of Groningen, *Kerklaan 30, 9751 NN Haren, The Netherlands*
[2]Biomembrane Structure Unit, Department of Biochemistry, University of Oxford, *South Parks Road, Oxford OX1 3QU, U.K.*

Abstract. The lactose transport protein of *Streptococcus thermophilus* catalyzes heterologous lactose/galactose exchange as well as lactose/proton symport. The properties of the binding sites for proton and sugar are discussed. Presented are (i) the differential structural requirements for sugar recognition at the inside and outside facing binding site, (ii) data from NMR, using sequence selective nitroxide labeling, and mutagenesis studies, indicating that the sugar binding site is located near helix XI and interhelix loop 10-11, and (iii) a model for proton binding at the N-terminal four amphipatic α-helices on the basis of site-directed mutagenesis and second-site suppressor studies.

1. Introduction

The lactose transport protein, LacS, of *Streptococcus thermophilus* belongs to a class of secondary transport systems comprising several dozen families to date. The LacS protein is a member of the galactosides-pentosides-hexuronides (GPH) family to which also the melibiose carrier (MelB) of *Escherichia coli* belongs [1]. Most members of the GPH family, for example MelB, have a structural fold that is composed of 12 transmembrane segments. LacS and some other members of the GPH family differ from these proteins by having an additional carboxyl-terminal cytoplasmic domain of about 180 amino acids (Fig. 1) [2]. The cytoplasmic domain can be phosphorylated at a histidine residue via the action of HPR~P, a central phosphoryl transfer component of the phosphoenolpyruvate:sugar phosphotransferase system [3, 4]. Phosphorylation of the cytoplasmic domain of LacS regulates the transport activity of the protein.

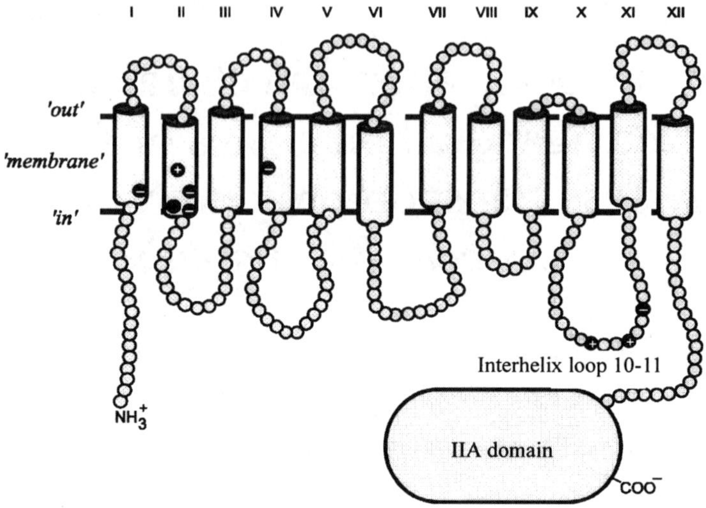

Figure 1: The fold of LacS protein predicted from hydropathy and reporter fusion analysis. Indicated are the 12 transmembrane helices, the regulatory IIA domain, and the positions of the residues mentioned in this chapter, that is: D22 in helix I; R64, E67, I70 and D71 in helix II; D133 in helix IV; K373, H376 and E379 in interhelix loop 10-11.

In *Streptococcus thermophilus* lactose is taken up via the lactose transport system, and intracellularly the disaccharide is hydrolyzed into glucose and galactose by the action of β-galactosidase. The glucose moiety is metabolized, and the galactose moiety is excreted into the medium by the action of the LacS protein. The overall *in vivo* reaction catalyzed by LacS is a lactose/galactose exchange, which is driven by the concentration gradients of both sugars across the membrane [5]. The LacS protein also catalyzes solute/ H^+ symport, but this transport reaction is one to two orders of magnitude slower than the exchange reaction and therefore less relevant *in vivo* [6]. Solute accumulation by the solute/ H^+ symport mode of transport is driven by the electrochemical gradient for protons across the membrane.

Any transport protein catalyzing an exchange or a proton symport reaction must oscillate between a minimum of two conformations, that is one in which the binding site faces towards the outside, and one where the binding site faces towards the inside of the cell (Fig. 2). A priori, one would expect that sugars bind with a higher affinity to the outside than to the inside facing binding site; the latter is the site from where the sugar taken up has to be released. However, this is not necessarily true for a system like LacS, where following release of lactose, galactose is bound to the inside facing binding site and subsequently released from the outside facing binding site. Indeed, the LacS protein displays a pronounced asymmetry in apparent binding affinities for lactose and galactose at the inside and outside facing binding sites, the apparent affinity constant at the inside being about one order of magnitude smaller [7, 8].

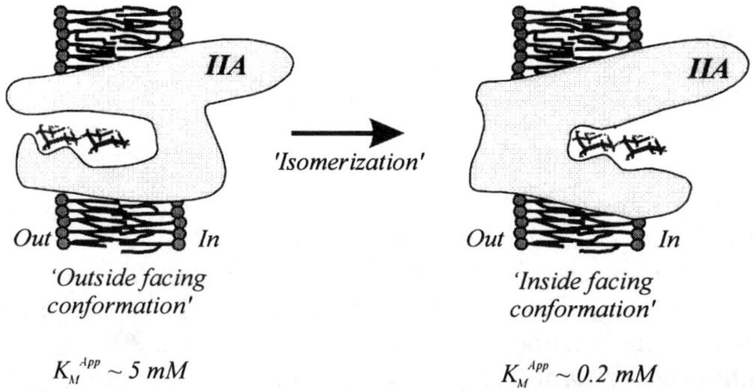

Figure 2: Two binding conformers of LacS, one exposing the binding site to the extracellular surface (left), and one where the binding site is accessible from the inside (right). The apparent affinity constants for lactose are approximately 5 mM and 0.2 mM at the outside and inside facing binding site, respectively. These constants are derived from transport kinetics rather than equilibrium binding.

A prerequisite for solute/H^+ symport is that reorientation of the binding site only occurs when the carrier is 'empty' or when both a sugar *and* a proton are bound [9]. The transporter would form a leak pathway when the binding site(s) reorient with either of the two ligands bound, that is the system would facilitate a proton or solute uniport [10]. Solute/ H^+ symport thus involves uptake of a proton and a sugar and reorientation of the empty carrier to the outside facing conformation. Kinetic analysis of transport has shown that not only in solute/H^+ symport but also in the exchange mode, protons are part of the translocation process [11]. A full turnover of the exchange mode involves the uptake of lactose together with a proton, which is followed by the exit of galactose with a proton. Thus, in both transport modes the conformational change from the inside to the outside facing conformation, and vise versa, only occurs when the carrier is 'empty' or when both sugar and proton are bound.

The aim of the work presented here is to find out where and how the two substrates, the sugar and the proton, are bound by LacS. Most biophysical and biochemical techniques to localize a ligand binding site within a protein rely on high affinity ligands, for example for monitoring fluorescence energy transfer between protein and ligand, or covalent linkage of a ligand to the protein. The same holds true for isothermal titration calorimetry or measurement of substrate retention, techniques used to characterize substrate binding in terms of stoichiometry, entropy, enthalpy, and affinity. Sugar transport proteins generally have a low affinity for their natural substrates, and high affinity substrates are not readily available or have properties (high hydrophobicity) that make them unsuitable for studies involving membranes.

The information described here on substrate binding by LacS comes from analyses of the aligned amino acid sequences of the GPH family members, site-directed mutagenesis studies, second-site suppressors, solid state NMR using sequence selective nitroxide labeling, and substrate specificity studies. The first part of this chapter relates to (i) the structural requirements of sugar recognition by the inside and outside facing binding sites of LacS, and (ii) the localization of the sugar binding site by mutagenesis and solid state NMR studies. In the second part, we discuss the localization and nature of the proton binding site on the basis of site-directed mutagenesis data and second-site suppressor analysis. Some general conclusions and perspectives end this chapter.

2. The sugar binding site

2.1 *Structural requirements for sugar recognition at the inside and outside facing binding site*

To obtain insight into the interactions between the sugar substrate and the inside and outside facing binding site of LacS, we determined transport rates and binding affinities at the inside and outside facing binding site for a range of natural and synthetic galactosides. For this purpose, the LacS protein was purified and reconstituted to form proteoliposomes with a unidirectional orientation of the protein [6]. The studies indicate that the binding site residues form a close fit around the galactose moiety of the sugar, as C-2 and C-6 methyl substituted galactosides are not bound, and, except for α- and β-D-galactose, none of the galactose epimers is bound. The binding site is thus specific for the spatial orientation of the C-2, C-3, C-4 and C-6 hydroxyl groups on the galactose ring, but not for the orientation of the C-1 hydroxyl group.

The LacS protein binds and transports galactosides with large hydrophobic or hydrophilic groups at the galactose C-1(α) or C-1(β) position, e.g., o-nitrophenyl-α-D-galactopyranoside, 5-bromo-4-chloro-3-indolyl-β-D-galactopyranoside, O-α-D-galactopyranosyl-[1,6]-a-D-glucopyranosyl-[1,2]-β-D-fructofuranoside or methyl-3-O-β-D-galactopyranosyl-β-D-galactopyranoside. A methoxy substituted C-4 position is also tolerated as methyl-4-O-lactose is bound and transported equally well as lactose. The binding site and translocation pathway are apparently spacious in the areas surrounding the galactose C-4 and even more so at the C-1 positions. In fact, galactosides with hydrophobic groups attached to the C-1 position have increased affinities at both the inside and outside facing binding site when compared to galactose.

The most important findings of this study concern the different interactions of the hydroxyl groups on the galactose moiety with the inside and outside facing binding site. Compared to galactose the affinities for 2- and 6-deoxy-galactose at the *in vivo* inside facing binding site are decreased 150- and 20-fold, respectively, whereas they are unchanged at the outside facing binding site. We speculate that the C-2-OH and C-6-OH contribute highly to the affinity for galactose at the *in vivo* inside facing binding site by forming hydrogen bonds with the protein, which does not take place when galactose is bound at the outside facing binding site. Clearly, the two binding conformers of LacS have binding sites with a distinct architecture.

In its normal conformation, the hydrogen atoms at the C-3, C-4, C-5, and C-6 of galactose can be considered to form a hydrophobic plane as illustrated in Fig. 3B. From the crystal structures of several sugar-binding proteins and sugar-binding toxins, it is known that sugar binding is usually accomplished by hydrophobic interactions with aromatic residues in addition to extensive hydrogen bonding [12, 13, 14, 15, 16]. The hydrophobic surface incorporating the hydrogen atoms of C-3, C-4, C-5, and C-6 of galactose would be a candidate for interaction with aromatic residues in the binding site. In contrast, bound [1-^{13}C]-galactose observed with a CP MAS (cross polarization magic-angle spinning) NMR showed no significant difference in chemical shift compared to galactose in free solution [17], which is indicative of a comparable chemical environment (also section 2.3). This would be consistent with C-1 being remote from the hydrophobic surface, as viewed in Fig. 3B, and existing in a more polar, or well hydrated region of the binding site.

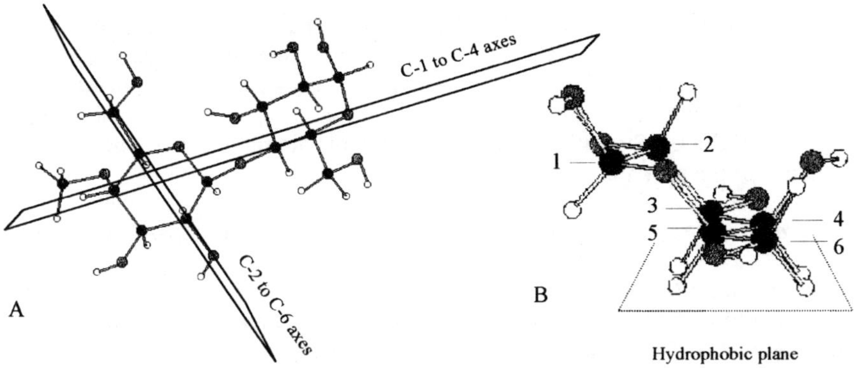

Figure 3: A: methyl-4-O-β-D-galactopyranosyl-β-D-glucopyranoside, highlighted are the C-1 to C-4 axes on the galactose moiety, that is where LacS tolerates substitution of the hydroxyl groups for larger groups, and the C-2 to C-6- axes which provides high affinity binding at the *in vivo* inside facing binding site.
B: β-galactose viewed along the C-6 to C-3 axes showing that the C-3, C-4, C-5, and C-6 hydrogen atoms can form a hydrophobic plane. Carbon atoms are black, oxygen gray, and hydrogen white.

The observations that (bulky) substituents are only tolerated at the galactose C-1 and C-4 positions, and that the C-2 and C-6 hydroxyl groups contribute highly to the affinity at the *in vivo* inside facing binding site, suggest that the binding site and translocation pathway are spacious along the C-1 to C-4 axes, and restricted along the C-2 to C-6 axes. Given these interactions and the structures of the galactosides (even tri-saccharides) transported, it seems reasonable to suggest that the sugars move through the protein along their galactose C-1 to C-4 axes. Fig. 3A shows the C-1 to C-4 axes and C-2 to C-6 axes in the structure of methyl-4-O-lactose.

2.2 *Localization of the sugar binding site*

Alignment of the sequences of the members of the GPH family shows that the sequence conservation is strongest in the regions encoding putative helices I and II, and the loop between helix X and XI. Analysis of the patterns of hydrophobicity and residue substitutions in the alignments of the transmembrane regions revealed that the helices II, IV and XI have a strong periodicity in terms of lipid facing and non lipid facing residues [1]. It seems that the packing of these helices is well conserved in each of the members of the GPH family, and that charged residues are excluded from the energetically unfavorable environment of the membrane. Although charged residues in regions of low dielectric constants such as the membrane could be stabilized by ion pairing, formed between residues in adjacent helices or within the same helix, the charged residues could also protrude in an hydrophilic channel in which substrate binding and/or translocation catalysis takes place.

Most of the sugar specificity mutants of MelB that have an altered sugar *and* cation recognition map in interhelix loop 10-11 and transmembrane segment XI [18]. Although none of these mutants has been studied in great detail, the observation that so many of these cluster together in the regions that are among the most conserved in the GPH family, suggests that they are in or near the substrate binding site of the protein.

Consistent with this notion is the observation that mutants carrying substitutions for His-376 or Glu-379 in interhelix loop 10-11 of LacS are highly defective in catalyzing coupled transport [3, 19]. Although the analysis of the alkaline phosphatase fusions of MelB suggest that interhelix loop X-XI is on the inner surface of the membrane [20, 21], the properties of the LacS position 376 and 379 mutants are more consistent with a location in the interior of the protein.

2.3 *Structural information from NMR using sequence selective nitroxide labeled LacS*

Solid-state NMR techniques can be used to selectively observe substrate that is immobilized within the binding site of sugar transport proteins [22, 23], whilst the proteins remain in a functional state within the native 'fluid' membranes. Discrimination of bound substrate by this approach is based on its dynamic properties and, therefore, this method is particularly useful for characterizing weak binding processes that are intractable to other means of physical analyses. The main advantage of this approach is in the possibility of combining labeled substrates with spectroscopic labels in the protein for the purpose of defining the binding center itself. For instance, nitroxide electron spin-labels can be used selectively at specific sites using single cysteine mutants [24]. The strong dipolar interaction from the electron spin circumvents the need for any detailed structural information to guide the selection of sites for labeling. The feasibility of the method is shown by the distance information obtained from the interaction between [1-^{13}C]D-galactose and a nitroxide label attached to a cysteine in interhelix loop 10-11 of LacS [17].

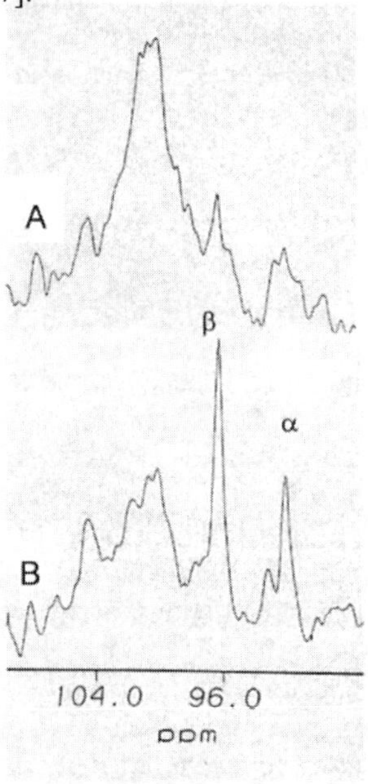

Figure 4: Expanded regions of CP MAS ^{13}C NMR spectra recorded from maleimide spin-labeled LacS membranes with 20 mM [1-^{13}C]D-galactose (A), and from Maleimide spin-labeled LacS membranes with substrate following chemical reduction of the spin label by incubating deoxygenated suspensions of the membranes with 100 mM sodium-ascorbate (over night; 4 °C) under nitrogen (B). The resonances from α- and β-[1-^{13}C]D-galactose are indicated with α and β. Figure taken from Spooner *et al.*, 1999 [17].

When LacS(K373C) protein in the native membrane is labeled with maleimide nitroxide spin label (MSL), the nitroxide spin-label at the Cys-373 appears to be essentially immobile on the time scale of both conventional ESR ($<10^{-8}$ s) and saturation transfer ($<10^{-3}$), which is important for the interpretation of its paramagnetic effect on the NMR observations. The MAS ^{13}C NMR spectrum of the membranes with MSL-LacS, equilibrated with 20 mM [1- ^{13}C]D-galactose, shows almost complete suppression of the high resolution signal of bound substrate (Fig. 4A). Chemical reduction of MSL-LacS(K373C) in the native membranes resulted in complete recovery of the high resolution signal from the substrate (Fig. 4B), demonstrating that substrate is within the binding center of MSL-LacS but that its NMR signal is evidently broadened beyond detection by paramagnetic interaction with the MSL. The MSL was calculated to be within 15 Å of the substrate binding site of LacS.

Despite the large upper limit placed on the electron-nuclear interaction observed here, the distance information together with the low mobility of the nitroxide label at Cys-373, suggests that interhelix loop 10-11 in LacS is folded into the membrane and likely located between the transmembrane helices involved in substrate binding.

3. The proton binding site

Transporters of the GPH family have several negatively and positively charged residues in the N-terminal helices, some of which are highly conserved, particularly in helix II. The catalytic role of several of these residues in the region has been examined by making various substitutions in both the *E. coli* MelB and the *S. thermophilus* LacS carrier proteins. The most important results can be summarized in the light of the hypothetical structure model of the amino terminal region of LacS (Fig. 5).

Neutral substitutions of the aspartate residues in MelB at positions 19, 55, 59 and 124 (corresponding to the positions 22, 67, 71 and 140 in LacS) lead to (i) loss of Na^+ linked TMG transport and (ii) retention of binding of α and β galactosides, but now independent of sodium ions [25, 26]. The corresponding Asp to Glu substitutions impair transport but sodium activation is still observed, albeit with lower efficiency. Similarly, neutral substitution of Glu-67, Asp71, and Asp-133 in LacS lead to an uncoupled transport activity, that is LacS still catalyzes exchange but galactoside/H^+ symport is either partially or completely impaired. The exchange reaction in these mutants most likely occurs without the accompanying movement of protons. Asparagine at position 58 in MelB (position 70 in LacS) is required for Na^+ coupled transport [27]. Again, analogous to MelB, the equivalent residue Ile-70 in LacS is required for proton coupling, as LacS(I70C) is impaired in galactoside/H^+ symport but can still catalyze exchange.

Overall, the data indicate that each of the acidic residues in the transmembrane segments I, II, and IV are important for cation binding and or translocation. In case of LacS, the bound and translocated cation may either be a proton or an hydronium ion.

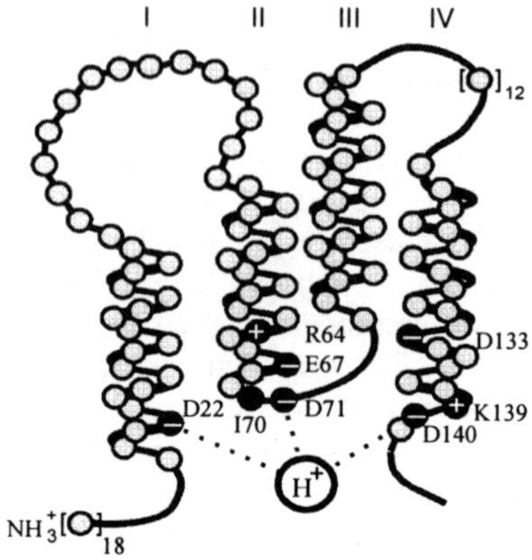

Figure 5: Hypothetical structure model of the amino-terminal region of LacS which has been implicated in proton binding and/ or translocation. Adapted from Poolman *et al.*, 1996 [1].

Mutants with a substitution in helix II of LacS or MelB that lead to inactivation of the carrier protein, that is D55S in MelB (D55 corresponds to position 67 in LacS) and R64C and D71C in LacS, have been used to search for second-site suppressor mutations. A suppressor mutation can be an indication for close approximation of two residues, especially when functional ion pairs within the protein are involved. The results of part of this analysis are discussed. The activity of MelB(D55S) is partly restored when Gly-117 is mutated to an aspartate, indicating that probably the negative charge on Asp-117 can compensate for the loss of negative charge in helix II [28]. Interestingly, in wild type LacS the equivalent of residue 117 is already an aspartate (Asp-133). Similarly, the activity of LacS(R64C) is restored when either Asp-133 is mutated to an asparagine or Ser-61 is mutated to a phenylanaline. Regained activity in LacS(R64C, D133N) suggests that in wild type LacS the opposite charges at positions 64 and 133 are stabilized by the formation of an ion pair, indicating a close approximation of helix II and IV. An unpaired charge on Asp-133 would thus inactivate the carrier protein, and neutralization of this residue restores activity. Along similar lines of reasoning one could explain the S61F suppressor mutation. The aromatic side chain of Phe-61, which is located one helix turn above Arg-64, might lower the polarity of the environment around the carboxylate of Asp-133. This would result in an increase in the pK_A of Asp-133, and the unpaired charge on the carboxylate could be neutralized at the prevailing pH.

Taken together these data strongly indicate that the polar faces of transmembrane helices II and IV of LacS are close together and make up at least part of the proton binding site. The aspartates and glutamate in these helices are either directly involved in proton binding or have a structural role, possibly by stabilizing different conformers of the protein through the formation of ion pairs, and thus have a more indirect role in catalysis.

4. Conclusions and future directions

The data presented in this chapter indicate that (i) the four amino terminal amphipatic helices are involved in proton binding and transport, and (ii) interhelix loop 10-11 and helix XI accommodate the sugar binding site. Obviously, proton and sugar transport by LacS are not separate events. Conformational coupling between sugar and proton binding and/or release will occur, and possibly the two substrates are even transported through the same translocation channel. Consistent with this idea is that neutral substitutions of residues in interhelix loop 10-11 (Glu-379 and His-376), and neutral substitution of residues in helix II and IV (Glu-67, Ile-70, Asp-71 and Asp-133) result in similar uncoupled transport proteins.

Evidence for conformational coupling in MelB comes from (i) proteolytic studies showing that cleavage in interhelix loop 4-5 is dependent on both cation and sugar binding [29], and (ii) fluorescence studies showing that the polarity of the environment of the fluorescent substrate changes in a sodium dependent fashion [30]. The observation that many MelB mutants exhibit simultaneous alterations in cation and substrate recognition [26, 30, 31] can also be interpreted in terms of conformational coupling. The only evidence for a close approximation of the N-terminal amphipatic helices II and IV and helix IX comes from fluorescence energy transfer data where the distance between tryptophans in interhelix loop 2-3 and in helix IX of MelB, were estimated to be 20 and 14 Å away from bound substrate, respectively [32, 33].

Currently, it is our task to bring together and complete the information on sugar and cation binding by LacS, and work towards a model where both are represented. Amongst the approaches taken will be those described in this chapter. Mutants impaired in proton and/or sugar binding will be studied with respect to their specificity at the inside and outside facing binding site. This may result in the identification of residues involved in substrate binding at either binding conformer. Extending our screen of second-site suppressor mutations might lead to the identification of neighboring residues in the C-terminal helices. Finally, single cysteine residues will be engineered at the ends of each of the transmembrane segments and in the putative transmembrane segments II, IV, XI of the LacS protein. Placing a nitroxide spin-label at each uniquely engineered site, and measurement of the spin label to ligand distance, will give a low resolution binding site map, which can be refined through triangulation.

5. Acknowledgment

This work was supported by funding from the European Community (Grants BIO-4-CT-960129 and 960439).

6. References

[1] Poolman, B., Knol, J., Does, van der, C., Henderson, P.J.F., Liang, W-J., Leblanc, G., Pourcher, T., and Mus-Veteau, I. (1996) *Mol. Microbiol.* **19**, 911-922

[2] Poolman, B., Royer, T.J., Mainzer, S.E., and Schmidt, B.F. (1989) *J. Bacteriol.* **171**, 244-253

[3] Poolman, B., Knol, J., Mollet, B., Nieuwenhuis, B., and Sulter, G. (1995) *Proc. Natl. Acad. Sci. USA.* **92**, 778-782

[4] Gunnewijk, M.G.W., and Poolman, B., manuscript in preparation

[5] Foucaud, C., and Poolman, B. (1992) *J. Biol. Chem.* **267**, 22087-22094

[6] Knol, J., Veenhoff, L.M., Liang, W-J., Henderson, P.J.F., Leblanc, G., and Poolman, B. (1996) *J. Biol. Chem.* **271**, 15358-15366

[7] Knol, J., Sjollema, K., and Poolman, B. (1998) *Biochemistry* **37**, 16410-16415

[8] Veenhoff, L.M., and Poolman, B., *J. Biol. Chem.* submitted

[9] Krupka, R.M. (1993) *Biochim et Biophys Acta* **1183**, 105-113

[10] Lolkema, J. S., and Poolman, B. (1995) *J. Biol. Chem.* **270**, 12670-12676

[11] Poolman, B., Knol, J., and Lolkema, J. (1995) *J. Biol. Chem.* **270**, 12995-13003

[12] Chaudhuri, B.N., Ko, J., Park, C., Jones, T.A., and Mowbray, S.L. (1999) *J. Mol. Biol.* **286** (5), 1519-1531

[13] Vyas, N.K., Vyas, M.N., and Quiocho, F.A. (1988) *Science* **242**, 1290-1295

[14] Vyas, M.N., Vyas, N.K., and Quiocho, F.A. (1994) *Biochemistry* **33** (16), 4762-4768

[15] Quiocho, F.A., Spurlino, J.C., and Rodseth, L.E. (1997) *Structure* **5**, 997-1015

[16] Merritt, E.A., Sixma, T.K., Kalk, K.H., Zanten, van, B.A.M., and Hol., W.G.J. (1994) *Mol. Microbiol.* **13**, 745-753

[17] Spooner, P., Veenhoff, L.M., Watts, A., and Poolman, B. (1999), *Biochemistry*, in press

[18] Botfield, M.C., and Wilson, T.H. (1998) *J. Biol. Chem.* **264**, 11649-11652

[19] Poolman, B., Modderman, R., and Reizer, J. (1992) *J. Biol. Chem.* **267**, 9150-9157

[20] Botfield M.C., Naguchi, K., Tsuchiya, T., and Wilson, T.H. (1992) *J. Biol. Chem.* **267**, 1818-1822

[21] Pourcher T., Bibi, E., Kaback, H.R., and Lablanc, G. (1996) *Biochemistry* **35**, 4161-4168

[22] Spooner, P.J.R., Rutherford, N.G., Watts, A., and Henderson, P.J.F. (1994) *Proc. Natl. Acad. Sci. USA* **91**, 3877-3881

[23] Spooner, P.J.R., O'Reilly, W.J. Rutherford, N.G., Henderson, P.J.F., and Watts, A. (1998) *Biophys.J.* **75**, 2794-2800

[24] Altenbach, C., Flisch, D.L., Khorona, H.G., and Hubell, W.L. (1989) *J. Biol. Chem.* **256**, 11804-11808

[25] Pourcher, T., Zani, M.L., and Leblanc, G. (1993) *J. Biol. Chem.* **268**, 3209-3215

[26] Zani, M.L., Pourcher, T., and Leblanc, G. (1994) *J. Biol. Chem.* **269**, 24883-24889

[27] Hama, H., Wilson, T.H. (1994) *J. Biol. Chem.* **269**, 1063-1067

[28] Wilson, D.M., Hama, H., and Wilson, T.H. (1995) *Biochem. Biophys. Res. Comm.* **209**, 242-249

[29] Gwizdek, C., Leblanc, G., and Bassilana, M. (1997) *Biochemistry* **36** (28), 8522-8529

[30] Shiota, S., Yazyu, H., and Tsuchiya, T. (1984) *J. Bacteriol.* **160**, 445-447

[31] Kawakami, T., Akizawa, Y., Ishikawa, T., Shimamoto, T., Tsuda, M., and Tsucchiya, T. (1988) *J. Biol. Chem.* **263**, 14276-14280

[32] Maehrel C., Cordat, E., Mus-Veteau, I., and Leblanc, G. (1998) *J. Biol. Chem.* **273**, 33192-33197

[33] Cordat E., Mus-Veteau, I., and Leblanc, G. (1998) *J. Biol. Chem.* **273**, 33198-33202

Protein, Lipid and Membrane Traffic: Pathways and Targeting
J.A.F. Op den Kamp (Ed.)
IOS Press, 2000

Homotypic Vacuole Fusion: A New Window on Interorganelle Traffick

Bill Wickner
Department of Biochemistry
Dartmouth Medical School
7200 Vail Building
Hanover, NH 03755-3844, USA

Correct subcellular compartmentation in eukaryotic cells requires both the selective delivery of newly made proteins into, and across, the membranes of specific organelles and highly regulated vesicular traffick between those organelles, sorting resident proteins from those which must move on to other intracellular destinations. While this was originally discovered and studied largely in mammalian systems, e.g. rat liver, the yeast S. cerevisiae has recently become the organism of choice, based on its advanced genetics and biochemistry. Vesicular trafficking between organelles consists of selective budding of vesicles, their movement along cytoskeletal elements, and their fusion at target organelles. Each step is complex, yet elegant. Budding is regulated by G proteins, and driven by the assembly of a coat on the nascent vesicle surface. The coat, via adapter proteins, selects cargo, and the cargo-enriched vesicle then pinches off from the donor organelle. The released trafficking vesicle then moves to its target, driven by a molecular motor along a cytoskeletal track. At the target organelle, it is recognized by "tethering" factors, which are often associated with GTPases. Priming by NSF (Sec18p) and _-SNAP (Sec17p) prepare the SNAREs on the vesicle and target organelle for pairing in *trans*, completing the process called "docking." Docking allows Ca flux, and this leads to membrane fusion and delivery of all vesicle contents to the target organelle. The complexity of this process is underscored by the estimate that encoding its catalysts occupies roughly 1/10th of the yeast genome.

Homotypic fusion is in some respects simpler, as it need not entail a selective budding step. Homotypic fusion is seen as a part of maintenance of low organelle copy number and is an integral part of organelle inheritance. Our studies of the inheritance of the vacuole (lysosome) or S. cerevisiae showed that vacuole-derived vesicles which are delivered into the bud will fuse to form the (often single) daughter-cell organelle [1]. This fusion has offered striking technical advantages for the study of membrane fusion in general, and thus has attracted substantial study.

To measure vacuole fusion [1,2], vacuoles are purified from two strains of yeast, one with normal vacuole proteases but a deletion in the PHO8 gene which encodes alkaline phosphatase and another with a deletion in the PEP4 gene, leading to accumulation of the catalytically-inactive pro-alkaline phosphatase. Though neither of these populations of purified vacuoles has alkaline phosphatase activity, their fusion allows the Pep4p protease access to the pro-alkaline phosphatase, and the resulting mature, catalytically-active phosphatase can then be assayed colorimetrically. The validity of this simple assay has been confirmed by the visual observation that vacuoles fuse and get larger during these same in vitro incubations and that each of the proteins identified as being required for this fusion in vitro are needed in the intact cell to maintain low vacuole copy number.

Our current understanding of this reaction is summarized in Figure 1. The reaction occurs in 3 successive steps, termed priming, docking, and fusion. Fusion has occurred when the organelle membranes and soluble contents mix, docking is the productive association of vacuoles prior to fusion, and priming is the reaction which is required for docking but which can be achieved prior to vacuole:vacuole contact. All the proteins which catalyze the overall reaction are found on the vacuole membrane as integral or peripheral membrane elements, though several of the latter can be removed by exposing the vacuoles to high salt concentrations. Two large, multisubunit complexes on the vacuole membrane have central roles in this reaction, the cis-SNARE complex and the complex of Ypt7p and Vam2/6p.

The cis-SNARE complex consists of the t-SNARE Vam3p, the SNAP-25 homolog (s-SNARE) Vam7p, and the v-SNAREs Nyv1p, Vti1p, and Ykt6p [3-5]. These are firmly associated with Sec18p, an ATPase chaperone which is the yeast equivalent of mammalian NSF, and Sec17p, the yeast equivalent of _-SNAP, which works with Sec18p. A second co-chaperone, LMA1 (low Mr activity #1), is bound to Sec18p [6]. Though the cis-SNARE complex has been highly purified [5,7], it has not yet been obtained in pure and fully functional form, and thus addtional elements may remain to be discovered. Priming is initiated by the hydrolysis of ATP by Sec18p. This leads to four reactions: 1. The 5 SNAREs are disassembled and can no longer be isolated in vacuole detergent extracts as a complex [8]. 2. The Sec17p is released from the vacuole [9]. 3. The t-SNARE Vam3p is activated for docking [8], though this activated state is inherently labile [9], and 4. LMA1 is transferred from Sec18p to the activated Vam3p [6], increasing its stability [10].

Four of the vacuolar SNAREs are integral membrane proteins, and thus remain firmly vacuole-bound after priming. However, the s-SNARE Vam7p does not [11], and remains vacuole-associated through interacting with Ypt7p and Vam2/6p, signaling to them that priming is complete and activating them for tethering. Tethering is a reversible stage of vacuole:vacuole association which is SNARE-independent and preceeds trans-SNARE pairing [12]. It requires prior priming and is mediated by the Ypt7p, a GTPase, and Vam2/6p [11, 12]. Vam2/6p is a large, multisubunit complex [13] which depends on Ypt7p for its stable vacuole association [11]. Vam2/6p is functionally similar to other large G-protein effector complexes of trafficking, such as the exocyst of yeast plasma membrane biogenesis [14], the TRAPP complex of yeast ER to Golgi traffick [15], and the EEA1 complex of mammalian endosome fusion [16]. Vam2/6p function is needed for the subsequent trans-SNARE pairing [11].

The pairing of t- and v-SNAREs in trans requires tethered vacuoles and the membrane proton gradient [17]. Trans-SNARE pairing completes the docking reaction, renders it irreversible, and triggers a Ca^{2+} efflux through a novel Ca^{2+} transport channel [18]. In successive steps which preceed fusion, Ca^{2+} then interacts with calmodulin, binding to an as-yet unidentified vacuole target, a protein phosphatase dephosphorylates a critical (but as yet unidentified) target [6], LMA1 is released, and the vacuoles fuse. Less is known of the molecular mechanisms of the fusion step per se, an area of obvious interest.

How was this knowledge of vacuole fusion gained? The soluble factors of the reaction--Sec18p, LMA1, Coenzyme A, and calmodulin--were either suspected by analogy with other trafficking reactions, and thence purified and tested in our system, or were discovered by fractionation of crude cytosol, assaying for fractions which are capable of supporting fusion [19]. Similarly, the membrane-bound elements, the SNAREs, Sec17p, Sec18p, Ypt7p, the Vam2/6p complex, were suspected by either analogy to other trafficking reactions or by the vacuole fragmentation phenotype seen in their mutants. In each case, they are tied to the reaction by inhibition of the reaction by specific antibodies, loss of the reaction when the vacuoles are prepared from deletion mutants, and specific (and, often, synthetic) mutant phenotypes.

Where do we go from here? Further components will be identified by the guess-and-test methods, but a systematic approach, yielding reconstitution of bona fide membrane fusion from all-purified components, will undoubtedly only come from systematic fractionation of vacuole detergent extracts, guided by the ability to reconstitute proteoliposomes from the extract or derived fractions which are themselves fully capable of physiological fusion reactions [7]. This approach, unique to vacuole homotypic fusion among all the trafficking reactions, is thus the central focus of my lab's efforts.

The vacuole fusion reaction has become a paradigm for biological fusion reactions, and offers far better technical tools for progress than virtually any other. This derives both from the rapid and colorimetric assay, from the advantages offered by yeast genetics and biochemistry, from the ability to perform the reactions with vacuoles and soluble components which can all come out of freezer stocks, and from the stimulating interactions within the yeast cell biology community. Forward!!

Figure 1. A working model-1999 of homotypic vacuole fusion. See text for details. Adapted from [17].

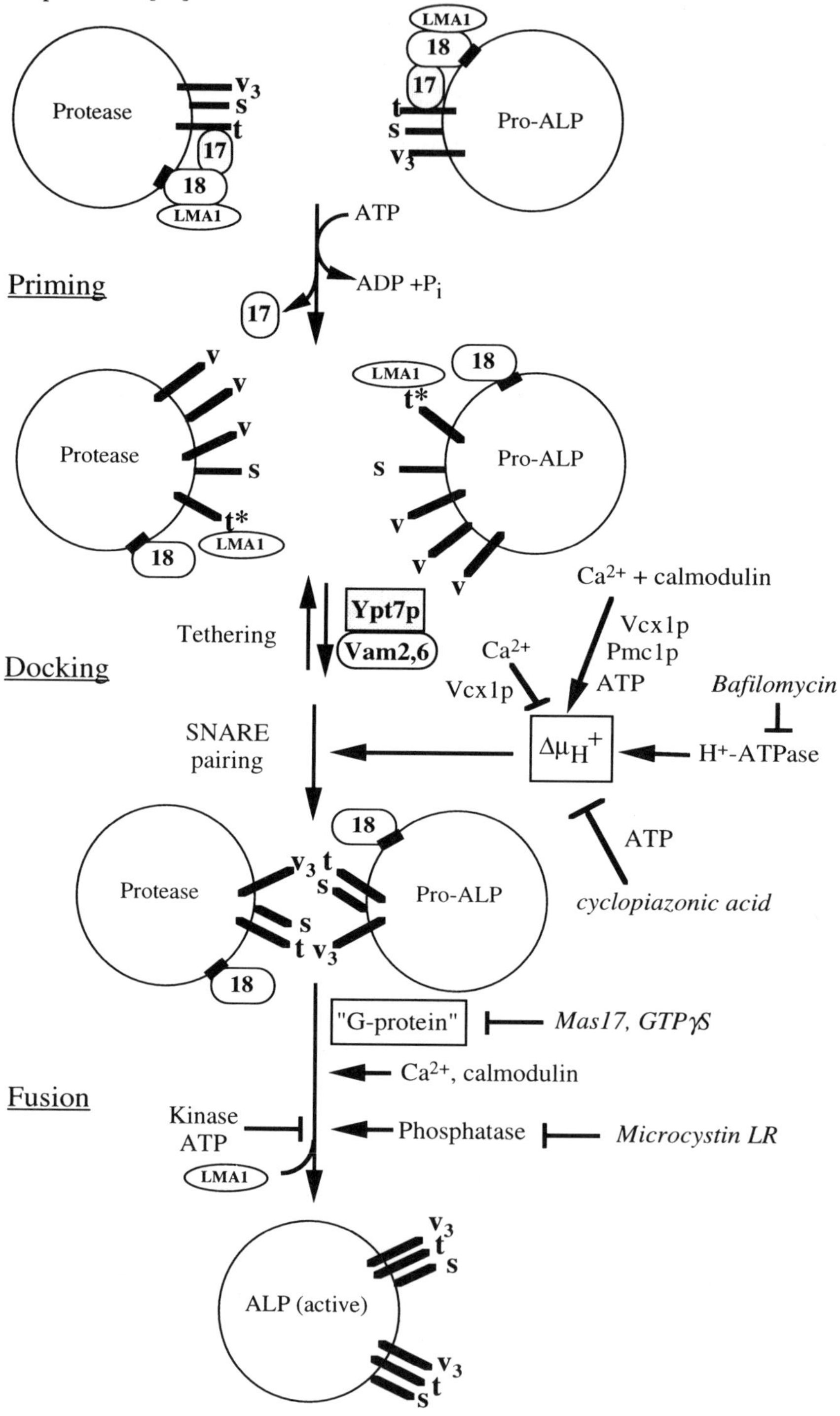

References

[1] B. Conradt, J. Shaw, T. Vida, S.D. Emr, and W. Wickner, In vitro reactions of vacuole inheritance. J. Cell Biol. **119** (1992) 1469-1479.
[2] A. Haas, B. Conradt, and W. Wickner, G-protein ligands inhibit in vitro reactions of vacuole inheritance. J. Cell Biol. **126** (1994) 87-97.
[3] B.J. Nichols, C. Ungermann, H.R.B. Pelham, W.T. Wickner, and A. Haas, Homotypic vacuolar fusion mediated by t- and v-SNAREs. Nature **387** (1997) 199-202.
[4] C. Ungermann and W. Wickner, Vam7p, a vacuolar SNAP-25 homolog, is required for SNARE complex integrity and vacuole docking and fusion. EMBO J. **17** (1998) 3269-3276.
[5] C. Ungermann, G. fischer von Mollard, O.N. Jensen, N. Margolis, T.H. Stevens, and W. Wickner, Three v-SNAREs and two t-SNAREs, present in a pentameric cis-SNARE complex on isolated vacuoles, are essential for homotypic fusion. Submitted (1999).
[6] Z. Xu, K. Sato, and W. Wickner, LMA1 binds to vacuoles at Sec18p (NSF), transfers upon ATP hydrolysis to a t-SNARE (Vam3p) complex, and is released during fusion. Cell **93** (1998) 1125-1134.
[7] K. Sato and W. Wickner, Detergent solubilization, purification, and functional reconstitution of an assembled vacuole v- and t-SNARE complex. Science **281** (1998) 700-702.
[8] C. Ungermann, B.J. Nichols, H.R.B. Pelham, and W. Wickner, A vacuolar v-t-SNARE complex, the predominant form in vivo and on isolated vacuoles, is disassembled and activated for docking and fusion. J. Cell Biol. **140** (1998) 61-69.
[9] A. Mayer, W. Wickner, and A. Haas, Sec18p (NSF)-driven release of Sec17p (α-SNAP) can precede docking and fusion of yeast vacuoles. Cell **85** (1996) 83-94.
[10] Z. Xu, A. Mayer, E. Muller, and W. Wickner, A heterodimer of thioredoxin and I2B cooperates with Sec18p (NSF) to promote yeast vacuole inheritance. J. Cell Biol. **136** (1997) 299-306.
[11] A. Price, W. Wickner, and C. Ungermann, Vam2/6p and Vam7p couple vacuole priming and docking. Submitted (1999).
[12] C. Ungermann, K. Sato, and W. Wickner, Defining the functions of trans-SNARE pairs. Nature **396** (1998) 543-548.
[13] N. Nakamura, A. Hirata, Y. Ohsumi, and Y. Wada, Vam2/Vps41p and Vam6/Vps39p are components of a protein complex on the vacuolar membranes and involved in the vacuolar assembly in the yeast saccharomyces cerevisiae. J. Biol. Chem. **272** (1997) 11344-11349.
[14] D.R. TerBush, T. Maurice, D. Roth, and P. Novick, The exocyst is a multiprotein complex required for exocytosis in Saccharomyces cerevisiae. EMBO J. **15** (1996) 6483-6494.
[15] M. Sacher, Y. Jiang, J. Barrowman, A. Scarpa, J. Burston, L. Zhang, D. Schieltz, Y.R. Yates, H. Abeliovich, and S. Ferro-Novick, TRAPP, a highly conserved novel complex on the cis-Golgi that mediates vesicle docking and fusion. EMBO J. **17** (1998) 2494-2503.
[16] S. Christoforidis, H.M. McBride, R.D. Burgoyne, and M. Zerial, The Rab5 effector EEA1 is a core component of endosome docking. Nature **397** (1999) 621-625.
[17] C. Ungermann, W. Wickner, and Z. Xu, Vacuole acidification is required for trans-SNARE pairing, LMA1 release, and homotypic fusion. Submitted (1999).
[18] C. Peters and A. Mayer, Ca2+/calmodulin signals the completion of docking and triggers a late step of vacuole fusion. Nature **396** (1998) 575-580.
[19] Z. Xu and W. Wickner, Thioredoxin is required for vacuole inheritance in S. cerevisiae. J. Cell Biol. **132** (1995) 787-794.

Protein, Lipid and Membrane Traffic: Pathways and Targeting
J.A.F. Op den Kamp (Ed.)
IOS Press, 2000

Membrane Targeting in Pea Root Nodule Cells

Eugenio BUTELLI and Nicholas J. BREWIN
John Innes Centre, Colney Lane, Norwich NR4 7UH, Great Britain

Abstract. Very little is known about how the Golgi apparatus is involved in the organisation of plant cell growth and development. Evidentally, it plays a key role in membrane biogenesis, providing many components that are targetted to the extracellular matrix and others that are targetted to intracellular compartments, e. g. storage bodies in seeds. The legume root nodule has many features that make it an excellent model in which to analyse the trans-Golgi network in plant cells. In the central tissues of pea nodules, each host cell is hyperinfected by endosymbiotic *Rhizobium* cells. Each bacterium is individually enclosed by a plant-derived "symbiosome membrane". The total surface area of this membrane is 30-100 times greater than for plasma membrane, making it a highly specialised, highly amplified membrane compartment. Each endosymbiotic bacteroid with its membrane envelope comprises a "symbiosome compartment" of the plant cell. The status of this compartment within the trans-Golgi network remains undefined. Using monoclonal antibody probes, we have defined many components of the peribacteroid membrane and peribacteroid fluid, including some antigens that distinguish between "early" and "late" stages of symbiosome development. We have also identified a symbiotically defective pea mutant (*sym31)* that is apparently arrested at an "early" stage of symbiosome development. Future studies of membrane biogenesis and targetting will be directed towards the analysis of differentially expressed transcripts in mutant and wild-type pea nodules

1. The Changing Role of the Golgi Apparatus

Nodulation, the symbiotic partnership between legume plants and nitrogen-fixing soil bacteria, begins with mutual signal exchange between plant hosts and *Rhizobium* bacteria, collectively called rhizobia [1]. Symbiosis leads to the production of a unique novel organ, the nodule, which contains new components that are not found in either free-living bacteria or uninfected plants [2]. Tissue and cell invasion by *Rhizobium* is an interesting developmental process involving the progressive reorganisation of the secretory pathway and the endomembrane system of host cells in order to accommodate the progressive internalisation of *Rhizobium* bacteria.

The association begins when the legume plants produce particular chemicals, flavonoids and isoflavonoids, that are recognized by *Rhizobium.* Different plants make different flavonoids, to which an appropriate strain of *Rhizobium* responds [3]. Each type of *Rhizobium* responds by inducing the expression of a unique molecule, called "Nod-factor". All Nod-factors are based on the same lipid-carbohydrate structure and various additions and substitutions to this structure generate host specificity [4]. Nod-factors are able to induce morphological changes to legume root hairs.

These changes are accompanied by the induction of several different nodulation genes, called nodulins, whose diverse functions are still not fully understood. Initially bacteria adhere to the root hair tip of the plant. This step is followed by root hair deformation and root hair curling, where the tip of the root hair cell curls round on itself, entrapping some bacteria [5]. The bacteria weaken the plant cell wall and the plant responds by forming a tubular invagination of the cell wall, called an "infection thread" [6], along which the bacteria can grow and divide [7]. Bacteria, through the production of Nod-factors, are also able to stimulate the cells of the root cortex to divide, inducing the

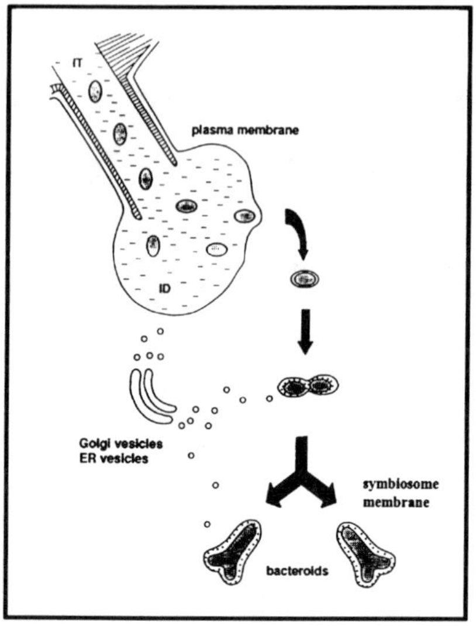

Fig. 1. An illustration of the infection process that leads to the formation of the symbiosome membrane. IT: infection thread. ID: infection droplet.

formation of a nodule primordium with peripheral vascular tissue [8]. The infection thread is propagated from the root hair cells into the dividing cells of the cortex, it branches, and delivers the bacteria into the host cytoplasm from unwalled infection droplets [9] (Fig. 1). Bacteria are released as single units and are enclosed in a membrane envelope, the symbiosomal membrane (SM), which undergoessubsequent transformation into a specialised compartment [10]. When the bacteria have been released, they enlarge and change their shape, such that in pea the endosymbionts have a characteristic Y conformation (Fig. 2). The final stage requires the coordinated development of the bacteria, now termed bacteroids, and the plant cells and leads to the formation of a mature N_2-fixing unit which has been called a "symbiosome", to underline its analogy with chloroplasts and mitochondria, prokaryote organelles that originally began their evolution as endosymbiotic bacteria.

2. Structure and Function of the Symbiosome Membrane

One of the amazing features of the process of symbiosis between legume plants and rhizobia is the biogenesis of the peribacteroid membrane (PBM), otherwise known as the symbiosome membrane (SM). Because each host cell contains several thousand individually enclosed bacteroids, the total extent of SM has been estimated to exceed that of the plasma membrane by 30 to 100 fold [11,12]. Therefore, it appears clear that the host cellular machinery must direct an intense process of membrane proliferation in order to establish the new subcellular compartment.

Considering the enormous amplification of this compartment, SM formation can be considered a good and fascinating system for the study of membrane biogenesis and vesicle targeting in plants. Despite the high rate of SM biogenesis, the nature of this structure and the targeting mechanism of the SM components are not fully understood.

Initially, SM can be considered as a plasmalemma-derived membrane. Bacteria are delivered from the infection thread inside the host cell by endocytosis [13]. In the following stages bacteria grow, divide and eventually differentiate into functional

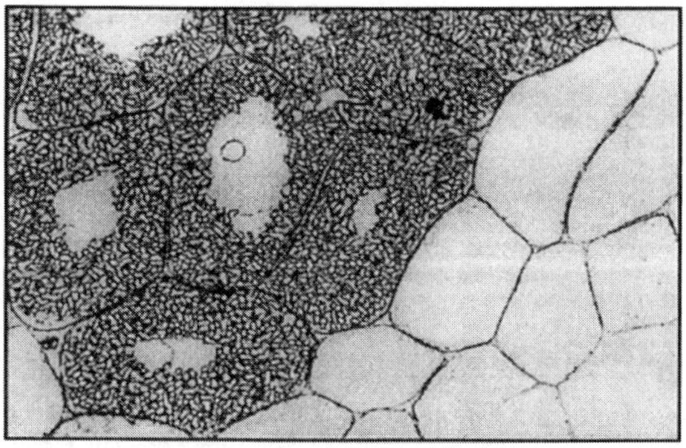

Fig. 2. Light micrograph showing infected and uninfected pea nodule cells, stained to reveal plasma membrane and symbiosomal membranes.

bacteroids. This process is accompanied by a rapid transformation of the surrounding plasma membrane into a specialized symbiotic compartment [14,15]. At the end of this process SM acquires new specialized features, some of which resemble those of the vacuolar membrane. For this reason SM is often considered as a mosaic membrane, sharing characteristic with both the plasma membrane and the tonoplast.

Identical to the plasma membrane is the thickness of SM (8-10 nm), while the composition presents a characteristically high lipid/protein ratio. Glycocalyx composition undergoes a transformation, too. This change has been used as a cytological marker for the differentiation of the SM [16]. K^+/H^+ -ATPase , a typical plasma membrane, is equally present in the SM, together with a Mg^{2+}/H^+ -ATPase, also present in tonoplast and Golgi membranes [12,17].

The SM apparently has two major functions [10]. First, it can be considered a physiological barrier excluding the endosymbiont from direct contact with the host cytoplasm, which might otherwise result in a pathogenic association by triggering plant defense responses. The second role is related to the control of the intense exchange of nutrients and signal molecules between the two symbionts [17]. In particular, SM excludes sugars and amino acids and controls the transport of dicarboxylic acids, which are probably the main carbon source provided to the bacteroids by the host, while an ammonium channel seems to be involved in the transport of fixed nitrogen [18] (Fig. 3).

3. Biogenesis of the Symbiosome Membrane

Although several SM-specific proteins and lipids have been identified, the targeting of these components to the SM is not well defined. SM proteins have been shown to be synthesised on membrane-bound ribosomes and co-translationally inserted into the endoplasmic reticulum [12]. Following synthesis, these proteins are modified and processed in the Golgi apparatus, packaged into vesicles and delivered to the SM. There is very convincing evidence for a specific SM localization of at least two plant proteins, nodulin-26 and nodulins-24. Nodulin-26, the most prominent and well characterised SM protein, is an intrinsic membrane protein with channel properties; nodulin-24 is a surface membrane protein, possibly involved in the physical interaction between the SM and the membrane of the bacteroids. Targeting of these nodulins does not follow the default pathway, because they are not found in the plasma membrane. However, no specific target information, either at the primary or secondary level, seems to be present in the different

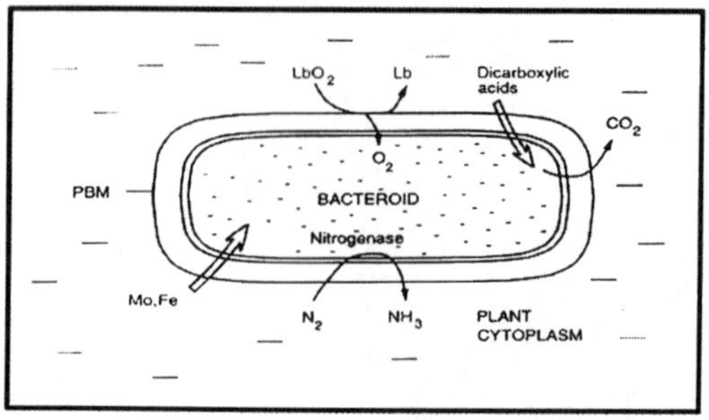

Fig. 3. Schematic representation of metabolite exchange between a nitrogen-fixing bacteroid and a plant cell. Lb: leghemoglobin.

nodulins analysed. By contrast, many proteins resident in the tonoplast and in other organelles have distinct protopeptidic targeting sequences that ensure the correct delivery of these proteins [19].

Assuming the existence of a common pathway for transport to the Golgi cisternae, there must be a specific sorting mechanism for vesicular transport of proteins from the *trans* Golgi network to different cellular compartments. A new vacuolar sorting receptor, BP-80, has recently been described [20]. BP-80 binds its ligands at the exit of the Golgi complex and releases them in the lytic vacuole or in a post-Golgi pre-vacuolar compartment. This kind of sorting receptor has not yet been identified for proteins delivered to the SM. Hence, regulatory mechanisms that allow the correct targeting to SM remain largely unexplored

It has been hypothesised that the SM uses a specialized targeting pathway, distinct from those to the plasma membrane or to the tonoplast. This targeting could follow more than one route and could be affected by unidentified proteins or signal molecules produced by the bacteroids [12]. Only recently, there has been a report that a single bacteroid-encoded polypeptide has been detected in the SM [21]. This study also suggests that some plant proteins are directly targeted to the symbiosome, without using a vesicle flow pathway. Both these interesting claims require further investigation.

Despite all these unresolved questions, it is generally accepted that the proliferation of the SM requires massive vesicular fusion with this membrane, and vesicular budding and fusion are clearly active in the proximity of the SM.

4. The "Snare" Hypothesis

Some key components of vesicular trafficking have been identified, especially in yeast and mammalian systems [22]. The basic machinery includes coat proteins, which are important for vesicle budding; vesicular- and target- specific proteins (v-SNARE and t-SNARE), which bind each other and thereby dock the vesicles to the acceptor membrane; and other proteins (SNAP and NSP) which bind to the SNARE complex and promote the final fusion step of the phospholipid bilayer. A group of monomeric small GTP-binding proteins are involved in the regulation of this complex reaction.

Recent evidence has indicated a role for Rab proteins and their effectors in vesicle targeting specificity and unidirectionality [23]. Rab proteins are a subfamily of small GTP-binding proteins. They have an average size of 200-230 amino acids and are localized in different subcellular compartments. They act as "molecular switches" and are known to cycle between donor/vesicular and acceptor membranes; between a GTP- and a

GDP-bound state; and between a membrane-associated and soluble form [24,25]. Several proteins belonging to the Rab/Ypt subfamily have been shown to have an important role during nodule biogenesis. Some of them showed elevated levels of expression in root nodules and inhibition of their expression severely impaired the functionality of the nodules resulting in the accumulation of aberrant structures [26].

5. Mutants and Molecular Genetics

As outlined in the previous section the legume nodule provides an excellent model system for studying membrane biogenesis in plants. The aim of our current research is to identify the genetic elements involved in the formation of the symbiosome compartment and the characterisation of the molecular steps that lead to the unique targeting of protein to the SM. Since nodules are not essential for the life of the plant, various mutations can be created experimentally in either symbiont [27]. The availability of several independent symbiotic defective mutants can provide a useful means to identify host plant genes affecting the development of symbiosis.

6. An Interesting Mutant Pea

An interesting case is Sprint 2 Fix$^-$ (*sym31*). Analysis of this pea mutant line suggests that *sym31* is blocked in the development of the specialised target pathway that leads to the formation of the SM [28]. The first steps of the infection process are not impaired; bacteria associate with the host and are normally delivered into the cortical cells by the infection thread. However, the plasma membrane in which they are engulfed is apparently incapable of differentiating into a mature SM and, as a consequence, bacteroids appear to be small, undifferentiated and not able to fix nitrogen. Another interesting feature is that in *sym31* several bacteroids are often enclosed in one membrane or are found in a vacuole-like compartment, suggesting an aberrant vesicle targeting pathway [29]. This peculiar morphology is similar to the one described in transgenic soybean nodules in which the amount of two legume proteins homologous to mammalian Rab1p and Rab7p was reduced using an antisense approach [26]. Our interest is therefore mainly oriented through the role of small GTP-binding proteins in specifying vesicular targeting pathways.

7. Differential Gene Expression

A comparative study of gene expression between *sym31* and wild type nodules could help us to identify which elements are involved in this process. Our experimental approach is designed to compare the patterns of gene expression in roots and in two different stages of nodule formation. Nodules formed by *sym31* are effectively arrested in a "juvenile" stage of development, compared to the mature symbiosomes found in the host cells of wild-type nodules.

Root ⟶ Juvenile Symbiosome ⟶ Mature Symbiosome
sym31 wt

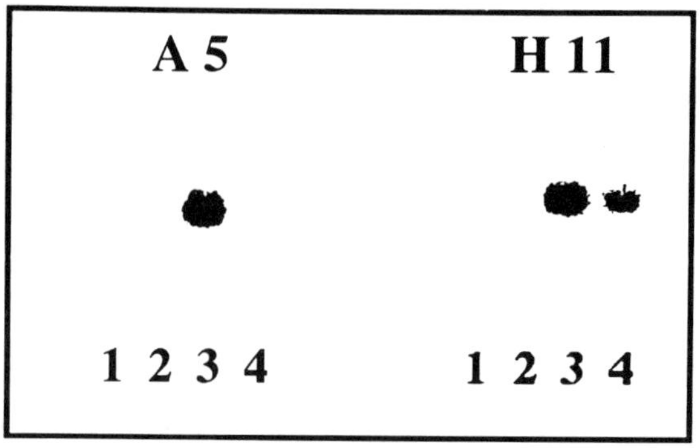

Fig. 4. Northern blot analysis confirming differential screening results.
A5: PsNlec1. H11: leghemoglobin. Each lane contains 10 μg of total RNA extracted from wild-type roots (1), *sym31* roots (2), wild-type nodules (3) and *sym31* nodules (4).

Suppression subtractive hybridization (SSH) is the current approach used to identify those transcripts with differential expression in these three stages [30,31]. This technique has already allowed the identification of several cDNA clones corresponding to genes with enhanced expression in one mRNA population compared to another. To confirm these preliminary results, three highly expressed clones were used in Northern blot analysis. Clones A5 and H11, corresponding to RNA with enhanced expression in wild type nodules strongly hybridised only with wild type nodules RNA, and gave weak or no signal in roots and *sym31* RNA (Fig. 4). These two clones were sequenced and turned out to be PsNlec1 (clone A5), an abundant nodule lectin previously identified by this group [32], and leghemoglobin (clone H11). On the other hand, the clone with a significantly enhanced level of expression in *sym31* nodules encoded a protein similar to the yeast and mammalian Sec15, a protein required for the fusion of secretory vesicles with the plasma membrane.

8. Future Work

Our preliminary trials with SSH have proved successful in the isolation of differentially expressed genes and have given a satisfactory rate. However, the technique needs to be further optimized before it can be used routinely. Once we have obtained a more reliable population enriched in differentially expressed cDNAs, it will be used to identify clones corresponding to small GTP binding proteins, taking advantage of degenerate PCR primers based on conserved sequences. At the same time, the clones isolated randomly by SSH will be sequenced and tested in Northern blot analysis. Due to our particular interest in small GTP binding proteins we can employ several available nodule-derived cDNA clones obtained from *Lotus japonicus* [33] and try to identify their counterparts in pea.

Acknowledgements

We are grateful to the UK Biotechnology and Biological Research Council for financial support.

References

[1] J. P. W. Young *et al.*, Diversity and phylogeny of rhizobia *New Phytology* **133** (1996) 87-94.

[2] P. Mylona *et al.*, Symbiotic nitrogen fixation *The Plant Cell* **7** (1995) 869-885.

[3] J. A. Downie, Signalling strategies for nodulation of legumes by rhizobia *Trends in Microbiology* **2** (1994) 318-324.

[4] R. Guerts *et al.*, Signal transduction in *Rhizobium*-induced nodule formation *Plant Physiology* **112** (1996) 447-453

[5] J.W. Kijne The *Rhizobium* infection process. In: G. Stacey, R.H. Burris H.J. Evans (eds.) Biological Nitrogen Fixation. Chapman & Hall, NY, London 1992, pp. 349-398.

[6] N.J. Brewin Tissue and cell invasion by *Rhizobium*: the structure and developments of infection threads and symbiosomes. In: H.P. Spaink, A. Kondorosi P.J.J. Hooykaas (eds.) The Rhizobiaceae. Kluwer Academic Publishers, Dordrecht 1998, pp. 417-429.

[7] N. J. Brewin, Development of the legume root nodule *AnnRevCell Biol* **7** (1991) 191-226.

[8] A-E. Hadri, H.P. Spaink, T. Bisseling N.J. Brewin, Diversity of root nodulation and rhizobial infection processes. In: H.P. Spaink, A. Kondorosi P.J.J. Hooykaas (eds.) The Rhizobiaceae. Kluwer Academic Publishers, Dordrecht 1998, pp. 347-360.

[9] P. Dart, Infection and development of leguminous nodules. In: R.W.F. Hardy W.S. Silver (eds.) A Treatise on Dinitrogen Fixation, Section III. Wiley and sons, New York 1977, pp. 367-472.

[10] N.J. Brewin, Plant membrane structure and function in the *Rhizobium*-legume symbiosis. In: M. Smallwood, J.P. Knox D.J. Bowles (eds.) Membranes: Specialised Functions in Plants. Bios Scientific Publishers, Oxford,U.K. 1996, pp. 507-524.

[11] J. G. Robertson *et al.*, The legume-*Rhizobium* symbiosis: a cell surface interaction *JCell SciSuppl* **2** (1985) 317-331.

[12] D. P. S. Verma *et al.*, Biogenesis of the peribacteroid membrane in root nodules *Trends Microb* **4** (1996) 364-368.

[13] J. G. Robertson *et al.*, Coated and smooth vesicles in the biogenesis of cell walls, plasma membranes, infection threads and peribacteroid membranes in root hairs and nodules of white clovers *JCell Sci* **58** (1982) 63-78.

[14] L. E. Roth *et al.*, Bacterium release into host cells of nitrogen-fixing soybean nodules: the symbiosome membrane comes from three sources *European Journal of Cell Biology* **49** (1989) 13-23.

[15] S. Perotto *et al.*, Molecular composition and development of the plant glycocalyx associated with the peribacteroid membrane of pea root nodules *Development* **112** (1991) 763-773.

[16] S. Perotto *et al.*, Differential expression of a glycosyl inositol phospholipid antigen on the peribacteroid membrane during pea nodule development *MolPlant-Microbe Interact* **8** (1995) 560-568.

[17] M. K. Udvardi *et al.*, Metabolite transport across symbiotic membranes of legume nodules *Annual Review of Plant Physiology and Plant Molecular Biology* **48** (1997) 493-523.

[18] S. D. Tyerman *et al.*, A channel-like transporter for ammonium ions on the symbiotic interface of nitrogen-fixing plants *Nature* **378** (1995) 629-632.

[19] A. Vitale *et al.*, The endoplasmic reticulum - Gateway of the secretory pathway *The Plant Cell* **11** (1999) 615-628.

[20] N. Paris *et al.*, Molecular cloning and further characterization of a probable plant vacuolar sorting receptor *Plant Physiology* **115** (1997) 29-39.

[21] A. C. W. Simonsen *et al.*, Origin of *de novo* synthesized proteins in the different compartments of pea-*Rhizobium* sp. symbiosomes *MolPlant-Microbe Interact* **12** (1999) 319-327.

[22] J. E. Rothman *et al.*, Protein sorting by transport vesicles *Science* **272** (1996) 227-234.

[23] S. Christoforidis *et al.*, The *Rab5* effector EEA1 is a core component of endosome docking *Nature* **397** (1999) 621-625.

[24] P. Novick *et al.*, The diversity of Rab proteins in vesicle transport *CurrOpinion Cell Biol* **9** (1997) 496-504.

[25] V. M. Olkkonen *et al.*, Role of rab GTPases in membrane traffic *International Review of Cytology* **176** (1997) 1-85.

[26] C. Cheon, N-G. Lee, A.B.M. Siddique, A.K. Bal, D.P.S. Verma Roles of plant homologues of Rab1p and Rab7p in the biogenesis of the peribacteroid membrane, a subcellular compartment formed *de novo* during root nodule symbiosis. *EMBO J* **12** (1993) 4125-4135.

[27] M. Schultze *et al.*, Cell and molecular biology of *Rhizobium*-plant interactions *International Review of Cytology* **156** (1994) 1-75.

[28] D. J. Sherrier *et al.*, Immunocytological evidence for abnormal symbiosome development in nodules of the pea mutant line Sprint2Fix⁻ (*sym*31) *Protoplasma* **199** (1997) 57-68.

[29] P. Dahiya *et al.*, Symbiotic gene *sym31* controls the presence of a lectinlike glycoprotein in the symbiosome compartment of nitrogen-fixing pea nodules *MolPlant-Microbe Interact* **11** (1998) 915-923.

[30] L. Diatchenko *et al.*, Suppression subtractive hybridization: a method for generating differentially regulated or tissue-specific cDNA probes and libraries *Proceedings of National Academy of Science, USA* **93** (1996) 6025-6030.

[31] O. D. von Stein *et al.*, A high throughput screening for rarely transcribed differentially expressed genes *Nucleic Acids Research* **25** (1997) 2598-2602.

[32] I. V. Kardailsky *et al.*, Identification of a new pea gene, *PsNlec1*, encoding a lectin-like glycoprotein isolated from the symbiosomes of root nodules *Plant Physiology* **111** (1996) 49-60.

[33] S. Borg *et al.*, Identification of new protein species among 33 different small GTP-binding proteins encoded by cDNAs from *Lotus japonicus*, and expression of corresponding mRNAs in developing root nodules *Plant Journal* **11** (1997) 237-250.

Subject Index

ABC transporter,
and glucose uptake in *Sulfolobus solfataricus,* 235-243
and protein secretion, 227-233
multidrug resistance proteins, 55-61
Acetyl-cholinesterase,
and GPI-anchoring, 32
Alkaline phosphatase,
and GPI-anchoring, 32
Aminopeptidase,
and GPI-anchoring, 32
Aminophospholipid translocator, 55-61
retention of transmembrane proteins, 73-84
Anterograde vesicle flow
and protein sorting in yeast, 63-71
12-(9-Anthroloxy)stearic acid,
in studies of lipid-peptide interactions, 85-94
Apical microvillar membranes, 73-84
retention of prominin, 73-84
Apical targeting, 211-219
and GPI-anchored proteins, 34
ARF proteins, 221-226
ARF-GTPase activating protein, 221-226
Atomic Force Microscopy,
and lipid domains, 39-54
of phase-separated lipid mixtures, 39-54
of biological membranes, 39-54
ATP-binding cassette transporters, 227-233
and glucose uptake in *Sulfolobus solfataricus,* 235-243
Bacteriorhodopsin,
crystallization, 137-150
Basolateral membranes,
and sorting of proteins, 73-84
lipid sorting, 55-61
Basolateral targeting, 211-219
Bilayer thickness,
and protein sorting in yeast, 63-71
Brefeldin,
and lipid translocation, 55-61
Carrier signature,
protein import in mitochondria, 151-159, 161-173, 161-173
Caveolin,
and apical sorting of lipids, 55-61
and membrane traffic, 176
CDP-choline pathway,
and regulation of phospholipid synthesis in yeast, 1-17
Chaperone membrane proteins,
in *Escherichia coli*, 95-104
Chaperones,
and protein folding, 227-233, 121-135
Cholesterol,
and rafts, 27-38
and protein sorting in yeast, 63-71
in sphingolipid and cholesterol-rich membrane domains, 27-38, 78
Choline,
and regulation of phospholipid synthesis in yeast, 1-17
Choline kinase,
in yeast, 1-17
Clathrin, 175-182, 211-219
and endocytosis, 211-219
and sorting signals, 175-182, 211-219
coated pits and vesicles, 8, 175-182, 211-219
Coat proteins,
of bacteriophage M13
Coated pits, 175-182
Coated vesicles, 8, 175-182
Cooperative phase separation of membrane lipids, 28
COPI , 175-182, 211-219, 221-226
COPI and COPII, 175-182, 221-226
coated vesicles and secretion, 8
COPI vesicles, 175-182
anterograde and retrogadc in Golgi , 8, 221-226
Co-translational protein import
in mitochondria, 151-159, 161-173
Co-translational translocation,
and GPI-anchored proteins, 31
Crystallization of membrane proteins, 137-150
Cytoskeleton,
and vesicular traffic, 255-259
Detergent resistant membrane (DRM), 27-38, 55-61, 78
and membrane traffic, 176
and multidrug resistance proteins, 55-61
and prominin, 78
visualised by AFM, 39-54
Detergents,
and membrane protein crystallization, 137-150
Diacylglycerol,
role in secretion, 8
Digalactosyl diacyl glycerol,
crystallization of plant light harvesting complex LH-II, 137-150

Domains in membranes, 27-38, 55-61, 86
and membrane traffic, 176
visualised by AFM, 39-54
Endocytosis, 183-193, 211-219
and clathrin, 211-219
and sphingosine base synthesis in yeast, 203-210
deficient mutants in yeast, 203-210
Endoplasmic Reticulum,
and protein targeting, 63-71
phosphatidylserine synthesis in yeast, 19-25
phospholipid synthesis, 55-61
synthesis of GPI-anchored proteins, 31
protein translocation, 211-219
phospholipid translocation, 55-61
vesicle budding, 183-193
Endoplasmic Reticulum to Golgi,
lipid transfer, 55-61
vesicle transport, 183-193, 63-71
Endosomal targeting signals, 211-219
Endosomes, 211-219
Escherichia coli,
membrane protein assembly, 95-104
protein translocation, 95-104, 195-201
signal recognition particle, 95-104, 195-201
Tat pathway of protein targeting, 95-104
Exocytosis, 183-193, 211-219
Fatty acylated proteins,
and rafts, 27-38, 55-61
Fluid phase endocytosis in yeast, 203-210
Fluorescence Resonance Energy Transfer,
and lipid-peptide interactions, 85-94
β-Galactosidase, 95-104
Galactosides-pentosides-hexuronides family,
and lactose transport, 95-104
Galactosylceramides, 55-61
Ganglioside GM1,
visualised by AFM, 39-54
Gel phase membrane domains, 27-38
visualised by AFM, 39-54
Galactoglycolipids,
and microdomains, 78
Glucose binding protein,
and glucose uptake in *Sulfolobus solfataricus,* 235-243
Glucosylceramides, 55-61
Glycosyl phosphatidylinositol,
and anchoring of proteins, 27-38, 55-61
Glycolipid translocator, 55-61
and multidrug resistance proteins, 55-61
Glycophorin,
three-dimensional structure, 105-119
Glycosphingolipids,
and membrane domains, 27-38, 44
biosynthesis and translocation, 55-61
Glycosylceramide, 55-61
Golgi apparatus,
and glycosphingolipids, 55-61
and protein targeting in plant cells, 261-268
and protein targeting in yeast, 63-71
phosphatidylserine decarboxylase, 19-25
protein targeting, 211-219, 221-226, 261-268
raft formation, 34, 55-61
vesiculation by rabGDI, 183-193
Golgi to ER vesicle transport, 183-193, 211-219, 63-71
retrograde vesicle transport, 221-226
GPI-anchored proteins,
and functions, 31-32
and intracellular sorting of membranes, 27-38
and signal transduction, 27-38
as apical targeting signal, 34, 74
in rafts, 27-38, 55-61
synthesis in the ER, 31
GTPase activating protein (GAP), 183-193
GTPases, 183-193
and retrograde transport, 221-226
and vacuole fusion, 255-259
Guanine nucleotide activating proteins (GAPs), 221-226
Guanine nucleotide exchange factors (GEFs), 221-226
Guanosine nucleotide dissociation inhibitors (GDIs), 183-193
and the early secretory pathway, 183-193
Halobacterium salinarum,
and protein crystallization, 137-150
Heat shock proteins,
and protein folding, 121-135
HSP60, HSP 70, HSP104, 121-135
Helical hairpins, 100
Helix-helix interactions in membrane proteins, 105-119
Heme binding protein,
secretion by *S. marcescens,* 227-233
Homotypic Vacuole Fusion, 255-259
Inositol,
and regulation of phospholipid synthesis in yeast, 1-17

Inositol-phosphate,
and membrane trafficking, 9
as second messenger, 8
Inositol-phosphate synthesis, 2
INO regulatory genes,
and phospholipid synthesis, 1-17
glucose response pathway, 10
Integral membrane proteins,
assembly and structure, 95-104
crystallization, 137-150
folding, 105-119
interaction with lipids, 105-119
structure, 105-119
topology mapping, 100
Internal targeting sequences, 151-159, 161-173
Interorganelle traffic, 175-182
vacuole inheritance, 255-259
Intracellular membrane trafficking, 175-182
Kennedy pathway,
phospholipid synthesis in yeast, 6
Lactose/galactose exchange, 95-104
Lactose/proton symport, 95-104
Lactose transport,
in *Streptococcus thermophilus*, 95-104
the galactosides-pentosides-hexuronides family, 95-104
Langmuir monolayers, v5
Lipid aggregation,
and protein interaction, 27-38, 55-61
visualised by AFM, 39-54
Lipid domains,
and lipid-protein interactions, 27-38, 55-61
and lipid sorting, 55-61
visualised by AFM, 39-54
Lipid intermembrane transport, 55-61
Lipid phase separation, 27-38
visualised by AFM, 39-54
Lipid phase transitions, 27-38, 85-94
visualised by AFM, 39-54
Lipid-peptide interactions, 95-104, 105-119
spectrofluorometric characterization, 85-94
Lipid sorting, 55-61
Lipid translocation, 55-61
and multidrug resistance proteins, 55-61
Lipidic mesophases,
and protein crystallization, 137-150
Lipidic cubic phases,
and protein crystallization, 137-150
Liquid-crystalline membrane domains, 27-38, 89
visualised by AFM, 39-54
Liquid-ordered membrane domains, 27-38
visualised by AFM, 39-54
Lysosomes,
and protein sorting, 211-219
Mannitol permease,
targeting in *E.coli*, 195-201
Mannose-6-phosphate receptors,
and protein sorting, 211-219
Matrix-targeting sequences,
mitochondrial protein import, 151-159, 161-173
Melting temperature of lipids,
and domain formation, 28
Membrane fusion protein, 227-233
Membrane lipid phase behaviour, 27-38,
and lipid-protein interactions, 85-94
Membrane potential,
protein import in mitochondria, 151-159, 161-173
Membrane proteins,
assembly and structure, 95-104
crystallization, 137-150
folding, 105-119
helix-helix interactions, 105-119
sorting, 73-84
structure, 105-119
Membrane targeting,
in pea root nodule cells, 261-268
Membrane trafficking, 175-182
and phospholipid metabolism in yeast, 7
Microdomains,
and cell polarity, 73-84
and multidrug resistance proteins, 55-61
in membranes, 27-38
visualised by AFM, 39-54
rafts, 27-38, 55-61
Microvilli,
and lipid domains, 73-84
Mitochondria,
of *Saccharomyces cerevisiae*, 19-25
phosphatidylserine decarboxylation, 19-25
phospholipid import and composition, 19-25
protein import, 151-159, 161-173
Mitochondrial processing peptidase,
protein import in mitochondria, 151-159, 161-173
Multidrug resistant P-glycoproteins,
and detergent resistant membranes, 55-61
and lipid translocation, 55-61

N-glycosylation of proteins,
and topology mapping, 100
Near Field Microscopy,
of artificial and biological membranes, 39-54
Neural cell adhesion molecule, 32
Neurospora crassa,
protein import in mitochondria, 151-159
Nodulins,
and symbiosis in plant cells, 261-268
5'-Nucleotidase,
and GPI-anchoring, 32
OmpA protein, 195-201
Organelle inheritance, 255-259
Pea root nodule cells, 261-268
Peptidyl-glycylleucine-carboxyamide,
and lipid-peptide interactions, 85-94
Phase transitions in the bilayer, 85-94
Phosphatidic acid,
and vesicle budding, 8
regulation of phospholipid synthesis in yeast, 1-17
Phosphatidylcholine,
via the Kennedy pathway, 6, 19-25
synthesis/translocation in yeast, 1-17
Phosphatidylethanolamine,
and GPI-anchor synthesis, 31
synthesis/translocation in yeast, 1-17, 19-25
Phosphatidylethanolamine-N-methyltransferases, 19-25
Phosphatidylinositol,
and GPI-anchor synthesis, 31
Phosphatidylinositol synthase in yeast, 1-17, 19-25
Phosphatidylserine,
synthesis/translocation in yeast, 19-25
Phosphatidylserine decarboxylase, 19-25
in*Saccharomyces cerevisiae*, 19-25
Phosphoinositides,
and vacuolar protein sorting, 8
Phospholipase D,
and turnover of phosphatidylcholine, 1-17
Phospholipid biosynthesis,
in yeast, 1-17, 19-25
and membrane trafficking, 7
and signal transduction, 1-17
and sorting, 55-61
structural and regulatory genes, 1-17
the CDP-choline pathway, 1-17, 19-25
Phospholipid localization, 55-61
Phospholipid sorting, 55-61
Phospholipid synthesizing mitochondria, 19-25
Phospholipid turnover,
in yeast, 1-17, 19-25
Phospholipid transfer proteins,
in yeast, 7
Phospholipid transport, 55-61
and multidrug resistance proteins, 55-61
between ER and mitochondria, 19-25
in yeast, 1-17
Phytic acid,
inositol synthesis, 7
Placental alkaline phosphatase,
and GPI-anchoring, 32
Posttranslational protein import,
in mitochondria, 151-159, 161-173
Posttranslational targeting of secretory proteins, 195-201
Preprotein translocase,
protein import in mitochondria, 151-159, 161-173
Prominin,
in apical microvillar membranes, 73-84
Protein anchoring, 27-38
Protein folding,
and chaperones, 121-135
and osmolytes, 121-135
and protein secretion, 227-233
and trehalose, 121-135
of integral membrane proteins, 105-119
two-stage model, 105-119
Protein glycosylation,
for topology mapping of membrane proteins, 100
Protein leader peptidase, 95-104
Protein import, 211-219
and membrane potential, 151-159, 161-173
ATP dependency, 151-159, 161-173
in mitochondria, 151-159, 161-173
Protein import receptor system, 151-159, 161-173
Protein insertion,
in *Escherichia coli*, 195-201
Protein misfolding, 125
Protein secretion, 211-219, 63-71
in *Serratia marcescens*, 227-233
Protein sorting, 211-219, 63-71, 255-259
and bilayer thickness, 63-71
and clathrin, 175-182, 211-219
and PIP3 kinase, 8
in COPI/COPII pathway, 175-182, 211-219

Protein sorting,
in membrane domains, 27-38, 73-84
in polarized cells, 73-84, 211-219
via rafts, 34
Protein targeting, 211-219
and signal sequences, 151-159, 161-173, 211-219
in *Escherichia coli*, 95-104, 195-201
into mitochondria, 151-159, 161-173
in plant cells, 261-268
of mannitol permease, 195-201
Tat pathway, 95-104
to rafts, 27-38
Protein topology,
studied by N-linked glycosylation, 100
Protein translocation, 211-219
and signal sequences, 211-219
ATP dependency, 151-159, 161-173
in *Escherichia coli*, 95-104, 195-201
in mitochondria, 151-159, 161-173
SEC-independent, 95-104
the SEC pathway, 95-104, 195-201
Protein transport channel,
in mitochondria, 151-159, 161-173
Proton motive force,
and glucose uptake in *Sulfolobus solfataricus,* 235-243
Rab cycle, 211-219
Rab escort protein (REP1), 183-193, 211-219
Rab proteins,
and the secretory pathway, 183-193, 211-219
and vesicle fusion, 183-193
Rafts, 27-38, 55-61
and detergent resistant membrane, 27-38
and lipid sorting, 55-61
and protein sorting, 73
and signal transduction, 35
formation in the Golgi, 34
visualised by AFM, 39-54
Receptor mediated endocytosis, 211-219
in yeast, 203-210
Receptor sorting, 211-219
by clathrin coated pits, 175-182
Retrograde transport vesicles, 221-226
and protein sorting in yeast, 63-71
Rhizobium,
and symbiosis with plant cells, 261-268
Saccharomyces cerevisiae,
and GPI-anchored proteins, 32
endocytosis, 203-210
intracellular vesicular transport, 183-193, 221-226
mutants in lipid biosynthesis, 1-17, 19-25
phosphatidylethanolamine synthesis, 1-17, 19-25
phosphatidylserine synthesis, 1-17, 19-25
phospholipid synthesis and turnover, 1-17
phospholipid translocation, 1-17
protein import in mitochondria, 151-159, 161-173
protein sorting, 63-71
sphingosine base synthesis and endocytosis, 203-210
vacuole fusion, 255-259
Scanning Force Microscope, 40
Scanning Tunneling Microscopy, 40
Scramblase, 55-61
Scrapie prion protein,
and GPI-anchoring, 32
SecB chaperone, 227-233
Sec-independent protein translocation, 95-104, 227-233
Sec protein translocation pathway, 95-104, 195-201, 227-233
SecY translocon, 195-201
Secretory pathways, 211-219, 227-233
and phospholipid metabolism in yeast, 7
in *Escherichia coli*, 95-104, 195-201
in *Serratia marcescens*, 227-233
Secretory proteins, 211-219, 227-233
and signal sequences, 211-219
and the early secretory pathway, 211-219
in *Escherichia coli*, 95-104, 195-201
Secretory vesicles, 175-182, 211-219, 221-226
Serine palmitoyl transferase,
in endocytosis deficient yeast, 203-210
Serratia marcescens,
protein export, 227-233
Signal-anchor sequences,
of membrane proteins, 195-201
Signal peptidase,
and secretion in *Escherichia coli*, 95-104,
Signal peptides,
in *Escherichia.coli*, 95-104
Signal recognition particle,
in *Escherichia coli*, 95-104, 195-201
Signal sequences,
and membrane receptors, 95-104

Signal transduction pathways,
and GPI-anchored proteins, 35
and lipid domains, 55-61
and phospholipid synthesis in yeast, 1-17
SNAP (soluble NSF Attachment) proteins, 175-182, 211-219
and protein targeting in plant cells, 261-268
and vacuole inheritance, 255-259
SNARE proteins, 175-182, 211-219, 221-226
and Golgi vesicle reassembly, 221-226
and protein targeting in plant cells, 261-268
and vacuole inheritance, 255-259
Solute/H^+ symport,
and lactose transport, 95-104
Sorting signals, 211-219
and clathrin pathway, 175-182, 211-219
of transmembrane proteins, 73
Spectrin,
retention of transmembrane proteins, 76
Sphingolipids,
and cholesterol rich membrane domains, 27-38, 44, 55-61
and GPI-anchored proteins, 27-38
and protein sorting, 63-71
in rafts, 27-38, 44, 55-61
translocators, 55-61
very-long-chain fatty acids, 28
Sphingomyelin,
and rafts, 27-38, 44, 55-61
synthesis and translocation, 55-61
Sphingomyelinase, 55-61
Sphingosine base synthesis,
and endocytosis in yeast, 203-210
Sterols,
and protein sorting in yeast, 63-71
Streptococcus thermophilus,
lactose transport, 95-104
Sulfolobus solfataricus,
glucose transport, 235-243
Symbiosis,
in plant cells, 261-268
Symbiosome membrane,
and protein targeting in plant cells, 261-268
Targeting of proteins, 211-219
into mitochondria, 151-159, 161-173
Targeting sequences, 211-219
and protein import in mitochondria, 151-159, 161-173
Tight junctions, 55-61, 73
Tim channel,
protein import in mitochondria, 151-159, 161-173
Tim proteins,
protein import in mitochondria, 151-159, 161-173
Tom channel,
protein import in mitochondria, 151-159, 161-173
Tom proteins,
receptors in protein import, 151-159, 161-173
Topology of proteins,
studied by N-linked glycosylation, 100
Trans-Golgi network, 211-219
protein sorting, 73, 211-219
raft formation, 34, 55-61
Transmembrane helices,
amino acid sequences, 95-104
interaction with lipids, 99
Transmembrane anchor sequences, 151-159, 161-173
Transmembrane proteins,
α-helices, 105-119
interaction with lipids, 99
retention by spectrin, 76
retention by the ankyrin/fodrin/uvomorulin complex, 76
sorting and targeting, 73, 95-104
structure, 105-119
Trehalase,
and GPI-anchoring, 32
Trehalose,
and protein folding, 121-135
Twin-arginine targeting (Tat) pathway of protein targeting, 95-104
Tyrosine sorting signals, 211-219
UAS_{INO}, containing genes,
and phospholipid synthesis, 1-17
Unfolded protein respons signal transduction pathway, 9
Vacuolar phosphatidylserine decarboxylase, 19-25
Vacuolar protein sorting, 255-259
and phosphoinositides, 8
in plant cells, 261-268
Vacuolar trafficking, 8
Vacuole fusion, 255-259
Vacuole inheritance, 255-259
Variant surface glycoprotein, 32

Vesicle budding, 175-182, 183-193, 221-226, 255-259
    and phosphatidic acid, 8
Vesicle fusion, 183-193
    and clathrin, 175-182
    vacuole inheritance, 255-259
Vesicle mediated transfer of lipids, 8, 55-61
Vesicle targeting, 211-219
    and clathrin, 175-182
Vesicle traffic, 183-193, 211-219, 255-259
    and clathrin, 175-182
    and protein sorting, 63-71
    and protein targeting in plant cells, 261-268
Vitronectin,
    and GPI-anchoring, 32
X-ray cristallography of membrane proteins, 137-150
Yeast,
    phospholipid synthesis, 1-17, 19-25
    protein sorting, 63-71
    retrograde transport, 221-226
    sphingosine base synthesis and endocytosis, 203-210
    vacuole inheritance, 255-259

Author Index

Albers, S.-V. 235
Almaguer, C.M. 1
Beck, K. 195
Birner, R. 19
Blondelle, S. 85
Bode, G. 183
Brewin, N.J. 261
Brown, D. 27
Bruno, V. 1
Bürgermeister, M. 19
Butelli, E. 261
Chang, H.J. 1
Corbeil, D. 73
Daum, G. 19
Delepelaire, P. 227
Dowd, S.R. 1
Driessen, A.J.M. 235
Elferink, M.G.L. 235
Engelman, D.M. 105
Fisher, L.F. 105
Friant, S. 203
Gangl, S. 85
Geissler, A. 161
Giocondi, M.-C. 39
Goudonnet, J.-P. 39
Grabner, G. 85
Graves, J.A. 1
Heitz, F. 39
Henry, S.A. 1
Hrastnik, C. 19
Huttner, W.B. 73
Kirchhausen, T. 175
Koch, H.-G. 195
Köhler, G. 85
Konings, W.N. 235
Laird, V. 211
Landau, E.M. 137
Lee, C.M. 151
Le Grimellec, C. 39
Lesniewska, E. 39
Levine, T.P. 63
Lindquist, S.L. 121
Lohner, K. 85
MacKenzie, K.R. 105
Mayer, T. 183
Müller, M. 195
Munn, A. 203
Munro, S. 63
Neupert, W. 151
Patton-Vogt, J. 1
Pfanner, N. 161
Poolman, B. 245
Popot, J.-L. 105
Raggers, R. 55
Riezman, H. 203
Röper, K. 73
Ruiz-Noriega, M. 1
Russ, W.P. 105
Sapriel, G. 227
Schäfer, U. 195
Schekman, R. 221
Sonoda, G.K. 105
Spang, A. 221
Spooner, P.J.R. 245
Sprong, H. 55
Sreenivas, A. 1
Staudegger, E. 85
Stuart, R.A. 151
Truscott, K. 161
Van Mau, N. 39
Van Meer, G. 55
Veenhoff, L.M. 245
Vié, V. 39
Villa-Garcia, M. 1
Von Heijne, G. 95
Watts, A. 245
Wickner, W. 255
Yang, J. 1
Zanolari, B. 203